Capturing Change in Science, Technology, and Innovation

IMPROVING INDICATORS TO INFORM POLICY

Panel on Developing Science, Technology, and Innovation Indicators for the Future

Robert E. Litan, Andrew W. Wyckoff, and Kaye Husbands Fealing, *Editors*

Committee on National Statistics
Division of Behavioral and Social Sciences and Education

and

Board on Science, Technology, and Economic Policy
Division of Policy and Global Affairs

NATIONAL RESEARCH COUNCIL
OF THE NATIONAL ACADEMIES

THE NATIONAL ACADEMIES PRESS
Washington, D.C.
www.nap.edu

THE NATIONAL ACADEMIES PRESS • 500 Fifth Street, NW • Washington, DC 20001

NOTICE: The project that is the subject of this report was approved by the Governing Board of the National Research Council, whose members are drawn from the councils of the National Academy of Sciences, the National Academy of Engineering, and the Institute of Medicine. The members of the committee responsible for the report were chosen for their special competences and with regard for appropriate balance.

The project that is the subject of this report was supported by grant number SES-0453930 between the National Science Foundation and the National Academy of Sciences. Support for the Committee on National Statistics is provided by a consortium of federal agencies through a grant from the National Science Foundation (grant number SES-1024012). Any opinions, findings, conclusions, or recommendations expressed in this publication are those of the author(s) and do not necessarily reflect the views of the organizations or agencies that provided support for the project.

International Standard Book Number-13: 978-0-309-29744-8
International Standard Book Number-10: 0-309-29744-3

Additional copies of this report are available from the National Academies Press, 500 Fifth Street, NW, Keck 360, Washington, DC 20001; (800) 624-6242 or (202) 334-3313; http://www.nap.edu.

Printed in the United States of America

Cover credit: Image of U.S. Capitol from Architect of the Capitol. All other images from iStock cover images. Cover designed by Jacqui Sovde of the National Research Council.

Suggested citation: National Research Council. (2014). *Capturing Change in Science, Technology, and Innovation: Improving Indicators to Inform Policy.* Panel on Developing Science, Technology, and Innovation Indicators for the Future, R.E. Litan, A.W. Wyckoff, and K.H. Fealing, Editors. Committee on National Statistics, Division of Behavioral and Social Sciences and Education. Board on Science, Technology, and Economic Policy, Division of Policy and Global Affairs. Washington, DC: The National Academies Press.

THE NATIONAL ACADEMIES

Advisers to the Nation on Science, Engineering, and Medicine

The **National Academy of Sciences** is a private, nonprofit, self-perpetuating society of distinguished scholars engaged in scientific and engineering research, dedicated to the furtherance of science and technology and to their use for the general welfare. Upon the authority of the charter granted to it by the Congress in 1863, the Academy has a mandate that requires it to advise the federal government on scientific and technical matters. Dr. Ralph J. Cicerone is president of the National Academy of Sciences.

The **National Academy of Engineering** was established in 1964, under the charter of the National Academy of Sciences, as a parallel organization of outstanding engineers. It is autonomous in its administration and in the selection of its members, sharing with the National Academy of Sciences the responsibility for advising the federal government. The National Academy of Engineering also sponsors engineering programs aimed at meeting national needs, encourages education and research, and recognizes the superior achievements of engineers. Dr. C. D. Mote, Jr., is president of the National Academy of Engineering.

The **Institute of Medicine** was established in 1970 by the National Academy of Sciences to secure the services of eminent members of appropriate professions in the examination of policy matters pertaining to the health of the public. The Institute acts under the responsibility given to the National Academy of Sciences by its congressional charter to be an adviser to the federal government and, upon its own initiative, to identify issues of medical care, research, and education. Dr. Victor J. Dzau is president of the Institute of Medicine.

The **National Research Council** was organized by the National Academy of Sciences in 1916 to associate the broad community of science and technology with the Academy's purposes of furthering knowledge and advising the federal government. Functioning in accordance with general policies determined by the Academy, the Council has become the principal operating agency of both the National Academy of Sciences and the National Academy of Engineering in providing services to the government, the public, and the scientific and engineering communities. The Council is administered jointly by both Academies and the Institute of Medicine. Dr. Ralph J. Cicerone and Dr. C. D. Mote, Jr., are chair and vice chair, respectively, of the National Research Council.

www.national-academies.org

PANEL ON DEVELOPING SCIENCE, TECHNOLOGY, AND INNOVATION INDICATORS FOR THE FUTURE

ROBERT E. LITAN (*Cochair*), Bloomberg Government, Washington, DC
ANDREW W. WYCKOFF (*Cochair*), OECD, Paris, France
CARTER BLOCH, Danish Center for Studies in Research and Research Policy, University of Aarhus, Denmark
NICHOLAS R. CHRISMAN, Geospatial Science, RMIT University, Melbourne, Australia
CARL J. DAHLMAN, Edmund A. Walsh School of Foreign Service, Georgetown University, Washington, DC
GEOFF M. DAVIS, Quantitative Insights Group, Google, Inc., Mountain View, CA
KATHARINE G. FRASE, Industries Research, IBM, Yorktown Heights, NY
BARBARA M. FRAUMENI, Muskie School of Public Service, University of Southern Maine
RICHARD B. FREEMAN, Department of Economics, Harvard University
FREDERICK D. GAULT, Economic and Social Research and Training Centre on Innovation and Technology, United Nations University, Maastricht, The Netherlands
DAVID GOLDSTON, Government Affairs, Natural Resources Defense Council, Washington, DC
MICHAEL MANDEL, Wharton School, University of Pennsylvania
JOHN E. ROLPH, Marshall School of Business, University of Southern California
LELAND WILKINSON, Data Visualization, Skytree, Inc., San Jose, CA

KAYE HUSBANDS FEALING, *Study Director*
ESHA SINHA, *Associate Program Officer*
ANTHONY S. MANN, *Program Coordinator*

COMMITTEE ON NATIONAL STATISTICS
2013-2014

LAWRENCE D. BROWN (*Chair*), Department of Statistics, Wharton School, University of Pennsylvania
JOHN M. ABOWD, School of Industrial and Labor Relations, Cornell University
MARY ELLEN BOCK, Department of Statistics, Purdue University
DAVID CARD, Department of Economics, University of California, Berkeley
ALICIA CARRIQUIRY, Department of Statistics, Iowa State University
MICHAEL E. CHERNEW, Department of Health Care Policy, Harvard Medical School
CONSTANTINE GATSONIS, Center for Statistical Sciences, Brown University
JAMES S. HOUSE, Survey Research Center, Institute for Social Research, University of Michigan
MICHAEL HOUT, Department of Sociology, New York University
SALLIE ANN KELLER, Virginia Bioinformatics Institute, Virginia Polytechnic Institute and State University
LISA LYNCH, The Heller School for Social Policy and Management, Brandeis University
COLM A. O'MUIRCHEARTAIGH, Harris Graduate School of Public Policy Studies, University of Chicago
RUTH D. PETERSON, Department of Sociology (emerita) and Criminal Justice Research Center, Ohio State University
EDWARD H. SHORTLIFFE, Departments of Biomedical Informatics, Columbia University and Arizona State University
HAL S. STERN, Donald Bren School of Information and Computer Sciences, University of California, Irvine

CONSTANCE F. CITRO, *Director*
JACQUELINE R. SOVDE, *Program Associate*

BOARD ON SCIENCE, TECHNOLOGY, AND ECONOMIC POLICY

Preface

Indicators are a scaffolding of statistics to which decision makers can relate other elements needed to make decisions. Indicators often are used to tell an end-to-end story on a policy-relevant topic. Science, technology, and innovation (STI) indicators are at a transition point in the formation of mainstream policy, especially in the current economic environment, in which a premium is placed on sustainable growth (e.g., growth that is not dependent on speculative bubbles). The United States has always played a leadership role internationally in the development of standards for statistical measurement of STI activities. This report is intended to offer the National Center for Science and Engineering Statistics (NCSES) at the U.S. National Science Foundation guidance that will help keep it at the forefront of this endeavor.

The primary audience for this report is the sponsor—NCSES—as well as similar statistical agencies that produce data and statistics on innovative activities worldwide. The report assesses the demand for STI indicators from different perspectives: national, international, subnational, and sectoral. Although STI indicators often are retrospective—measuring stocks, flows, and networks within the system—some bellwethers can show prospective trends in science, technology, engineering, and mathematics (STEM) talent; new areas of scientific exploration; potentially vibrant regions of innovation activity; and arcs toward the international rise or decline of countries based on their technological capabilities. STI indicators, therefore, can be used to report macro-level STI activities, outputs, outcomes, and linkages, as well as micro-level metrics related to actors and intermediate outputs in the system. New indicators can be developed by layering, linking, or blending existing data and indicators.

The users of NCSES data and statistics also are expected to be an important audience for this report. The user community for NCSES's STI indicators is diverse and includes the National Science Board; staff at other statistical agencies in the United States and abroad; local, state, federal, and international policy makers and government administrators; academic researchers and administrators; and business managers and consultants. Users want indicators that relay timely information about drivers, trends, advances, vulnerabilities, culture/climate, and distributions related to the STI system. Users also are eager to explore nontraditional sources of data from which indicators are derived and want to be informed about the reliability of those data. Lastly, qualitative or descriptive information should accompany quantitative measures to explain the economic tenets, social norms, regulatory constructs, and political atmosphere with which the STI system engages. Qualitative information and even case studies allow for deeper insights into not only what happened but also why.

The user community relies on STI indicators for answering key questions regarding the global science and technology enterprise, including: What are direct measures of innovation, and what drives innovation? Where is leadership in science and engineering knowledge generation trending? What is the status of STEM talent around the world? What is the portfolio of spending and other support by governments and private firms and organizations for STI activities, particularly those at universities? What institutions, networks, and regulations facilitate or impede advances in science, technology, and entrepreneurship? What is the trend in online learning in the United States and abroad, and what impact will that have on university finance, operations and recruitment? What are important subnational collaborative activities that promote innovation and economic growth? What are the general perceptions about science and the public value of science in the general population in the United States and abroad? Dozens more such questions are enumerated in this report.

Given its broad disciplinary scope, the study was conducted by a panel of experts that collectively represent more than a dozen fields, including computer science, economics, education, engineering, finance, geography, mathematics, physics, political science, psychology, statistics, and visual analytics. The panel also reflects the international nature of the topic, with members from Canada, Denmark, France, and the Netherlands.

In undertaking this study, the panel first relied on users, experts, and written reports and peer-reviewed articles to establish current and anticipated user needs for STI indicators. Second, the panel recognized that no one model informs the types of indicators NCSES needs to produce. Policy questions served as an important guide to the panel's review, but the study was also informed by systems approaches and international comparability. Third, it was important to identify data resources and tools NCSES could exploit to develop its indicators program. Understanding the network of inputs—including data from NCSES surveys, other federal agencies, international organizations, and the private sector—that can be tapped in the production of indicators gave rise to a set of recommendations for working with other federal agencies and public and private organizations. Fourth, the panel did not limit its recommendations to indicators but also addressed processes for prioritizing data development and the production of indicators in the future, because it was clear that the changing environment in which NCSES operates is a key determinant of the agency's priorities from year to year. Internal processes that are observant, networked, and statistically and analytically balanced are important for NCSES's indicators program.

On request of the sponsor, an interim report was published in February 2012, summarizing the panel's early findings and recommendations. The recommendations offered in this report expand on those of the interim report. They are intended to serve as the basis for a strategic program of work that will enhance NCSES's ability to produce indicators that capture change in STI to inform policy and optimally meet the needs of its user community.

Robert E. Litan and Andrew W. Wyckoff, *Cochairs*
Panel on Developing Science, Technology,
and Innovation Indicators for the Future

Acknowledgments*

It is with extreme gratitude that the panel thanks the many people who made contributions to this study on science, technology, and innovation (STI) indicators. The staff of the National Center for Science and Engineering Statistics (NCSES) at the National Science Foundation, under the directorship of John Gawalt and formerly Lynda Carlson, gracefully provided invaluable input and insights, including clear direction on what they wanted to learn from the study, as well as useful sources of information from their division and other resources. Robert Bell, Lawrence Burton, John Jankowski, Nirmala Kannankutty, Beethika Khan, Rolf Lehming, Francisco Moris, Jeri Mulrow, Christopher Pece, and Emilda Rivers all contributed their knowledge and expertise to answer our questions.

The panel's work benefited greatly from presenters and attendees at our open meetings. The insights of the following individuals were critical for the framing of policy issues that are relevant to this study: Jeff Alexander (SRI International), Patrick Clemins (formerly of the American Association for the Advancement of Science), Mark Doms (U.S. Department of Commerce), Matthew Gerdin (U.S. State Department), Kei Koizumi (U.S. Office of Science and Technology Policy), Christine Matthews (Congressional Research Service), Amber Hartman Scholz (President's Council of Advisors on Science and Technology), Dahlia Sokolov (U.S. House of Representatives), and D. Greg Tassey (formerly of the National Institute of Standards and Technology). Conceptual frameworks for STI indicators were presented by Michelle Alexopoulos (University of Toronto), Bronwyn Hall (University of California, Berkeley), and Adam Jaffe (Brandeis University).

Opportunities for advances in STI data collections and statistics, particularly among U.S. federal agencies, were discussed with panel members by B.K. Atrostic, Cheryl Grim, Richard Hough, Dave Kinyon, Erika McEntarfer, and Mary Potter (U.S. Census Bureau); Ana Aizcorbe, Maria Borga, and Carol Robbins (Bureau of Economic Analysis); Laurie Salmon, Jim Spletzer, and David Talan (Bureau of Labor Statistics); David McGranahan and Tim Wojan (U.S. Department of Agriculture); Daniel McGrath, Jessica Shedd, Matthew Soldner, and Tom Weko (National Center for Education Statistics); Stuart Graham (U.S. Patent and Trademark Office); and George Chacko and Walter Schaffer (National Institutes of Health). We also thank Rochelle (Shelly) Martinez and her colleagues and Katherine Wallman at the Office of Management and Budget for an engaging discussion regarding synergies in the federal statistical system with respect to measures of STI activities.

Because international comparability is an important aspect of this study, the panel convened two workshops of international researchers and practitioners who use STI indicators. The first workshop, in July 2011, covered STI measures and described opportunities and obstacles that NCSES should anticipate as it further develops its STI indicators program. We thank the workshop presenters: Shinichi Akaike (Hitotsubashi University, Tokyo), Howard Alper (Canada's Science Technology and Innovation Council), Jayanta Chatterjee (Indian Institute of Technology), Gustavo Crespi (Inter-American Development Bank), Matthieu Delescluse (European Commission), Changlin Gao (Chinese Academy of Science and Technology), Jonathan Haskel (Imperial College Business School), Hugo Hollanders (United Nations University-Maastricht Economic and Social Research Institute on Innovation and Technology), Brian MacAulay (National Endowment for Science, Technology and the Arts, London), and Philippe Mawoko (The New Partnership for Africa's Development).

The second workshop, in June 2012, included participants from the OECD-National Experts on Science and Technology Indicators (NESTI) representing nations, economic regions, and international institutions, including the African Union, Australia, Austria, Belgium, Brazil, Chile, the Czech Republic, Estonia, the European Union, Germany, Hungary, India, Israel, Italy, Japan, Korea, La Red

*All listed affiliations are as of February 2014.

Iberoamericana e Interamericana de Indicadores de Ciencia y Tecnología (RICYT), Luxembourg, the Netherlands, Norway, Poland, Portugal, Russia, the Slovak Republic, South Africa, Spain, Sweden, Switzerland, Turkey, the United Kingdom, and the United States. The panel chair, study director, and panel member Fred Gault, who is also chair of the NESTI advisory board, discussed the interim report from the panel and solicited comments from attendees, primarily on which indicators would be most useful to policy makers in their nations or regions. We thank the presenters, who discussed prioritization methods and specific indicators that are important for international comparisons: Alessandra Colecchia, Dominique Guellec, and Joaquim Oliveria Martins (OECD); Matthieu Delescluse (European Commission); Almamy Konté (African Union); Leonid Gokhberg (National Research University, Russia); and Veijo Ritola (Eurostat).

During the course of its work, the panel also obtained input from several other science and technology policy experts, including Aaron Chatterji (Duke University, formerly of the Council of Economic Advisers); Bhavya Lal and Stephanie Shipp (Institute for Defense Analyses–Science and Technology Policy Institute); Donna Ginther (University of Kansas); and Alessandra Colecchia, Gili Greenberg, and Fernando Galindo-Rueda (OECD). The panel explored the use of microdata, particularly administrative records and web tools, to create STI statistics. We heard from several experts in this diverse field of study at the July 2011 workshop, including Carl Bergstrom (University of Washington), Stefano Bertuzzi (National Institutes of Health and the STAR METRICS Program), Erik Brynjolfsson (Massachusetts Institute of Technology), Lee Giles (Pennsylvania State University), John Haltiwanger (University of Maryland), Richard Price (Academia.edu), and Alicia Robb (Kauffman Foundation).

The development of STI indicators at subnational levels was also an important topic for this study. At the July 2011 workshop, Rob Atkinson (Information Technology and Innovation Foundation), Maryann Feldman (University of North Carolina), Andrew Reamer (George Washington University), and Robert Samors and David Winwood (Association of Public and Land-grant Universities) presented options for measuring STI activities at a variety of geographic scales.

Nicholas Donofrio (IBM) participated in a roundtable discussion with panel members during the workshop. We greatly appreciate his insights from a business perspective on measuring research and development and innovation. His comments reminded us that the role of multinational corporations in the global STI system should be examined carefully and that entrepreneurial activities at firms of various sizes deserve careful measure.

Because the National Science Board (NSB) is a primary user of NCSES's STI indicators—the biennial *Science and Engineering Indicators* volumes are published by NSB—the panel conducted two rounds of interviews with board members. We thank Ray Bowen, Kelvin K. Droegemeier, José-Marie Griffiths, Arthur Reilly, and Arnold F. Stancell for in-depth and insightful responses to our questions, as well as the board's staff members Jennie Moehlmann, Michael Van Woert, and Matthew Wilson, for facilitating the meetings.

This report has been reviewed in draft form by individuals chosen for their diverse perspectives and technical expertise, in accordance with procedures approved by the National Research Council's Report Review Committee. The purpose of this independent review is to provide candid and critical comments that will assist the institution in making its published report as sound as possible and to ensure that the report meets institutional standards for objectivity, evidence, and responsiveness to the study's charge. The review comments and draft manuscript remain confidential to protect the integrity of the deliberative process. We wish to thank the following individuals for their review of this report: Henry Brady, Office of the Dean, Goldman School of Public Policy, University of California, Berkeley; Michael Conlon, Clinical and Translation Science Institute, University of Florida; Martin Fleming, chief economist and vice president, Business Performance Services, International Business Machines Corporation; Jacques S. Gansler, Center for Public Policy and Private Enterprise, School of Public Policy, University of Maryland; Christopher T. Hill, Emeritus Department of Public Policy and Technology, School of Public Policy, George Mason University; Graham G. Kalton, Westat, Inc., Rockville, Maryland; Jason Owen-Smith, Barger Leadership Institute, University of Michigan; Georgine M. Pion, Department of Psychology and Human Development, Vanderbilt University; Hal Salzman, Bloustein School of Planning and Public Policy and Heldrich Center for Workforce Development, Rutgers University; Phillip Swagel, School of Public Policy, University of Maryland; Albert H. Teich, director, Science and Policy Programs, American Association for the Advancement of Science; and Ward Ziarko, Scientific and Technical Information Service, Belgian Federal Science Policy.

Although the reviewers listed above provided many constructive comments and suggestions, they were not asked to endorse the report's conclusions or recommendations, nor did they see the final draft of the report before its release. The review of this report was overseen by Chuck Phelps, provost emeritus, University of Rochester, and John Haltiwanger, Department of Economics, University of Maryland. Appointed by the National Research Council, they were responsible for making certain that an independent examination of this report was carried out in accordance with institutional procedures and that all review comments were carefully considered. Responsibility for the final content of this report rests entirely with the authoring panel and the institution.

We extend special thanks to staff of the Committee on National Statistics. The study director, Kaye Husbands Fealing, provided invaluable assistance to the panel in organizing the meetings and preparing this report. Connie Citro, Tom Plewes, and Michael Cohen gave excellent guidance to the panel. Stephen Merrill, director of the Board on Science, Technology, and Economic Policy, contributed to panel meetings. Esha Sinha assembled and developed a public use databank of STI indicators from several international sources (see http://sites.nationalacademies.org/DBASSE/CNSTAT/Science_Technology_and_Innovation_Indicators/index.htm [May 2014]. In collaboration with panel member Leland Wilkinson, she also conducted the cluster analysis and heat map exercises for this study and cowrote the accompanying data appendix (Appendix F). Anthony Mann provided outstanding administrative and logistical support to the panel. Our Mirzayan fellow, Daniel Grady, drew on his expertise in systems dynamics and web tools to benefit the panel's work.

Robert E. Litan and Andrew W. Wyckoff, *Cochairs*
Panel on Developing Science, Technology,
and Innovation Indicators for the Future

Acronyms, Abbreviations, and Descriptions

ACRONYMS AND ABBREVIATIONS

America COMPETES Act	America Creating Opportunities to Meaningfully Promote Excellence in Technology, Education, and Science
APLU	Association of Public and Land-grant Universities
BEA	Bureau of Economic Analysis, U.S. Department of Commerce
BERD	business enterprise expenditure on research and development
BLS	Bureau of Labor Statistics, U.S. Department of Labor
BOP	balance of payments
BRDIS	Business Research and Development and Innovation Survey
CICEP	Commission on Innovation, Competitiveness, and Economic Prosperity
CIS	Community Innovation Survey
ERS	Economic Research Service, U.S. Department of Agriculture
ETA	Employment and Training Administration, U.S. Department of Labor
EU	European Union
FTE	full-time equivalent
GBAORD	government budget appropriations or outlays for research and development
GDP	gross domestic product
GERD	gross domestic expenditure on research and development
GO→SPIN	Global Observatory on Science, Technology and Innovation Policy Instruments
GOVERD	government intramural expenditure on research and development
HERD	higher education expenditure on research and development or Higher Education Research and Development Survey
ICT	information and communication technology
IMF	International Monetary Fund
IPEDS	Integrated Postsecondary Education Data System
MEP	Manufacturing Extension Partnership
MIST	Microbusiness, Innovation Science, and Technology
MOOC	massive open online course

NAICS	North American Industry Classification System
NCES	National Center for Education Statistics, U.S. Department of Education
NCSES	National Center for Science and Engineering Statistics, U.S. National Science Foundation
NEPAD	New Partnership for Africa's Development
NESTI	National Experts on Science and Technology Indicators
NIH	National Institutes of Health, U.S. Department of Health and Human Services
NIST	National Institute of Standards and Technology, U.S. Department of Commerce
NLM	National Library of Medicine, U.S. Department of Health and Human Services
NSCG	National Survey of College Graduates
NSF	U.S. National Science Foundation
NSRCG	National Survey of Recent College Graduates
OECD	Organisation for Economic Co-operation and Development
OMB	U.S. Office of Management and Budget
OSTP	U.S. Office of Science and Technology Policy
PISA	Programme for International Student Assessment
R&D	research and development
S&E	science and engineering
S&T	science and technology
SBIR	Small Business Innovation Research
SDR	Survey of Doctorate Recipients
SED	Survey of Earned Doctorates
SEI	*Science and Engineering Indicators*
SESTAT	Science and Engineering Statistical Data System
SIBS	Survey of Innovation and Business Strategy
STAR METRICS	Science and Technology for America's Reinvestment: Measuring the Effect of Research on Innovation, Competitiveness and Science
STEM	science, technology, engineering, and mathematics
STI	science, technology, and innovation
STTR	Small Business Technology Transfer
UNESCO	United Nations Educational, Scientific and Cultural Organization
UNU-MERIT	United Nations University's Maastricht Economic and Social Research Institute on Innovation and Technology
USDA	U.S. Department of Agriculture

DESCRIPTIONS

BERD: Research and development expenditure in the business enterprise sector in a given year at the regional level.

GBAORD: Government budget appropriations or outlays for research and development, a way of measuring government support for research and development activities.

GERD: Gross domestic expenditure on research and development, defined as total intramural expenditure on research and development performed on the national territory during a given period.

GO→SPIN: Global Observatory on Science, Technology and Innovation Policy Instruments. Contains not only indicators but also an inventory of science, technology, and innovation (STI) national system descriptions; STI legal frameworks (with access to text of the acts and decrees); and an inventory of operational policy instruments that promote STI activities in a country. Developed by a group at the United Nations Educational, Scientific and Cultural Organization (UNESCO) Regional Bureau for Science for Latin America and the Caribbean.

HERD: Research and development expenditure in the higher education sector in a given year at the national and different subnational geographic scales.

Human capital: The ability, knowledge, and skill base that are typically acquired or enhanced by an individual through education and training.

Innovation: The implementation of a new or significantly improved product (good or service) or process; a new marketing method; or a new organizational method in business practices, workplace organization, or external relations (OECD-Eurostat, 2005, p. 46). A common feature of an innovation is that it must have been *implemented*. A new or improved product is implemented when it is introduced on the market. New processes, marketing methods, or organizational methods are implemented when they are brought into actual use in a firm's operations (OECD-Eurostat, 2005, p. 47).

NLSY: National Longitudinal Survey of Youth, conducted by the Bureau of Labor Statistics.

Research and development (R&D): Comprises creative work undertaken on a systematic basis to increase the stock of knowledge, including knowledge of man, culture, and society, and the use of this stock of knowledge to devise new applications (OECD, 2002a, p. 30).

Science and technology (S&T): A broad concept that includes science and technology activities, defined as follows: "For statistical purposes, Scientific and Technological Activities (STA) can be defined as all systematic activities which are closely concerned with the generation, advancement, dissemination, and application of scientific and technical knowledge in all fields of science and technology, that is the natural sciences, engineering and technology, the medical and the agricultural sciences (NS), as well as the social sciences and humanities (SSH)" (United Nations Educational, Scientific and Cultural Organization, 1984, p. 17). Also included are Scientific and Technological Services (STS) and Scientific and Technological Education and Training (STET), the definitions of which are found in UNESCO (1978). Research and development is included in STA. The OECD *Frascati Manual* (OECD, 2002a, p. 19) notes that "R&D (defined similarly by UNESCO and the OECD) is thus to be distinguished from both STET and STS." In the *Frascati* definition, R&D includes basic research, applied research, and experimental development.

SIBS: Survey of Innovation and Business Strategy, conducted by Statistics Canada.

Statistic: A numerical fact or datum, especially one computed from a sample.

Statistical data: Data from a survey or administrative source used to produce statistics (OECD, 2002b, pp. 205-230).

Statistical indicator: A statistic, or combination of statistics, providing information on some aspect of the state of a system or of its change over time. For example, gross domestic product (GDP) provides information on the level of value added in the economy, and its change over time is an indicator of the economic state of the nation. The decline of GDP for two quarters is indicative of a recession. The ratio of gross domestic expenditure on research and development (GERD) to GDP is an indicator of the formal generation of new knowledge and is used both for international comparisons and for the setting of targets, such as the Lisbon target for the European Union of 3 percent. There are also composite indicators involving many component indicators. A single indicator is indicative but not definitive in its description of the system. As an example, GDP per capita provides one piece of information about an economy and may be indicative of wealth or productivity, but the income distribution for the country, another indicator summarized in a Gini coefficient, provides complementary information on income inequality. Employment is yet another indicator of the state of the economy.

Statistical information: Statistical data, or a statistic, placed in a context. As an example, the number of people making less than a dollar a day in a country is a statistic populated by statistical data that may result from estimation based on a sample. The context is the analysis of poverty, and in that context, the statistic provides information on poverty, but it is only one dimension.

Contents

Summary

The availability of relevant, accurate, timely, and objective information on the health of the science, technology, and innovation (STI) enterprise in the United States is critical to addressing vital policy questions for the nation. Just some of these questions are

- How is the contribution of STI to productivity, employment, and growth in the broader U.S. economy changing in a world of economic globalization?
- What are the drivers of innovation that benefit the economy and society?
- Does the United States have the STI knowledge capital it needs to move the nation forward, address its social challenges, and maintain competitiveness with other countries?
- What effect does federal expenditure on research and development (R&D) and on science and engineering education have on innovation, economic health, and social welfare, and over what time frame?
- What characteristics of industries and geographic areas facilitate productive innovation?

Since the 1950s, under congressional mandate, the U.S. National Science Foundation (NSF)—through its National Center for Science and Engineering Statistics (NCSES) and predecessor agencies—has produced regularly updated measures of R&D expenditures, employment and training in science and engineering, and other indicators of the state of U.S. science and technology. A more recent focus has been on measuring innovation in the corporate sector. NCSES collects its own data on STI activities and also incorporates data from other agencies to produce indicators that are used for monitoring purposes—including comparisons among sectors, regions, and with other countries—and for identifying trends that may require policy attention and generate research needs. NCSES also provides extensive tabulations and microdata files for in-depth analysis.

Changes in the structure of the U.S. and global economies and in NCSES's mandate and budget environment pose not only significant challenges but also opportunities for its efforts to monitor STI activities in the United States. On the challenge side, what used to be the relatively simple task of tracking domestic R&D spending by a small number of U.S. manufacturers has evolved into the need to monitor STI activities across the globe and across a wide range of industrial and commercial sectors. Similarly challenging are the increasing velocity and changing character of the innovation system. Yet another challenge is the constrained budget environment, which means that NCSES, like other federal agencies, is trying to do more with less.

Offering both challenge and opportunity is the recent broadening of NCSES's statistical mission by the America Creating Opportunities to Meaningfully Promote Excellence in Technology, Education, and Science (America COMPETES) Reauthorization Act of 2010. Section 505 of the act expanded and codified NCSES's role as a U.S. federal statistical agency charged with collecting, acquiring, analyzing, reporting, and disseminating data on R&D trends; the science and engineering workforce; U.S. competitiveness in science, technology, and R&D; and the condition and progress of U.S. science, technology, engineering, and mathematics (STEM) education. The act also charged NCSES with supporting research that uses its data and with improving its statistical methods.

Further affording both challenge and opportunity is the emergence of new types of information with which to track innovation, R&D, and the STEM workforce. Historically, statistical agencies such as NCSES have relied on sample surveys and censuses to collect consistent and unbiased information in these and other areas. In recent years, however, the amount of raw data available online has soared, creating possibilities for new STI indicators. Microdata from administrative records and other sources are increasingly being used to produce measures of capacities and trends in the global STI system. Also, frontier methods are emerging for monitoring phenomena such as the number of new product introductions through sophisticated web-scraping algorithms; for tracking innovation activities through help-wanted ads; and for tracing networks of scientists engaged in research through textual analysis of grant abstracts, patent applications, and online working papers and publications. At present, such data sources, although promising, are largely untested and hence of highly uncertain quality, which means they will require careful evaluation to determine those that may be suited for statistical use.

These new challenges and opportunities raise questions about whether NCSES's current STI statistical activities are properly focused to produce the information needed by policy makers, researchers, and businesses. Such questions have become especially acute given tightening fiscal measures and the importance of innovation for economic growth and other aspects of social well-being.

STUDY PURPOSE AND METHODOLOGY

In response to a request from NCSES, the Committee on National Statistics and the Board on Science, Technology, and Economic Policy of the National Research Council convened the Panel on Developing Science, Technology, and Innovation Indicators for the Future. The panel was charged to assess and provide recommendations regarding the need for revised, refocused, and newly developed indicators of STI activities that would enable NCSES to respond to changing policy concerns. In carrying out its charge, the panel reviewed STI indicators from the United States as well as Australia, Canada, China, India, Japan, Russia, South Korea, and several countries in Africa, Europe, and Latin America. It consulted with a broad range of users of STI indicators in the academic, public, and private sectors and with statistical agencies and organizations that produce relevant information in the United States and abroad. Although its focus was on indicators, the panel also believed it important to identify and assess both existing and potential data resources and tools that NCSES could exploit to further develop its indicators program. Finally, the panel considered strategic pathways for NCSES to move forward with an improved STI indicators program.

The panel recognized that no one model informs the types of indicators NCSES needs to produce. Policy questions for the United States served as an important guide to the panel's review, but the study was also informed by considerations of international comparability. In addition, the panel recognized the need to balance the introduction of new indicator series against the need to maintain long-standing bellwether indicators that require continual monitoring over long periods of time—for example, to ascertain the rise or decline of countries based on their technological capabilities.

KEY FINDINGS

The United States has historically played a leadership role internationally in the development of STI indicators, particularly in the areas of human capital and R&D expenditures. NCSES specifically has displayed areas of strength and innovation in its data collection and indicator development, outreach to the user community, and collaboration with statistical agencies in the United States and abroad. This report is intended to offer NCSES guidance that will help keep it at the forefront of providing accurate, relevant, timely, and objective STI indicators that can inform policy makers about the health of the STI enterprise, signal trends of policy concern, and suggest questions for in-depth research.

In reviewing STI indicators around the globe, the panel found a depth and breadth of indicator programs that is truly remarkable—many countries are placing a high priority on collecting information on innovation and related activities, and they are gathering high-quality data. Nevertheless, after hearing presentations from different countries ranging across the African, Asian, and European continents, the panel was unable to identify any proven STI indicators or methodologies used by other countries that NCSES lacks and could easily and inexpensively adopt for its own program. Given the global nature of STI, however, it is essential for NCSES to remain aware of the experimentation currently under way and to benefit from the lessons learned from these experiments.

The panel also identified a number of ways in which NCSES could improve its current STI indicators program with relatively little new investment in original data collection. Examples include increasing comparability with international classifications and concepts and improving the usefulness of the agency's Business Research and Development and Innovation Survey (BRDIS) to policy makers and researchers.

Finally, changes in the economy have made it necessary to develop new concepts and measures of STI and its economic and social impacts. Economic changes also have given rise to new methodologies for data collection and analysis that have the potential to lower costs and expand the usefulness of STI indicators while enabling NCSES to maintain and enhance its global leadership. NCSES may find it difficult to fund and supervise the development of new STI measures and methodologies, especially while continuing its current program of STI indicators. Nonetheless, continued production of only the traditional STI measures will provide an incomplete and possibly misleading indication of how well or poorly the economies of the United States and other countries are performing in generating the innovations in products, services, and production and delivery chains that lead to improved living standards.

STRATEGIC RECOMMENDATIONS

To help NCSES deal with the implications of the above findings, the panel offers five recommendations that together form a strategy for moving forward with an improved STI indicators program. The first strategic recommendation entails according priority to data quality, broadly defined, which is a principal objective of a statistical agency. The remaining recommendations offer four strategic pathways for development: working with other agencies to share data and link databases to produce indicators that would not be possible if the agencies worked independently; using existing grants and fellowship programs to support relevant methodological research; making data holdings available to researchers on a timely basis for substantive work on new and revised STI indicators, in addition to methodological advances; and establishing a position of chief analyst to link policy and research needs more effectively with analytical concepts and data collection methods.

NCSES will first need to assign relative priorities to the above four strategic pathways and then use the other 26 more specific recommendations in this report to flesh out a program of work along each pathway. NCSES has already made significant strides in each of these areas. Strengthening its program of work by incorporating the panel's recommendations should ensure that NCSES is well positioned to maintain and enhance its domestic and international leadership in the production and interpretation of STI indicators needed to inform policy.

The panel's five strategic recommendations are briefly outlined below, with examples of selected specific recommendations (numbered according to the chapter of the main text in which they appear). The panel's specific recommendations also are summarized and linked to the strategic recommendations in Chapter 8.

Data Quality (Recommendation 8-1)

The panel recommends that NCSES, as a statistical agency, place data quality at the top of its priority list. The panel defines data quality to include accuracy, relevance, timeliness, and accessibility. To make data quality central, NCSES should review its data quality framework, establish a set of quality indicators for all of its surveys, and publish the results of its review at least annually.

The overriding need with respect to data quality for NCSES's STI indicators program is for the indicators, and the data used to generate them, to accurately reflect policy-relevant dimensions of the underlying processes of science, technology, and innovation that contribute to economic growth and societal well-being. To be relevant, indicators should also be available on a timely basis and widely accessible to the data user community. A concern with quality means that indicators should be carefully evaluated for their conceptual soundness, and all feasible steps should be taken to minimize measurement error. Quality considerations should inform decisions to revise, drop, or add indicators. Indicators that are highly relevant but do not meet measurement standards may be published as experimental or research series, accompanied by clear qualifications.

Data Linkage and Sharing (Recommendation 8-2)

The panel recommends that NCSES work with other federal agencies bilaterally and in interagency statistical committees to share data, link databases where feasible, and produce products that would not be possible if the agencies worked independently. The use of data from outside the federal system, where appropriate, should be part of this process. This strategic pathway is central to enable NCSES to develop selected new policy-relevant, internationally comparable indicators that are based on existing NCSES survey data and on data collections of other statistical agencies and organizations, both within and outside the government. Selected recommendations for implementation via this pathway include the following:

- NCSES has many data series that have not been fully analyzed but have great potential to help answer questions posed by users. **Recommendation 3-1** stresses the importance of developing new STI indicators from existing data.
- In particular, existing BRDIS data should be exploited by, for example,
  - developing innovation-related tabulations from BRDIS data for comparison purposes using the same cutoffs for firm size used by other OECD countries and cross-tabulations from BRDIS data that link innovation indicators to a variety of business characteristics, including the amount of R&D spending by U.S.-based companies outside of the United States (**Recommendation 4-2**);
  - matching existing BRDIS data to ongoing surveys and administrative records at the U.S. Census Bureau and Bureau of Labor Statistics to create measures of activities by high-growth firms and of firm dynamics, such as births and deaths of businesses, linked to innovation outputs (**Recommendation 4-4**);
  - making greater use of BRDIS data to provide indicators of payments and receipts for R&D services purchased from and sold to other countries, an effort that would require continued collaboration with the U.S. Bureau of Economic Analysis on the linked dataset (**Recommendation 5-2**); and
  - developing a suite of indicators that track the development and diffusion of general-purpose technologies, including information and communication technologies, biotechnology, nanotechnology, and green technologies, using BRDIS data

as well as patent and bibliometric data (**Recommendation 5-4**).
- Better access to BRDIS data by NCSES staff is imperative for the timely distribution of new and improved BRDIS-based indicators (**Recommendation 4-3**).
- NCSES should also draw on longitudinal datasets on occupations and education levels to create indicators of labor mobility. Data from the Survey of Doctorate Recipients, the Longitudinal Employer-Household Dynamics Study, and the Baccalaureate and Beyond Longitudinal Study would be particularly useful for understanding the correspondence between supply and demand for skill sets in science and technology sectors and worker mobility (**Recommendations 6-1 and 6-2**).
- An important prerequisite for linking data from different sources is the development of a consistent taxonomy of science and engineering fields and occupations (including the health and social sciences). NCSES should engage with other statistical agencies, including but not limited to the Bureau of Labor Statistics, the U.S. Census Bureau, the National Center for Education Statistics, and the National Institutes of Health, to develop this taxonomy and to establish a process for updating it as needed (**Recommendation 2-2**).

Methodological and Substantive Research Through a Community of Practice (Recommendations 8-3 and 8-4)

The panel recommends that NCSES build a community of practice around existing and emerging methodological issues so it can update its data acquisition and analysis techniques to support new and revised STI indicators. Such a community should include not only NCSES staff but also, given constrained staff resources, outside researchers. NCSES should leverage its existing grants and fellowship programs to support methodological research that addresses its specific needs.

Relatedly, the panel recommends that NCSES make its data holdings available to external researchers on a timely basis to facilitate their research while protecting confidentiality and privacy. The timeliness with which NCSES delivers indicators to user communities depends on its own access to data resources, primarily surveys and censuses, but increasingly other sources as well, such as databases that involve text processing.

Outside researchers can help NCSES address methodological issues entailed in improving the accuracy and timeliness of new and revised STI indicators and evaluating the usefulness of new kinds of data for indicator production. External researchers can also help NCSES develop new and improved indicators that are relevant to policy on such topics as the following (**relevant recommendations are in Chapters 4 through 7**):

- organizational and market innovations, as well as innovations in training and design;
- hindrances to the innovation process;
- new measures of innovation based on business practice data obtained through administrative records and web-based data;
- knowledge networks that contribute to innovation;
- improved measures of labor mobility, career paths, stay rates for students at various levels of education, wages and salaries by skill set, and demand and supply of skill sets in various industries;
- international trade in technological goods and services;
- emerging regions for entrepreneurial activity in science and technology; and
- precommercialized inventions, to shed light on the early stages of the innovation process.

Chief Analyst (Recommendation 8-5)

The panel recommends that NCSES establish a unit to manage data quality assurance, cooperation with other organizations, and focused analysis to support the development of its indicators program and underlying datasets. Establishment of such a unit would enable NCSES to more fully embody an important practice for federal statistical agencies, which is to have an active research program on the substantive issues for which the agency compiles information and to understand how that information is used for policy and decision making (National Research Council, 2013b, p. 22). NCSES should include within this unit a new position of chief analyst, whose role would be to interface with users of NCSES data, including indicators; provide other NCSES staff with periodic updates on areas of change that may affect the agency's statistical operations; and assess the utility of new types of datasets and tools that NCSES could use either in house or by contractual arrangement.

The panel believes a chief analyst would better enable NCSES to keep up to date with changing demand for STI indicators and other data products; with socioeconomic changes that have implications for STI indicators; and with new ways of collecting, acquiring, analyzing, and disseminating the most relevant, accurate, and timely indicators for policy and research use. Through adopting this recommendation, along with the other strategic and specific recommendations in the panel's report, NCSES will be well positioned to carry out its role from the America COMPETES Reauthorization Act of 2010 to "collect, acquire, analyze, report, and disseminate statistical data related to the science and engineering enterprise in the United States and other nations that is relevant and useful to practitioners, researchers, policymakers, and the public."

1

Introduction

The National Science Foundation's National Center for Science and Engineering Statistics (NCSES) is one of 13 major statistical agencies[1] in the federal government, about half of which collect relevant information on science, technology, and innovation (STI) activities in the United States and abroad.[2] The America Creating Opportunities to Meaningfully Promote Excellence in Technology, Education, and Science (America COMPETES) Reauthorization Act of 2010 expanded and codified NCSES's role as a U.S. federal statistical agency (see Box 1-1).[3] Aspects of the agency's mandate most relevant to this study include collecting, acquiring, analyzing, reporting, and disseminating data on (1) research and development (R&D) trends; (2) the evolution of the science and engineering workforce in the United States and abroad; (3) U.S. competitiveness in science, technology, and R&D; and (4) the condition and progress of U.S. science, technology, engineering, and mathematics (STEM) education.[4] NCSES is also explicitly charged with supporting research on the data it collects and its methodologies for data collection, analysis, and dissemination.

NCSES's day-to-day activities serve two main functions: data curation and the provision of statistics, including indicators that distill underlying data into metrics that track inputs, processes, outputs, and outcomes of interest to policy analysts and others. Clients worldwide look to NCSES for basic measurements of STI activities, as well as aggregations, blended statistics, or indices designed to inform specific issues. NCSES's clientele is diverse. Some use a wide range of its data on a variety of STI actors and activities, while others require more processed, analytically salient statistics derived from those data. Users of NCSES data and statistics include staff of other statistical agencies in the United States and abroad; local, state, federal, and international policy makers and government administrators; academic researchers and administrators; and, to a much lesser extent, managers of businesses, consultants, and nonprofit organizations.[5] As it works to meet these commitments, NCSES also is focused on deepening its core competencies in anticipation of future demands. As it further develops its indicators program, the agency must in particular consider the synergies between its R&D and human resources statistics programs.

The purpose of this report is to make recommendations for what STI indicators NCSES should produce in the near term and over time, and what process changes or innovations are needed to help the agency produce high-utility indicators in a timelier fashion.

[1]The Office of Management and Budget-chaired Interagency Council on Statistical Policy has 14 members, 13 of which are recognized as major statistical agencies. See the full listing of Principal Statistical Agencies in Federal Statistics (2007).

[2]Agencies collecting such information include six statistical agencies—the Bureau of Economic Analysis, Bureau of Labor Statistics, Census Bureau, Economic Research Service (U.S. Department of Agriculture), National Center for Education Statistics, and NCSES—and also the National Institute of Standards and Technology (U.S. Department of Commerce).

[3]The act also gave the agency its new name; it was previously called the Science Resources Statistics Division.

[4]In this report, STEM occupations include engineers, mathematical and computer scientists, life scientists, physical scientists, social scientists, technicians, programmers, and science and engineering managers, but not health care practitioners and technicians. This is the definition of STEM occupations used by NCSES (see Regets, 2010).

[5]". . . NCSES has responsibility for statistics about the science and engineering enterprise. NCSES designs, supports, and directs a coordinated collection of periodic national surveys and performs a variety of other data collections and research, providing policymakers, researchers, and other decision makers with high quality data and analysis on R&D, innovation, the education of scientists and engineers, and the S&E workforce. The work of NCSES involves survey development, methodological and quality improvement efforts, data collection, analysis, information compilation, dissemination, web access, and customer service to meet the statistical and analytical needs of a diverse user community. It also prepares two congressionally mandated biennial reports—*Science and Engineering Indicators (SEI)* and *Women, Minorities, and Persons with Disabilities in Science and Engineering*. The data collected by NCSES also serve as important tools for researchers in SBE's Science of Science and Innovation Policy (SciSIP) Program" (National Science Foundation, 2012a).

BOX 1-1
Role of the National Center for Science and Engineering Statistics Under the America COMPETES Reauthorization Act

SEC. 505. NATIONAL CENTER FOR SCIENCE AND ENGINEERING STATISTICS.

(a) Establishment—There is established within the Foundation a National Center for Science and Engineering Statistics that shall serve as a central Federal clearinghouse for the collection, interpretation, analysis, and dissemination of objective data on science, engineering, technology, and research and development.

(b) Duties—In carrying out subsection (a) of this section, the Director, acting through the Center shall—

(1) collect, acquire, analyze, report, and disseminate statistical data related to the science and engineering enterprise in the United States and other nations that is relevant and useful to practitioners, researchers, policymakers, and the public, including statistical data on—

(A) research and development trends;

(B) the science and engineering workforce;

(C) United States competitiveness in science, engineering, technology, and research and development; and

D) the condition and progress of United States STEM education;

(2) support research using the data it collects, and on methodologies in areas related to the work of the Center; and

(3) support the education and training of researchers in the use of large-scale, nationally representative datasets.

(c) Statistical Reports—The Director or the National Science Board, acting through the Center, shall issue regular, and as necessary, special statistical reports on topics related to the national and international science and engineering enterprise such as the biennial report required by section 4(j)(1) of the National Science Foundation Act of 1950 (42 U.S.C. 1863(j)(1)) on indicators of the state of science and engineering in the United States.

CHALLENGES

In carrying out the mandate outlined above, NCSES confronts three key challenges: (1) the increasing complexity of STI activities, including the rapid technological changes that continue to reshape product markets and demand for human resources; (2) the globalization of technology and economic activities; and (3) the budget constraints under which NCSES (along with most other federal agencies) must operate.

Rapid technological change is a healthy feature of modern economies, and indeed decades of economic research have confirmed that it is the dominant driver of growth in gross domestic product (GDP) over the long run for developed countries such as the United States (Barro and Sala-I-Martin, 1997; Romer, 1990; Solow, 1956, 1957). At the same time, the pace of change today in areas such as information and communication technologies, medical sciences, materials science and engineering, and the organization of educational institutions present fundamental problems for government statistical agencies charged with measuring innovation activities. Although certain inputs into the innovation process can be counted using traditional methods,[6] continued rapid change makes it difficult to capture well the nature of innovation *outcomes,* as well as many specific characteristics of the nation's innovation ecosystem (or ecosystems, because innovation in particular industries often tends to cluster in certain geographic regions). Particularly challenging as well is to use traditional methods to develop metrics that link changes in inputs to changes in outputs and societal impacts, all as institutions are evolving driven in part by the same underlying forces. Yet these are the types of questions that users are looking to answer with NCSES's STI indicators.

At the same time, NCSES must deal with an STI system that is globalizing at breathtaking speed. Virtually all companies with major domestic R&D spending also have growing R&D operations in foreign countries, especially emerging economies such as Brazil, China, and India or oil-rich countries such as the United Arab Emirates (see, e.g., MADAR Research and Development [2006] and the chapter on "The Strategic Vision for 2030" in *The Abu Dhabi Economic Vision 2030* [The Government of Abu Dhabi, 2008]). The same is increasingly true for major domestic research universities, which are reaching out to create research centers around the world.[7] Research projects increasingly are the result of cross-border collaboration, accomplished either in person or by electronic connection. In addition, the delivery of education and training in science and engineering fields increasingly is distributed geographically via electronic media—as illustrated by the emergence of massive open online courses (MOOCs)—supplementing or substituting for traditional, local, bricks-and-mortar classroom facilities.

The recent economic crisis, coupled with the high cost of innovating at the technological frontier, has led to the growing adoption of "open" innovation strategies based on enhanced collaboration, sharing, and acquisition of knowledge from external sources (Chesbrough et al., 2008). The emergence of global networks of innovation, as well as global markets for knowledge as contract R&D and the buying and selling of intellectual property become more prevalent, has further fueled globalization.

The combination of rapid technological change and globalization of the innovation ecosystem is placing increased stress on traditional STI indicators, such as expenditures

[6]For example, R&D expenditures, patents, and broad measures of human capital (such as undergraduate and graduate degrees awarded in science and technology, as well as on-the-job training).

[7]Stanford, for example, has opened a new research center at Peking University; see http://news.stanford.edu/news/2011/march/scpku-china-peking-033111.html [January 2014]. For a broad overview of the globalization of universities, see Wildavsky (2010).

on research and development without linking expenditures to innovative outputs. Continued use of the traditional STI measures will give an incomplete and possibly misleading indication of how well or poorly the U.S. economy and other economies are doing in generating the improvements in products, services, and processes that lead to improved living standards. This stress is exacerbated by the budget constraints faced by federal agencies in fiscal year (FY) 2013 and beyond. The challenges posed by NCSES's broadening of responsibilities and constrained budget resources were reflected in the FY 2013 budget request to Congress for the agency (National Science Foundation, 2012a, p. SBE-3):

> For FY 2013, NCSES will accelerate efforts to rely more heavily on data from the National Survey of College Graduates, built from the American Community Survey, which will reduce overall survey costs while still continuing to meet the needs of policy makers, researchers, and the general public for data on the overall science and engineering workforce. NCSES will develop plans for a project to utilize federal agency administrative records to measure research and development activity and to explore new methods to enhance data collection, analysis and data sharing capabilities better to serve all its customers interested in the science and engineering enterprise.

Adding to the challenges outlined above are rapid changes in the ways statistical information is obtained, data are generated, and statistics are estimated. In his July 2012 presentation to the Senate Subcommittee on Federal Financial Management, Government Information, Federal Services, and International Security,[8] Robert Groves identified these challenges and suggested solutions that the U.S. Census Bureau and other U.S. statistical agencies should consider: (1) engage university researchers and technology firms in the development of more efficient processes for data extraction and manipulation; (2) use new data sources (including "organic data" and data from quick-response surveys and administrative records) to complement traditional survey-based data; and (3) link or bridge datasets across federal agencies to produce a richer set of timely, high-quality statistics at reduced cost. These solutions may apply to NCSES as it confronts the challenges it faces today.

CHARGE TO THE PANEL

NCSES has a global reputation for excellence in collecting and reporting STI-related information. Essential indicators are available in the National Science Board's biennial volume *Science and Engineering Indicators*, in *InfoBriefs* on a wide variety of topics, in statistics published in the Scientists and Engineers Statistical Data System and *National Patterns of R&D Resources, Business R&D and Innovation: 2008-2010* (NSF 13-332), and in statistics available on WebCASPAR and IRIS. Nonetheless, NCSES recognizes the challenges described above, and in July 2010 reached out to the National Academies to request a study that would lead to recommendations for revised, refocused, and new STI indicators reflecting the fundamental and rapid changes that have occurred in the domestic and global STI system.[9] In response to that request, the Committee on National Statistics of the National Research Council, in collaboration with the Board on Science, Technology, and Economic Policy, convened the Panel on Developing Science, Technology, and Innovation Indicators for the Future. The panel's detailed statement of task is presented in Box 1-2.

NCSES sought the panel's assessment of the types of data, metrics, and indicators that would be particularly influential in evidentiary policy and decision making for the long term. NCSES also charged the panel with recommending indicators that would reflect the fundamental and rapid changes in the global STI system while having practical resonance with a broad base of users in the near, medium, and long terms.

ORGANIZING FRAMEWORK

Given the panel's charge and the backdrop of a changing data environment, three principles guided this study.

First, the panel endeavored to map existing STI indicators against the known and anticipated high-level issues the indicators are used to address—a *policy-driven framework*. For example, the current set of indicators could be improved to better assist policy makers in making basic science and technology funding decisions, such as the mix of government support to business R&D, or in understanding the position of the United States relative to that of various other nations in the production of a STEM workforce. STI indicators can provide early warning that a critical investment or activity is waning (or succeeding) or that important advances are being made overseas. STI indicators also offer useful measures of total factor productivity, a key element of the national income and product accounts. NCSES and other international organizations[10] produce a vast array of STI indicators, which

---

[8]Available: http://www.hsgac.senate.gov/subcommittees/federal-financial-management/hearings/census-planning-ahead-for-2020 [October 2012].

[9]To some degree, this study can be seen as a complement to three previous studies of NCSES programs by the National Research Council's Committee on National Statistics: *Measuring Research and Development Expenditures in the U.S. Economy* (National Research Council, 2005), which focuses on survey redesign to improve measurement of R&D and innovation; *Data on Federal Research and Development Investments: A Pathway to Modernization* (National Research Council, 2010), which addresses improvements to surveys of federal funds for R&D and federal science and engineering support for universities, colleges, and nonprofit institutions; and *Communicating Science and Engineering Data in the Information Age* (National Research Council, 2012), which highlights data dissemination techniques using Internet-based tools.

[10]Other organizations include but are not limited to: Eurostat; OECD; Statistics Canada; and United Nations Educational, Scientific and Cultural Organization (UNESCO).

BOX 1-2
Statement of Task

An ad hoc panel, convened under the Committee on National Statistics, in collaboration with the Board on Science, Technology, and Economic Policy, proposes to conduct a study of the status of the science, technology, and innovation indicators (STI) that are currently developed and published by the National Science Foundation's National Center for Science and Engineering Statistics (NCSES). Specifically, the panel will

1. Assess and provide recommendations regarding the need for revised, refocused, and newly developed indicators designed to better reflect fundamental and rapid changes that are reshaping global science, technology and innovation systems.
2. Address indicators development by NCSES in its role as a U.S. federal statistical agency charged with providing balanced, policy relevant but policy-neutral information to the President, federal executive agencies, the National Science Board, the Congress, and the public.
3. Assess the utility of STI indicators currently used or under development in the United States and by other governments and international organizations.
4. Develop a priority ordering for refining, making more internationally comparable, or developing a set of new STI indicators on which NCSES should focus, along with a discussion of the rationale for the assigned priorities.
5. Determine the international scope of STI indicators and the need for developing new indicators that measure developments in innovative activities in the United States and abroad.
6. Offer foresight on the types of data, metrics and indicators that will be particularly influential in evidentiary policy decision-making for years to come. The forward-looking aspect of this study is paramount.
7. Produce an interim report at the end of the first year of the study indicating its approach to reviewing the needs and priorities for STI indicators and a final report at the end of the study with conclusions and recommendations.

can be classified into the following subtopics: human capital (including education and workforce statistics); R&D (including business, academic, and nonprofit expenditures on R&D and government intramural and extramural appropriations or outlays on R&D); outputs (including patents, articles and citations, and innovation); technology balance of payments and trade in R&D-intensive industries; venture capital in science and technology sectors; and public attitudes on science and technology. One often used indicator of a country's preeminence in STI activities is the ratio of R&D to GDP. A full discussion of the sources and types of STI indicators is included in Chapter 3 and Appendix F of this report.

The second principle guiding this study was the priority of *international comparability* for key STI indicators. Indeed, policy relevance and international comparability were, to a large extent, complementary elements of the panel's focus. Discussions with U.S. policy makers and international experts on STI indicators throughout this study made clear the similarities in the types of policy questions asked worldwide and the interest in having internationally comparable data at the national and subnational scales. This principle also comports with NCSES's mandate to produce statistics on U.S. competitiveness in science, engineering, technology, and R&D.

The third key principle was *efficiency*, which entailed identifying those STI indicators that are essential and those that can be eliminated. NCSES requested that the panel deliver a prioritized list of STI indicators. The panel interpreted this request as necessitating efficiency gains that would require limiting the number of measures that can be considered key national STI indicators and establishing time- and resource-saving processes for creating those measures. The panel also used this opportunity to recommend approaches that NCSES could use to develop new or reinterpret existing indicators. Moreover, because STI indicators draw on data from the R&D and human resources statistics programs at NCSES, the panel identified synergies among all three programs as prospects for efficiency gains.

STUDY APPROACH

The panel's approach to this study comprised five components (see Figure 1-1).

First, the panel consulted users, experts, and written reports and peer-reviewed articles to establish current and anticipated user needs for STI indicators. Users interviewed included policy makers, government and academic administrators, researchers, and corporate decision makers in high-tech manufacturing and service industries. The panel also sought input from developers of STI indicators and from individuals who are called on by policy makers and other decision makers to conduct assessments of high-tech sectors in the United States and abroad. To this end, the panel held two workshops, one at the National Academies in Washington, DC, July 2011, and the other in Paris following the meeting of the OECD-National Experts on Science and Technology Indicators (NESTI) Working Group in June 2012. These discussions collectively yielded a wealth of information on policy questions and specific measures considered of high priority by users of STI data and indicators.[11]

[11]See Appendix B for the full list of users and providers of STI indicators who informed the panel about key measures that NCSES should develop or continue to produce. That appendix also includes an updated list of policy issues published in the panel's interim report.

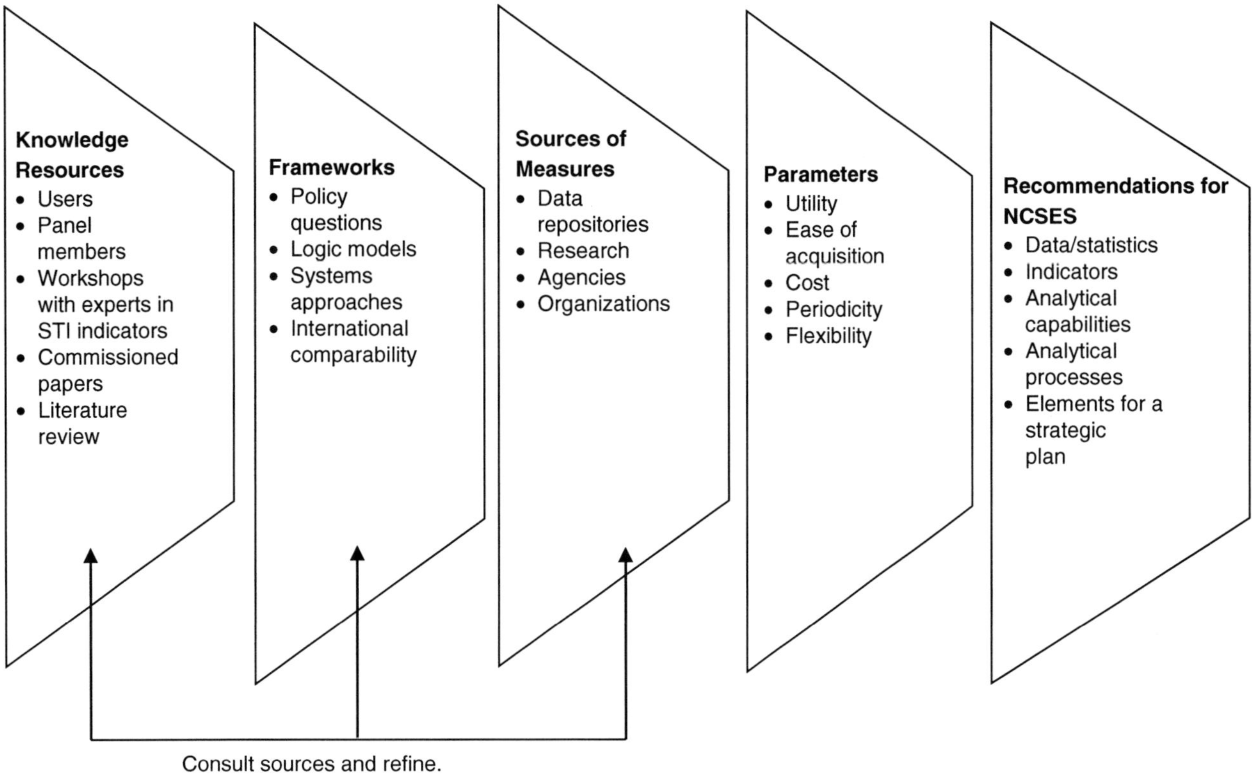

FIGURE 1-1 Panel's organizing framework.

Second, recognizing that no one model informs the types of indicators NCSES needs to produce, the panel looked to policy questions, logic models, systems approaches, and international comparability to refine the suite of indicators it would recommend as priorities. Although policy relevance is an important principle for prioritizing STI indicators, the panel notes that an indicator rarely can answer a policy question in and of itself, given that policy issues are usually highly complex and depend on multiple, interrelated factors.[12] At the same time, indicators have a major role in informing policy debates.

Third, it was important to identify data resources and tools that NCSES could use to further develop its indicators program. Understanding the network of inputs—including data from NCSES surveys, other federal agencies, international organizations, and the private sector—that currently and should in the future feed into the production of indicators was important for this study.

Fourth, the panel applied several parameters in refining its recommendations to NCSES. Primary among these was utility but also considered were ease of acquisition, cost, periodicity, and flexibility.

Fifth, the panel's recommendations were not limited to indicators, but also encompassed processes for prioritizing data development and the production of indicators in the future, because it was clear that the changing environment in which the agency operates is a key determinant of its priorities from year to year. Internal processes that are observant, networked, and statistically and analytically balanced are important to NCSES's indicators program. The panel also outlined elements of a prioritization strategy for producing timely, reliable STI indicators.

NCSES did not ask the panel to recommend new survey designs. It also did not request the panel's attention to the formulation of a much-needed updated data taxonomy or to the development of theoretical foundations of measurement

[12]In a paper commissioned for this study, Hall and Jaffe (2012, p. 39) make this point as well: "In our view, indicators and other data will never, in and of themselves, provide answers to questions of this kind. The questions posed are, for the most part, research questions. Answering them requires not just data but modeling and analysis. Indicators and data more generally are necessary inputs to the research process that can provide answers to these questions, but they do not provide the answers without that analysis. . . . There is one category of normative questions that does, in principle, relate to indicators, and that is the allocation of public resources across different disciplines or areas of research."

for indicators derived from web sources or administrative records.[13] These are excellent topics for companion studies.

REPORT ORGANIZATION

Chapter 2 of this report establishes the systems framework that STI indicators are designed to inform and then identifies the types of measures that users collectively indicated to the panel are essential or of high priority. Chapter 3 reviews NCSES's current suite of indicators and those indicators the agency should improve or create to meet high-priority user needs. Chapters 4-6 are the core chapters of the report, focusing, respectively, on the panel's recommendations for measuring innovation, knowledge capital and networks, and human capital. These chapters, therefore, cover innovation activities and related outcomes extensively. Other measures, related to actors and various linkages in the STI system, as well as to market conditions, culture, and socioeconomic impacts, span these three core areas and are taken up in Chapters 4-6 where appropriate. Chapter 7 expands on frontier methods for data extraction and manipulation that are introduced in the three core chapters. In Chapters 2-7, the panel's recommendations on the respective topics are presented at appropriate points in the text. Finally, Chapter 8 presents the panel's recommendations for processes NCSES could use to make strategic decisions about data acquisition for indicators and methods for indicator development, and links these processes to the recommendations in Chapters 2-7 to form a strategic program of work.

Eleven appendixes follow the main chapters of the report. In brief biographical sketches, Appendix A conveys the expertise of panel members conducting this study. Because the organizing framework for prioritizing STI indictors was informed by user needs, Appendix B goes in depth into questions posed by users and specific measures they would like to see NCSES produce in the future. Workshops were one means by which the panel gleaned information on demand for STI indicators and the types of measures that are found most useful in various countries at the national level and at smaller geographic scales. Appendixes C and D present the agendas and lists of participants for the panel's two internationally focused workshops: the Workshop on Developing Science, Technology, and Innovation Indicators for the Future, which was held in Washington, DC, July 2011, and the OECD-NESTI Workshop, which was held in Paris, June 2012. The next five appendixes highlight the research- and data-driven functions of NCSES's indicators program. Appendix E gives abstracts of grants funded by NCSES's Grants and Fellowships program. Appendix F goes into depth on STI indicators produced by NCSES and other organizations around the world; this appendix includes background material for Chapter 3 of the report. Appendixes G through K show statistical measures derived from NCSES's Business Research and Development and Innovation Survey (BRDIS), which are important for the analysis of innovation activities reported by the indicators program. Appendixes B through K are all referenced in the main text of the report where relevant.

[13]It should be noted that *Science and Engineering Indicators 2012* has an appendix titled "Methodology and Statistics" that describes data sources and potential biases and errors inherent in the data collected (National Science Board, 2012a). OECD-Eurostat (2005, pp. 22-23), in the *Oslo Manual*, also has an extensive published methodology on the development of indicators. In addition, a previous study of the National Research Council (2005, pp. 144-151) has a chapter on sampling and measurement errors entailed in surveys and the use of federal administrative records. The economics literature is also replete with articles examining the existence of and correction for measurement errors that occur owing to survey methods (see, e.g., Adams and Griliches, 1996; Griliches, 1974, 1986). Further study of measurement theory as it relates to STI indicators is beyond the scope of this study. As detailed later in this report, however, the panel recommends that NCSES fund research to explore biases introduced in data that are gathered using web tools or administrative records.

2

Concepts and Uses of Indicators

The purpose of this chapter is to introduce the concept of "indicators" as distinct from raw data and basic science and engineering statistics. It is also useful for the reader to understand how the production of science, technology, and innovation (STI) indicators is informed by the precepts of conceptual or logic models that attempt to reflect the actors, actions, dynamics, systems, and resulting outcomes that analysts try to capture. Entailed in this process are assumptions, reasoned inferences, and a "black box" (see Rosenberg, 1982) where tangible and intangible inputs become measurable outputs.

One difficulty encountered in constructing the STI indicators users want is that users are not monolithic. The specific types of indicators users need depend on the types of decisions they must make and the networks to which they belong. User diversity was therefore an important consideration as the panel deliberated on which indicators the National Center for Science and Engineering Statistics (NCSES) should produce in the future. Also considered was the expected diversity of the future user base—for instance, the potential for more business users and more users who must make decisions in regional contexts within the United States and abroad.

At the same time, however, all users want reliable values and to varying degrees wish to have the "black box" mechanisms exposed and detailed to the extent possible. As discussed in Chapter 1, users of STI indicators share the need for high-quality, accessible, and timely observations on the rapidly changing global STI system. In addition, they expect those measures to be based on fundamentals and not merely on ad hoc relationships.

After defining the term "indicators" for this volume, the main task of this chapter is to demonstrate the utility of STI indicators, specifying those that address specific policy issues. In the process, the chapter establishes the characteristics of and user priorities for these indicators, many of which are already satisfied by NCSES's publications and data tables (as discussed in Chapter 3).

DESIRABLE ATTRIBUTES OF INDICATORS

Generally, indicators point toward or foreshadow trends, turning point patterns, expectations, and intentions. They are often things one should know about issues of interest to a variety of users. Indicators have analytical qualities such that they typically go beyond raw data. As a result, they are usually rough proxies for activities that are difficult to observe or measure directly. They are like baseball statistics: a single statistic is unlikely to tell the whole story; instead, users often rely on a collection or suite of indicators. Furthermore, indicators should not be used in isolation; they require contextual information to be useful. Indicators can be composite indices of other statistics, designed to smooth out volatility in contributing factors. Indicators also provide input for the construction of econometric models used to evaluate the key determinants in systems and guide policy development.

Most familiar indicators are related to the weather or the economy. For example, The Conference Board publishes leading, coincident, and lagging economic indicators. The index of leading economic indicators comprises 10 individual measures, each of which is a leading indicator. These leading indicators are designed to signal coming peaks and troughs in the economic business cycle. Leading indicators inform prospective analyses, while coincident and lagging indicators facilitate contemporaneous or retrospective analyses.

This report focuses specifically on statistical indicators of STI activities—their composition, uses, and limitations—and hence the statistical measurement of activities that fall within the mandate of NCSES. To discuss measurement, the report defines a statistical indicator as a statistic, or combination of statistics, providing information on some aspect of the state or trends of STI activities. International comparability of these indicators is an important quality because it provides a benchmark against which to judge the performance of one system relative to others. STI indicators often substitute for direct measures of knowledge creation, invention, innovation, technological diffusion, and science and engineering talent, which would be difficult if not impossible to obtain. For example, economic growth in a given nation is linked to the ingenuity of residents in science and engineering sectors. Since it is difficult to measure ingenuity directly, proximate measures that are more readily observed are used, such as numbers of master's or Ph.D. degrees produced in a given nation within a given time period. Following trends in the number of degrees also enables users of indicators to develop projections of future economic growth. Notably, these indicators do not precisely measure ingenuity, but they are arguably reasonable proxies. Techniques for obtaining data that directly measure innovation activities are improving, however,[1] and these data are already being used to complement indicators derived with traditional methods.

Some indicators—those derived from modeling—can answer certain policy questions. Indicators also can reveal potential issues that require exploring, for example, the impact and cost-effectiveness of research and development (R&D) tax credits in stimulating incremental business R&D. Moreover, indicators can help refine and perhaps usefully narrow the policy question being asked. For example, is it innovation by large businesses or small, young firms that yields faster and more lucrative breakthroughs?

A comprehensive review of the use of STI indicators for policy decisions is provided by Gault (2010), who outlines four ways indicators are used for policy purposes: monitoring, benchmarking, evaluating, and forecasting or "foresighting":[2]

1. monitoring—the international innovation system, linkages within and between national innovation systems, regional innovation systems and industrial clusters, the implementation of national science and technology (S&T) projects, the selected quantitative indicators in the S&T development goals;
2. benchmarking—international and interprovincial (or interstate) benchmarking;
3. evaluating—the performance of public investment in S&T, the performance of government research institutes and national laboratories, national S&T programs, specialization of S&T fields, advantages versus disadvantages, emerging industries (e.g., information technology, biotechnology, energy, health, knowledge-based services); and
4. forecasting—the latest data not available in gathered statistics.

These categories are widely accepted as functional characteristics of STI indicators. For instance, at the panel's July 2011 workshop, Changlin Gao reported that they are being used by China to target its STI indicators program.

At the same workshop, several other presenters suggested attributes that NCSES should keep in mind as it develops new STI indicators and improves existing indicators. One such attribute is low sensitivity to manipulation. During the workshop, Hugo Hollanders of UNU-MERIT[3] stated that composite indices have both political and media appeal,[4] although caution is essential in interpreting such indices, which may be readily understood but may not be adequate for conveying complex information. Other desirable characteristics of indicators mentioned by workshop participants included being scientifically derived and evidence based, comparable across regions, powerful for communication, affordable, accessible, scalable, sustainable, and policy and analytically relevant. STI indicators also should be policy neutral, even though the particular indicators selected may reflect the preferences of the stakeholders who request them.

IN SEARCH OF A FRAMEWORK

During its deliberations, the panel encountered several challenges in seeking a single framework or model for the STI system.

First, all the known elements of the STI system are not necessarily measurable; some are "unknown knowns." Because poor measurement often leads to poor decision making, recommending the development of new but perhaps mismeasured indicators could leave users worse off than they are now.

Second, although linkages among the elements in a representation of the STI system are important to measure, such linkages often are difficult to identify quantitatively. In these instances, case studies or qualitative representations may be preferable to indicators. Moreover, spillover effects—for example, when two or more elements in the system contribute to synergistic or configural outcomes—are difficult to disentangle, and developing indicators that measure such effects is therefore a difficult task. Thus, linkages and spillovers often are "unknown unknowns"; that is, developing reliable indicators of these important components of the STI

[1]See Eurostat's Community Innovation Statistics in European Commission (2010) and NCSES's Business Research and Development and Innovation Survey statistics in U.S. Department of Commerce (2011).

[2]Wehrmeyer and colleagues (2002) give extensive definitions of foresighting as the term is used differently in business consulting and in government decision-making practices. Citing Coates (1985, p. 30), Wehrmeyer and colleagues give the generic definition of foresighting as follows: "Foresighting is a process by which one comes to a fuller understanding of the forces shaping the long-term future which should be taken into account in policy formulation, planning and decision-making. . . . Foresight involves qualitative and quantitative means for monitoring clues and indicators of evolving trends and developments and is best and most useful when directly linked to the analysis of policy implications."

[3]UNU-MERIT—the United Nations University Maastricht Economic and Social Research Institute on Innovation and Technology—is a research and training center of the United Nations University and works in close collaboration with the University of Maastricht.

[4]To clarify, the panel is not advocating that NCSES develop one composite indicator, or as it is often termed, a "headline indicator." A suite of key STI indicators should be more informative for users.

system is difficult. For example, developing valid measures of intangible assets is problematic precisely because they are typically intermediate inputs with realized values that depend on the values of other outputs over some time horizon.

Third, models of the STI system or its components are plentiful and typically are shaped by the user's goal (see, e.g., Cornell University, INSEAD, World Intellectual Property Organization, 2013, p. 6; Crépon et al., 1998; Department of Business Innovation and Skills, 2011, p. 30; European Union, 2013, p. 4; Griliches, 1998, pp. 17-45; Hall and Jaffe, 2012; Hall et al., 2010; National Science Board, 2012c, p. 3; OECD, 2011; Sampat and Lichtenberg, 2011; Shanks and Zheng, 2006, pp. 105 and 288; Tassey, 2011). For example, some models list elements that have been shown to matter, either alone or in combination with other elements, to generate new ideas, products, processes, and other outcomes of the STI system, while other models are functional, in that the stocks (boxes) and flows (arrows connecting the boxes) represent estimable elements. Many of these models identify the same key elements that should be measured (or at least assessed), while offering added dimensionality depending on the model's utility.

Economists, policy analysts, information scientists, material scientists, physicists, statisticians, and geographers (all represented on the panel) have different predilections for how to develop a representative model of the STI system. The identification of one common model by the panel was unlikely and could have appeared to be ad hoc or arbitrary. Therefore, instead of choosing a single model for the STI system, the panel used aspects of several models to inform its decisions about which elements of the system are most important to assess. Box 2-1 shows seven of the "models" that informed the panel's guiding framework of key STI indicators that NCSES should produce. Since its charge was to focus on identifying policy-relevant, internationally comparable STI indicators, the panel also decided to use a policy-driven approach. This approach was informed by the published work of leading academicians and practitioners who map the STI system, as well as experiences in the international S&T policy arena. The resulting policy-driven framework, depicted in Figure 2-1, identifies key activities that should be measured, as well as the links among these activities and the actors and outcomes in the STI system.

A POLICY-DRIVEN FRAMEWORK

The panel's policy-driven framework provides a useful rubric for identifying the key policy issues and the indicators that can support analysis of these issues. These issues can range from highly aggregate (e.g., What is the contribution of STI to growth?) to highly granular (e.g., What is the supply of individuals with science, technology, engineering, and mathematics [STEM] skills by gender and ethnicity?). The issues tend to change over time (e.g., geographic interest has shifted from Japan to China, while sectoral interest has shifted from space to nanotechnologies). In some cases, the indicators needed to examine these issues are quite advanced, in other cases they are being developed, and in still other cases they are still in an embryonic state and require that NCSES partner with other organizations in their development. In nearly all cases, indicators offer only partial insight into the issue; gaining insight into the key determinants often requires empirical analysis involving econometric or growth accounting techniques or qualitative analysis that makes use of stylized facts or case studies. In any event, high-quality, policy-relevant data are needed to construct the indicators, support the econometric analysis, and create the stylized facts.

Policy makers, policy analysts, and the greater user community have an almost inexhaustible supply of questions they would like to have indicators to inform. Statistical agencies therefore are challenged as they seek to address user demands within the practical limits of available resources and expertise. With this tension in mind, the panel sought to populate its framework with a set of policy questions it believes are enduring and can serve as part of a strategic plan going forward.

As shown in Figure 2-1, the key question on which almost all users want bedrock statistics is: What are the social returns to public and private expenditures on STI activities? The follow-on question is: Given expenditures on STI activities, what is the impact on economic growth, competitiveness,[5] and jobs? These questions are nuanced in several ways. Users want to know the drivers of innovation that could be encouraged through funding mechanisms and creative organizational structures. For example, indicators are sought not only for static measures of stocks of human capital, but also for trends as to which countries will be generating the most scientific research that can be commercialized or which countries are likely to attract the most R&D investments in the near future. Users have questions about advances in science on the horizon or vulnerabilities in the innovation ecosystem that could impede the commercialization of new ideas. They want quantitative measures to inform these questions, but they also need stories or case studies to provide a full understanding of the issues. Users are interested in the most fertile organizational structures or networks that foster creativity and the transfer of technology from bench to market. They also are interested in the nature of cooperative relationships that foster collaboration while protecting intellectual property rights and downstream profits and mitigating risks. Distributional questions are

[5]The term "competitiveness" as used here denotes relative standing. Users of STI indicators often want to know the U.S. position relative to other nations on factors that are critical to U.S. preeminence in STI outcomes. Users are also interested in the standing of certain demographic groups and economic regions vis-à-vis other groups and geographic regions, respectively. The term "competitiveness" here does not relate to low cost or high profitability (as it often does in economics), and it does not necessarily have a strategic basis (as it does in the business literature).

BOX 2-1
Conceptual and Functional Models of the Science, Technology, and Innovation System (Synthesized by the Panel to Create Figure 2-1)

Innovation Systems: The National Science Board has used a systems model to illustrate what its data and statistics attempt to measure. The purpose of this diagram (see Figure Box 2-1A) is to show key elements in the global innovation system and the relationships between elements in the system. As important as it is to measure variables in the boxes or to develop scenarios that explain those elements, it is as important to measure or explain the linkages (arrows) between the boxes. This diagram has several "black boxes" or "assumptions" that require further explanation, and government expenditures on research and development (R&D) at universities and firms, and public sector R&D are not explicitly shown in this diagram.

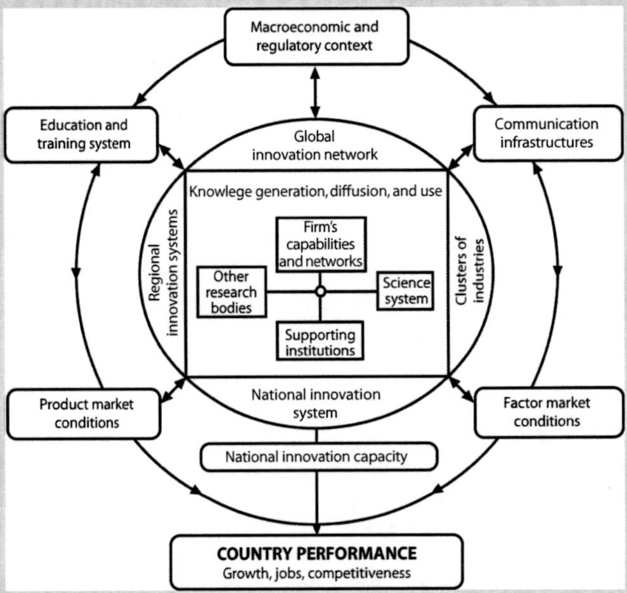

FIGURE BOX 2-1A
SOURCE: National Science Board (2012b).

Knowledge Capital: With a focus on measuring innovation and knowledge assets, Cornell University, INSEAD, and the World Intellectual Property Organization collaborated on a representation of the innovation system (see Figure Box 2-1B). The European Commission framework has similar elements, with human and financial capital inputs, linkages and employment outputs specifically identified (see Figure Box 2-1C). Together these frameworks capture many important dimensions of the STI system.

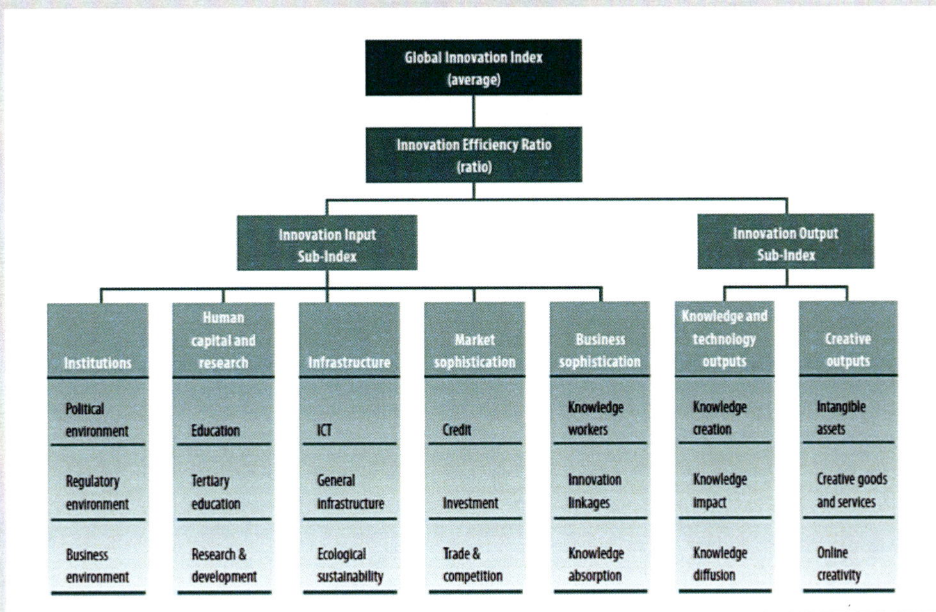

FIGURE BOX 2-1B
SOURCE: Cornell University, INSEAD, and World Intellectual Property Organization (2013).

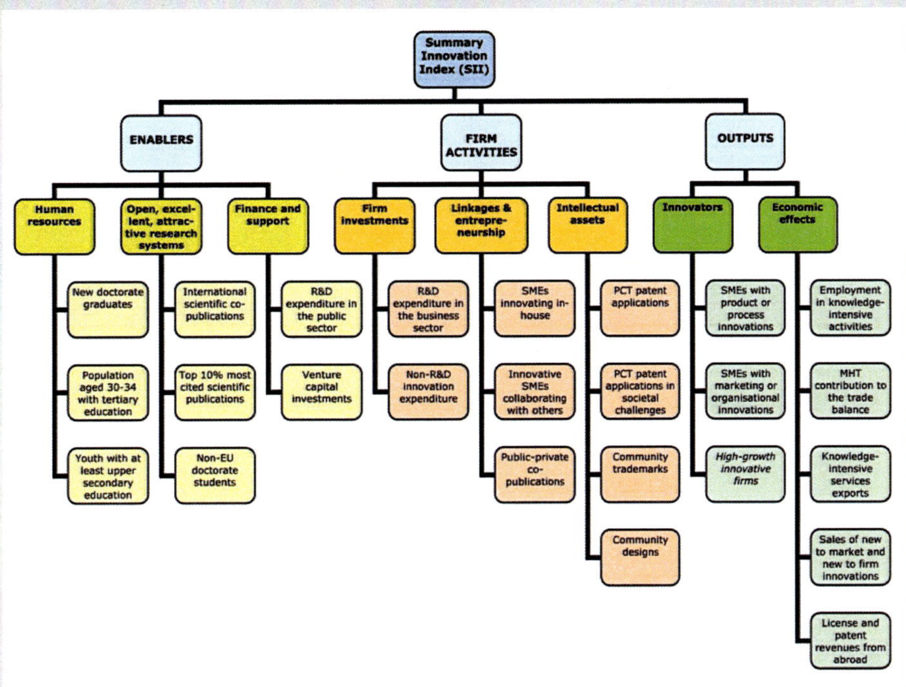

FIGURE BOX 2-1C
SOURCE: European Union (2013).

BOX 2-1 Continued

Return on Expenditure: Growth accounting models are also used to describe the STI indicators system. STI indicators are often used to relate knowledge inputs to outputs, outcomes, or impacts. At a very basic level, knowledge inputs include years of schooling, level of degree, and the amount of training an employee receives on the job. Outputs are specific products, processes, or services. Outcomes and impacts are the near-term and long-term effects and ramifications to the economy or society in which the technological ecosystem operates. Productivity and returns on expenditures are often used to measure economic outcomes of STI activities. Other social outcomes, such as improvements in health outcomes associated with cleaner water or more effective medical treatments, are important to assess. For example, scientific advancement in detecting and removal of pathogenic microorganisms leads to technological mechanisms that in turn lead to cleaner water, thereby increasing productivity (through a healthier workforce) and hence increasing inputs in the production of goods and services, as well as increased welfare of citizens. Indicators are relied on for both post-activity evaluations and analysis prior to an activity, although there are major limitations in using STI indicators for predictive exercises. [See Abramovitz (1956); Carson et al. (1994); Fraumeni and Okubo (2005); Jorgenson and Griliches (1967); Solow (1957).] Other models focus on returns to R&D, where the framework is similar to the traditional production function/total factor productivity model. It is important to note that the second diagram calls out obsolescence of knowledge, making it important to measure not only the depreciation of capital stock but also the depreciation of knowledge and human capital. [See Crépon et al. (1998), see Figure Box 2-1D below; David (2010); Furman et al. (2002); Griliches (1958, 1998); Hall-Jaffe (2012); OECD (2009); Jorgenson and Gollop (1992); Mairesse and Mohnen (2010); Nelson (1993); Rogoff (2012); Shanks and Zheng (2006), see Figure Box 2-1E below; Solow (1994); and Soete (2012).]

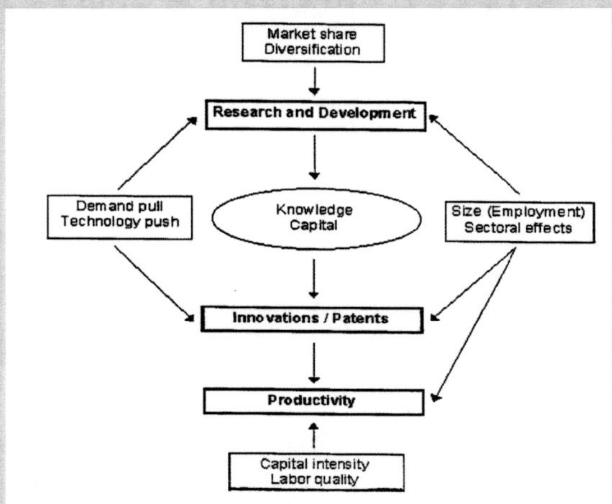

FIGURE BOX 2-1D
SOURCE: Crépon et al. (1998).

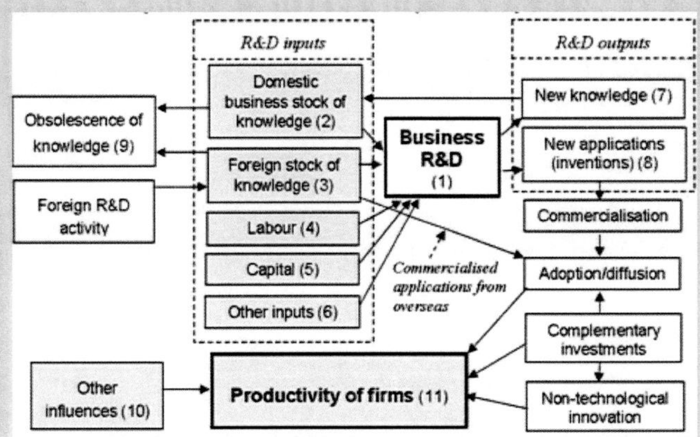

FIGURE BOX 2-1E
SOURCE: Shanks and Zheng (2006).

Specific Outcomes: There are models that look at specific outcomes of scientific and innovation activities, such as Acemoglu et al. (2012), David (1990), David et al. (2011), Mowery (2010), Popp (2010), Rogoff (2011), and Sampat (2011). Sampat's logic model of publicly funding R&D shows the pathways to new medical products and eventually health outcomes resulting from initial R&D expenditures. This model (see Figure Box 2-1F) shows the importance of measuring nodes and connections between nodes, which show the influence on health outcomes of various elements in the system.

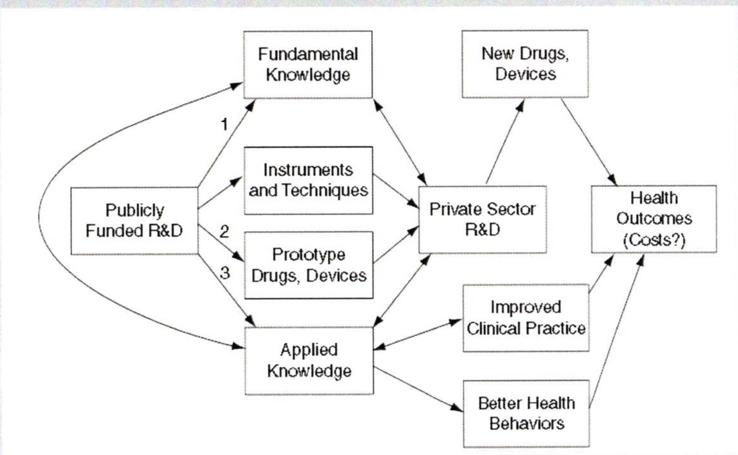

FIGURE BOX 2-1F
SOURCE: Sampat (2011).

Linkages: There are frameworks that identify specific relationships between actors in the STI system (see Figure Box 2-1G). Tassey (2011) highlights the coexistence of government and private funding for technological development. The blue shows contributions from government agencies while the red shows funding from private firms and organizations. This is a relational model but not necessarily predictive.

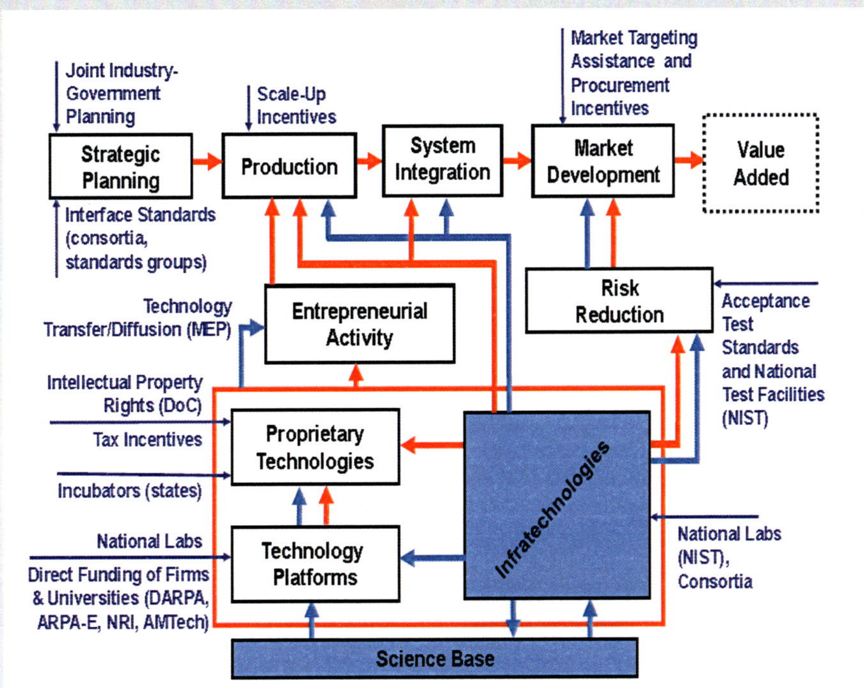

FIGURE BOX 2-1G
SOURCE: Tassey (2011).

Key Questions

Social Returns on Public and Private Expenditures on STI Impact on Economic Growth, Competitiveness, and Jobs

STI Indicators

Drivers, Trends, Advances, Vulnerabilities, Culture/Climate, and Distribution

Actors
- Individuals
- Collectives
  - Teams
  - Governments
  - Education and research institutions
  - Businesses
  - Private nonprofit organizations

Activities
- Research
- Invention
- Development
- Engineering/design
- Innovation
- Diffusion
- Education
- Training
- Capital investment
- Job mobility
- Firm dynamics
- Policy, regulation, and governance

Linkages
- Grants
- Contracts
- Collaboration
- Partnerships
- Codevelopment
- Copublication
- Social networks

Outcomes
- Knowledge stocks
- Social capital
- Intangibles
- Products and services
- Productivity
- Product life cycles
- Trade in S&T products
- Trade in R&D services
- Job mobility
- Firm dynamics
- Socioeconomic impacts/ well-being

FIGURE 2-1 A policy-driven framework for STI indicators.
NOTE: R&D = research and development; S&T = science and technology; STI = science, technology, and innovation.
SOURCE: Panel's own work.

important in many areas, including geospatial hot spots for entrepreneurial activities; potential avenues for broadening the participation of women, minorities, and people with disabilities in STEM fields; contributions to S&T breakthroughs from the social and behavioral sciences; the uptake of ideas for innovation from consumers; and the inclusivity of growth for various rungs of society. All of these top-level issues—drivers, trends, advances, vulnerabilities, relationships, and distributions—have underlying metrics that users want.[6]

At the panel's June 2012 workshop, representatives of the OECD-National Experts on Science and Technology Indicators (NESTI) Working Group implicitly described the STI system.[7] A system consists of actors, engaged in activities, with linkages to other actors and activities, giving rise to short-term outcomes and long-term impacts. Actors are people who are observed individually or as collectives, such as teams or organizations. In a high-level analysis, the actors are governments, institutions of education and research, businesses, and others such as private nonprofit organizations. The activities in which the actors engage include research, invention, development, design and other engineering tasks, innovation, diffusion of technologies and practices, education and training, and capital investment. Examples of linkages are grants and contracts, collaboration, partnerships, codevelopment, copublication, and social networks.

Mapping the system, understanding it, and explaining it to policy makers—all themes that emerged in the workshop—require data linkage and microdata analysis. The result of addressing these themes would be new and better indicators on linkages in addition to existing indicators on activities such as R&D, trade in R&D services, and the production and mobility of highly qualified people. Workshop participants also stressed that a system exists in space and time, and looking in more detail at regions is important, as is providing policy-relevant indicators to policy makers in a more timely manner.

USER PRIORITIES[8]

This section summarizes the priorities of two key groups of users of NCSES products: users of microdata and users of STI indicators.

[6]A list of policy issues and related questions appears in Appendix B and is referenced in Chapters 4-6.

[7]See Appendix D for the workshop agenda and the list of attendees. NCSES staff were represented at the workshop.

[8]The panel was unable to obtain a full list of users from NCSES. Identifying its full customer base is difficult for NCSES because the vast majority obtain information anonymously via the World Wide Web. Therefore, the panel derived information about key users from NCSES, panel members who are experienced data users, and some of the users who were interviewed for this study.

Users of Microdata

Although data development and access were not the focus of this study, it is important to mention here the primary request of a significant number of data users. Academic researchers, as well as data analysts at government agencies and private organizations, uniformly requested greater access to microdata (in this case, disaggregated data from NCSES surveys), not just aggregates or other statistics derived from those data. This type of user relies on NCSES for expertly curated datasets.[9]

Over the years, for example, NCSES has collected longitudinal data,[10] administering follow-up surveys to students after graduation from college. These data are useful to researchers who, for instance, want to determine the propensity of STEM-educated students to persist in STEM occupations. Another example relates to data from the Business Research and Development and Innovation Survey (BRDIS). Since the BRDIS pilot was conducted in 2008, NCSES has published only one *InfoBrief* reporting statistics on innovation based on that survey. The 2009, 2010, and 2011 BRDIS data were collected. Users of NCSES's datasets are eager to analyze the data on innovation from these surveys. However, only researchers with access to U.S. census data at census research data centers can work with BRDIS data. Upon request, NCSES makes statistics derived from these data available in tabular form.[11] The tables include the incidence of innovation in the United States, measures of R&D expenditure, performance and employment domestically and worldwide, and measures of projected R&D costs and intellectual property (e.g., see Borousch, 2010, p. 5).[12]

NCSES has long had means through which data users can gain access to microdata, with some stipulations. From time to time, the agency has invited proposals from researchers for use of its datasets, and in 2012 it revived its grants program under the title "Research on the Science and Technology Enterprise: Statistics and Surveys." This program increases data access for academic and other researchers, potentially yielding dividends in improved S&T indicators, as well as improved methodologies for analyzing and disseminating data and statistics.

Abstracts from the NCSES research awards are given in Appendix E. They show a wide array of topics, including the impact of tax incentives on increasing R&D, factors that affect time to degree for doctoral students, the impact of firms' economic activities on economic outcomes, differences in employment rates for women and other groups underrepresented in STEM, differences in promotion and retention rates for women and minority professors, experimental methods for assessing ways of mitigating the survey nonresponse problem, and experimental methods for improving recall accuracy on questionnaires. Just this small group of grants shows that NCSES data can be used effectively by researchers to (1) examine the questions the panel heard were critically important to users of STI indicators, especially those that indicators cannot fully address and therefore require analytical research; and (2) inform data extraction and statistical practices that could enhance NCSES's survey development, its data collection and analysis activities, and ultimately its productivity. Compared with the contract vehicle used by NCSES to produce some of its analytical products, the grants program allows for greater breadth of content through an investigator-initiated research agenda.

RECOMMENDATION 2-1: The National Center for Science and Engineering Statistics should continue its Grants and Fellowships Program for using its datasets, maintaining the high National Science Foundation standards for peer-reviewed award decisions.

One additional issue raised by data users was the need for more up-to-date taxonomies. For example, there is some discrepancy between the Science and Engineering Indicators 2008 figures and those calculated by staff at the National Institutes of Health (NIH) on the number of postdoctoral employees in the medical sector. Several factors could account for this discrepancy, including differences in the data taxonomies used at NCSES and NIH.

A previous National Research Council study (National Research Council, 2005) sponsored by the National Science Foundation (NSF) identified the need for collaborative efforts among federal agencies to review ways in which multiple classifications of science and engineering fields could be reconciled. The study recommended that a follow-on data taxonomy study be conducted to produce consistent definitions of fields and occupations across the sciences,

[9]Data curation is the active and ongoing management of data through their life cycle of interest and usefulness to scholarship, science, and education. Data curation enables data discovery and retrieval, maintains data quality, adds value, and provides for reuse over time through activities including authentication, archiving, management, preservation, and representation. (See http://www.lis.illinois.edu/academics/programs/specializations/data_curation [June 2013].)

[10]The Survey of Doctorate Recipients (SDR) has been conducted every 2 to 3 years since 1973. It follows a sample of doctorate recipients in science, engineering, and health fields throughout their careers up to age 75. See National Science Foundation (2012c) for more detail on this dataset.

[11]From the 2009 BRDIS Table Notes: "There are two types of tables in the table set. Most tables classify survey items by detailed industry, company size, or business code. Table 1 is a different kind of table. It contains aggregate totals for a variety of survey items that may not be tabulated at detailed levels. Since there is a large number of data items in BRDIS, tabulating all of them at a detailed level at this time is impractical and would unduly delay release of the statistics. Consequently, only selected items have been tabulated at the detailed industry, company size, or business code level. Most of the rest of the items are included in Table 1, but only at the aggregate (all industry) level. In the future, NSF intends to add more tables to cover some of the most requested data items that are not currently tabulated at the detailed level."

[12]The panel was able to obtain data tables from NCSES. Chapter 3 and Appendix F of this report provide more detail on NCSES's datasets and comparable data at other organizations worldwide.

including the health sciences, the social sciences, and engineering. Specifically:

> The panel recommends that it is now time for the U.S. Office of Management and Budget to initiate a review of the Classification of Fields of Science and Engineering, last published as Directive 16 in 1978. The panel suggests that OMB appoint the Science Resources Statistics office of the NSF to serve as the lead agency for an effort that must be conducted on a government-wide basis, since the field classifications impinge on the programs of many government agencies. The fields of science should be revised after this review in a process that is mindful of the need to maintain continuity of key data series to the extent possible (Recommendation 6-1). (National Research Council, 2005, p. 127)

Such consistency is particularly important if NCSES data are to be linked with data from other statistical agencies, as is recommended later in this report.

A subsequent NSF-funded NRC study (National Research Council, 2010) included extensive analysis on the issue of data taxonomies for science and engineering statistics. That study found a need for harmonization of data taxonomies on R&D across federal agencies. Below are two relevant excerpts from the report on that study:

> The importance of updating the taxonomy to better incorporate interdisciplinary research is widely recognized by policy makers, funding agencies, professional organizations, and across academia. The growing role of research involving more than one discipline is a serious challenge to any taxonomy of fields and therefore to gathering, analyzing, and using federal funds data based on a single-field taxonomy. (National Research Council, 2010, p. 22)

> . . . No single taxonomy will satisfy all. However, for purposes of collecting data on research and development statistics in a consistent manner across federal government agencies, it is necessary to establish a common taxonomy that will be useful to the largest number of data providers and users. In the longer term, a provision can be made for tailoring structures that meet the specific needs of providers and users by flexibly categorizing administrative records. (National Research Council, 2010, p. 32)

The panel that produced the 2010 NRC report did not recommend, however, an immediate broad updating of the science and engineering taxonomy because of concerns about breaks in time series and the difficulties of the task. It limited its formal recommendation to a call for NCSES to "in the near term . . . make the changes necessary to improve the comparability of the federal funds taxonomy and the taxonomy for the academic research and development expenditures survey" (National Research Council, 2010, Recommendation 3-1, p. 44).

Recognizing the problems of developing classifications that can satisfy a variety of user needs and the need for historical continuity to the extent possible, this panel nonetheless concludes, consistent with the 2005 NRC report, that a broader effort to revise the existing classifications for science and engineering fields and occupations is long overdue. Changes in the U.S. economy led to a government- and continent-wide effort to develop what became the North American Industry Classification System in 1997, which is regularly updated every 5 years and has enabled federal statistics to take cognizance of the growth of the service sector and other changes in the economy. So, too, is it important to update the science and engineering taxonomy, given the evolution of new areas of science and the growth of interdisciplinary research. Indeed, NCSES has undertaken some work along these lines, which it is important to continue and to step up to the extent feasible. It is also important for these efforts to include development of a process for performing updates as needed.

> **RECOMMENDATION 2-2: The National Center for Science and Engineering Statistics should engage with other statistical agencies, including but not limited to the Bureau of Labor Statistics, the U.S. Census Bureau, the National Center for Education Statistics, and the National Institutes of Health, to develop a consistent taxonomy of science and engineering fields and occupations (including the health and social sciences). There should also be an established process for performing updates of this taxonomy as needed.**

Users of STI Indicators

Issue-driven requests for analytical measures are legion, and NCSES does not have the capacity to develop indicators for all issues. As discussed in Chapter 1, moreover, indicators address but cannot fully answer most of the policy questions posed regarding the functioning of the STI system. Therefore, priorities need to be set regarding (1) which issues indicators can address, at least in part; (2) which users' needs will be met; (3) where high-quality indicators will be obtained absent processing in house; and (4) where additional dollars will be spent and what will be produced less frequently or not at all should funds decrease. The remainder of this chapter focuses on the first two questions; the latter two relate to make-buy decisions for NCSES and strategic expenditures on activities within the agency, which the panel believes are for NSF staff to determine.

User-identified high-priority STI indicators are listed in Box 2-2. The list is broken down by the categories in Figure 2-1—STI activities, outputs and outcomes, and linkages; metrics related to actors and intermediate inputs in the system appear under activities and outcomes. Although the list is extensive, it represents only a subset of the measures users said they wanted to have, either specifically from NCSES or in general. The panel used its collective expertise to determine which indicators from that total set should be deemed priorities for NCSES, focusing in particular on the utility of indicators in the policy context. The panel also considered other factors that could be used to prioritize STI indicators produced by NCSES, including ease of acquisition, cost, flexibility, and periodicity (see Figure 1-1 in Chapter 1). These factors often were difficult for the panel to ascertain. For example, the cost of producing any given indicator depends on the type of data (survey, unstructured, administrative) required, the need to link to other datasets that NCSES might have to acquire, the possibility that NCSES would have to invest in new skill sets in house to use new data acquisition and manipulation techniques, and so on. Therefore, the panel focused primarily on high-utility indicators; suggestions are offered later in the report on how to acquire new types of data and the skill sets needed to work with those data.

This report reflects the systems framework of STI activities shown earlier in Figure 2-1. However, the lines between activities and outcomes (inputs and outputs) are imprecise, and attempts to draw them are imperfect. Therefore, linkages between various elements of the system are highlighted throughout the report to help the reader appreciate why a suite of indicators is important for a fuller understanding of drivers, trends, advances, and vulnerabilities in the global STI system.

International Comparability

Because one of the primary goals of this study was to determine how to improve the comparability of STI indicators in the United States and abroad, the panel discussed priorities for indicators with internationally known experts in the field during its two workshops. Several recurring themes emerged from these discussions.

First, mapping and understanding the innovation system (or system of systems) is important. However, priorities for indicator development are driven by policy makers' queries.

Second, understanding and producing indicators on the commercialization of publicly funded knowledge is a priority. Linkage indicators need to show the flow of knowledge from public institutions to businesses and from businesses to the market, leading to social and economic impacts. This flow of knowledge includes highly qualified people as well as licenses for intellectual property and R&D services. Thus, it is important to have good measures of the STEM workforce and other talent employed in S&T sectors. In addition, measures of knowledge assets (including those considered intangible) and innovation are a high priority for improved measurement.

Third, the development of STI indicators at different geographic scales and for a variety of sectors is a priority. Other distributional elements of the STI system also are important, including characteristics of people—their gender, age, level of education and experience, and willingness to take risks and be entrepreneurial, as well as their employment and mobility. Measures of social capital that affect the development and diffusion of knowledge capital are important indicators of STEM talent. The characteristics of firms matter, too—their location, size, rate of employment and revenue growth, entrepreneurial characteristics, and complexity (multinational corporations are different from small and medium-sized firms).

Fourth, outcome measures are necessary but not well developed. It is important to have internationally comparable measures of innovation and of its social and economic impacts.

Finally, there is a need to measure innovation that is not the direct result of R&D expenditures. This class of indicators could begin to answer questions about what governments obtain in return for their funding and how their expenditures affect productivity, economic growth, and job creation.

Dahlman (2012) discusses the "changing geography of innovation," with a focus on the engineered emergence of technological capabilities in Brazil, China, and India. In addition to measures of innovation activities that are used in the developed-country context, Dahlman presents several indicators that are particularly useful early warning signals of the potential for ascendency of a developing country in the international innovation system. These indicators include (1) human capital (enrollments in higher education outside the home country, number of back-expatriated citizens, number of top foreign scientists recruited by local universities and industry); (2) R&D expenditure (information-enabled service industries and knowledge services); (3) learning (investments from transnational corporations, including locations of research, development, and engineering functions; exports and imports with the rest of the world; technology licensing and S&T cooperative agreements at home and abroad); (4) institutions (regulations, including protection of property rights, trade restrictions at home and abroad); (5) intermediate and final outputs (share of total world R&D);[13] (6) domestic demand for high-technology products (including alternative energy technologies);[14] and (7) social outcomes (income inequality; measures of children's access to education, health care, and food; programs that support product, process, and service innovations that address the needs of low-income populations).

NCSES publishes many STI indicators that are comparable to those available in OECD's 2011 Science, Technology, and Industry Scoreboard; the Cornell-INSEAD-World Intellectual Property Organization (WIPO) Global Innovation Index 2013; and the European Union's Innovation Union Scoreboard 2013. NCSES is most prolific in international

[13]Dahlman (2012, p. 6) states that ". . . .there has been concern about significant fraud and cheating in research and scientific publications and that many patents are of little value. The argument is that this activity has proliferated because promotions and salary increases in universities and research labs have been reoriented to be based on publications and patenting."

[14]"Domestic demand conditions" is one of Porter's (1990) four determinants of national competitive advantage.

BOX 2-2
Key Indicators Suggested by Major Users of STI Indicators

Activities

Research and Development (R&D)

- National R&D expenditures
  - Federal and state funds for basic research
  - Public-sector R&D (focus on advanced manufacturing, green technologies, energy-related R&D, nanotechnology, agriculture, weapons)
  - Public R&D spending as a share of gross domestic product (GDP)
  - Business R&D spending
  - Business R&D as a share of GDP
  - Industry support for R&D in universities
  - Social science R&D
- National R&D performance (by type of industry and source of funds)
- Trends in grant size to universities
- Number of R&D centers in the United States and other countries

Innovation

- Direct measures of innovation (data similar to those from the Community Innovation Survey)
  - Ratings for propensity to innovate
  - Subject matter experts (SMEs) innovating in house as a share of SMEs
  - Firms (<5, 5+, 10+, 20+ employees) introducing new or significantly improved products or processes as a share of all firms
  - Firms (<5, 5+, 10+, 20+ employees) introducing new or significantly improved goods or services as a share of all firms
  - Firms (<5, 5+, 10+, 20+ employees) introducing marketing or organizational innovations as a share of all firms
- Numbers and types of new products per year, by region (Thomasnet.com)
- Drug and other approvals per year, by region
- Sale of new-to-market and new-to-firm innovations as a share of turnover
- Non-R&D expenditures on innovation activities and non-R&D innovation spending as a share of turnover
- Inclusive innovation for development (case studies)
- Capital expenditures related to the introduction of new processes
- Marketing expenditures related to new products
- Expenditures on design and technical specifications
- Expenditures on service-sector innovation
- Investment in new information and communication technology (ICT) hardware and software
- Innovation inhibitors (case studies)

Market Capital Investments

- Venture capital investments in science and technology (S&T) (early-stage, expansion, and replacement) and venture capital investments in S&T as a share of GDP
- Number of initial public offerings (IPOs) in S&T
- Number of S&T spinoffs
- Expenditures in later phases of development/testing that are not included in R&D

Outputs and Outcomes

Commercial Outputs and Outcomes

- Performance of high-growth small and large firms
- High-growth enterprises as a share of all enterprises
- Medium- and high-tech manufacturing exports as a share of total product exports
- Knowledge-intensive service exports as a share of total service exports
- Value added in manufacturing

- Value added in technical services
- Trade flows of S&T products and services
- ICT outputs and sales (intermediate and final)
- Other intermediate inputs
- Technology balance of trade (especially intellectual property)
- Contracts to S&T firms
- Advanced manufacturing outputs (information technology-based processes)
- Market diffusion activities
- Emerging industries (based on universities, government laboratories, firms, value chains, key occupations, and individuals)
- Help-wanted ads, "how to" books, and other derivative STI activities
- Use and planned use of general-purpose technologies

Knowledge Outputs

- U.S. receipts and royalty payments from foreign affiliates
- U.S. patent applications and grants by country, technology
- U.S. trademark applications and grants by country, technology
- Patent citations
- License and patent revenues from abroad as a share of GDP
- Triadic patent families by country
- Percentage of patent applications per billion GDP
- Percentage of patent applications related to societal challenges per billion GDP (e.g., climate change mitigation, health)
- Intangible assets
- Average length of a firm's product life cycle or how often the firm usually introduces innovations
- Births and deaths of businesses linked to innovation outputs; firm dynamics by geography, industry, business size, and business age
- Knowledge depreciation
- Knowledge stocks and flows in specific sectors, including nanotechnology; information technology; biotechnology and agriculture research (local foods, organic foods, biofuels, environment, nutrition, health); oil and gas production; clean/green energy; space applications; weapons; health care technologies; educational technologies (massive open online courses [MOOCs]); and mining

Science, Technology, Engineering, and Mathematics (STEM) Education

- Expenditures, direct and indirect costs, investments, revenues, and financing for STEM education
- Percentage of faculty in nonteaching and nonresearch roles at universities
- Enrollment data by STEM field at various levels (e.g., associate's, bachelor's, master's, doctorate) and for various types of institutions
- New degrees (e.g., associate's, bachelor's, master's, doctorate); new doctoral graduates per 1,000 population aged 25-34
- Stock of degrees (e.g., associate's, bachelor's, master's, doctorate)
- Share of population aged 30-34 having completed tertiary education
- Share of youth aged 20-24 having attained at least upper-secondary-level education
- Persistence and dropout rates in education, by geographic and demographic distinctions
- Number of high school students pursuing associate's degrees and implications for the workforce and the cost of higher education
- Disciplines in which community colleges have a relative advantage
- Foreign-born STEM-educated individuals—country of birth, immigration visas, etc.
- Stay rates of foreign students
- Trends in online learning and MOOCs

STEM Workforce/Talent

- Postdoctoral levels and trends in various STEM fields, by country of birth and country of highest degree
- Number of postdoctorates in health, but specific fields
- STEM employment
- Labor mobility and workforce migration
- Demographic composition of people who would enter specific occupations (e.g., clean energy, ICT, biotechnology, health services)
- Fraction of STEM degree holders that hold STEM jobs
- Earnings by degree type and occupation

BOX 2-2 Continued

- Feeder fields in agricultural science
- On-the-job training activities in S&T, manufacturing, and services
- STEM demand
- Employment in knowledge-intensive activities (manufacturing and services) as a share of total employment

Socioeconomic Impacts/Well-Being
- Economic growth
- Productivity
- Other measures of impact on GDP and jobs
- Agricultural preparedness
- Energy preparedness
- Return on investment (ROI) on grants to universities, by type of S&T
- National security/defense
- Environment
- Energy use
- Geographic hot spots

Linkages

Organizations/Institutions/Infrastructure
- Public-private copublications per million population
- University-industry research collaborations
- Number and value of international collaborations
- Business structure dynamics
- Technology transfer between academic institutions and businesses, including mechanisms
- Technology transfer (Manufacturing Extension Partnership [MEP])
- Technology transfer from national laboratories

indicators of human capital stocks and flows,[15] and it has many essential indicators of firm activities in R&D, innovation, and knowledge-intensive services; financial expenditures; patent grants; and international trade in high-technology products. As discussed in Chapters 4 through 6 of this report, however, there are elements of innovation, knowledge generation, knowledge networks and flows, and even human capital for which NCSES should consider expanding its portfolio of indicators. Doing so would improve the comparability of STI indicators internationally, thereby improving the utility of these measures for a variety of users.

Subnational Statistics[16]

Users want more disaggregated STI information on multiple levels. They want STI comparisons across U.S. regions and between U.S. and foreign regions. Cooke and

[15]For example, NCSES's *InfoBriefs* and the National Science Board's *Science and Engineering Indicators* volume (for which NCSES provides statistical indicators) include the following statistics: enrollments in master's and Ph.D. science and engineering programs in the United States by countries or economic regions of origin; baccalaureate origins of U.S.-trained science and engineering doctorate recipients; number of science and engineering degrees earned in foreign countries; international mobility and employment characteristics of recent U.S. doctorates, including stay rates; employment in R&D activities worldwide, with specifics on R&D workers in multinational corporations; and international collaborations of scientists and engineers in the United States.

[16]This report uses the term "subnational" instead of "regional" to denote geographic areas that are defined within a nation's boundaries. While the term "regional" is used extensively in the literature to denote states or provinces, standard metropolitan statistical areas (SMSAs), or even well-defined industry clusters, the term is also used in reference to clusters of countries (e.g., the North American region or the Pacific Rim). Cooke and Memedovic (2003, p. 5) give useful criteria for defining a region or subnational area: "(1) a region must not have a determinate size, (2) it is homogeneous in terms of specific criteria, (3) it can be distinguished from bordering areas by a particular kind of association of related features, and (4) it possesses some kind of internal cohesion. It is also important to mention that the boundaries of regions are not fixed once for all; regions can change, new regions can emerge and old ones can perish."

- Bilateral S&T agreements (including international)
- Collaboratories
- Industry clusters
- Incubators
- Consortia (Defense Advanced Research Projects Agency [DARPA], Advanced Research Projects Agency-Energy [ARPA-E], Technology Innovation Program [TIP])
- Intellectual property rights and policies
- Standards
- Market planning assistance (Department of Commerce [DoC], Bureau of Labor Statistics [BLS], Small Business Administration [SBA])
- Research and experimentation (R&E) tax credits (federal and state)
- Innovative SMEs collaborating with others as a share of SMEs
- Alumni contributions to R&D
- Communications linkages (including broadband)

Culture
- Public value of S&T
- Business climate
- Entrepreneurial activities
  - Mappings of entrepreneurial density
  - All establishments and firms with at least one employee, including start-ups, 1976 to the present
  - All nonemployer firms and integrated-with-employer firms, 1994 to the present
  - All employer-employee matches and transitions (hires, separations, job creation, and job destruction) 1990 to the present
  - Information on innovation policy and its outcomes (contexts; national, regional, sectoral levels)
  - Data on the existence of dealmakers and entrepreneurs and their connections in a given market
- Risk tolerance
- Social networks
- Social capital

Memedovic (2003, p. 31) surmise that "global economic forces have raised the profile of regions and regional governance not least because of the rise to prominence of regional and local business clusters as vehicles for global and national economic competitiveness." Hollanders (2013, p. 79) states that "regions are increasingly becoming important engines of economic development." Drawing on comparisons at the subnational level, Hollanders finds that understanding various components of the innovation system at this level yields useful insights into performance outcomes at the country level. A key caveat, however, is that subnational STI statistics are scarce relative to comparable national statistics. Furthermore, Hollanders asserts that comparing small countries (such as Bahrain and Cyprus) to large countries such as (China and India) clouds the ability to determine what he terms "best practices." Hollanders (2013, p. 84) states:

> Applying best practices from these large countries to smaller ones will be difficult because of the differences in scale. We need to be able to compare smaller countries with regions of larger countries that are similar to the smaller countries in size or in industrial structure. Such a comparison requires a breakdown of country-level statistics into regional statistics, where regions should not be defined as static administrative regions, . . . but rather as economic regions that can be distinguished from bordering regions and that should have a certain degree of internal cohesion. There are no guidelines for determining the "ideal" region, but large metropolitan areas seem to emerge as a natural category.

Research shows that innovation depends on many factors that together contribute to the success or failure of a given idea and therefore is highly local. The particular arrangement of R&D facilities, industry concentration, labor force skills, and labor mobility makes the local setting productive (see, e.g., Belleflamme et al., 2000; Braunerhjelm and Henrekson, 2013; Clark et al., 2003). A substantial body of evidence shows that these local settings have been highly influential in creating concentrations of innovation over the past century (see, e.g., Audretsch et al., 2005). For example, Route 128 around Boston saw the emergence of new industries and relationships in the 1950s and 1960s characterized by a complex interaction among venture capital, real estate promoters, and major research universities; Silicon Valley subsequently emerged as the focal point for a range of new products and services (these two cases are considered in great detail by Saxenian [1996]; see also Gertler et al., 1995; Link and Scott, 2003). Some of the most successful clusters of innovation have far exceeded any original plans, arising from particular combinations that others have tried to replicate with varying degrees of success. State laws vary on postemployment covenants (also known as noncompete agreements), which can have differential effects on entrepreneurial outcomes (Marx and Fleming, 2012). Many cities have explicit policies for incubators, including some that have produced tangible results for the region and the economy as a whole (see, e.g., Tödtling and Trippl, 2005). Yet such policies are not panaceas and at times produce measured successes or failed outcomes. Therefore, it is critically important to have a variety of information, including subnational STI indicators that can inform judgment on whether such expenditures should continue and whether the portfolio of programs is optimal.

Countries other than the United States have more explicit regional strategies for fostering innovation, established over longer time frames (Cooke and Memedovic, 2003; Falck and Heblich, 2008). In recent years, for example, much of the funding for innovation in the European Union has been channeled through regional initiatives. The federal structure of the United States obscures some of the explicit regional strategies across states and in a complex collection of regional bodies, some spanning multiple states. Yet many U.S. policies that support innovation are at the local level, where industries are created and incubated.

Thus, local decision makers need indicators for their specific region in comparison with others; no single geographic subdivision scheme will serve all needs. Demographic statistics often are tabulated by legislated entities (e.g., states, counties, cities) because of the constitutional relationship to representation. These entities, though easily recognizable, may not well represent the economic realities of market areas and regional variation. Consider the vast commercial and demographic variability within some states, such as California, compared with the relative homogeneity of a state like Delaware. These two states are not at the same scale analytically, yet each needs indicators to manage its investments. A confounding factor is that economic information has another hierarchy—from the industry to firm to plant or establishment level. Only the finest level has a spatial expression, and therefore raises questions of subnational policy relevance. Some states, particularly those dependent on sales tax revenues and other highly cyclical sources, expend great effort in operating their own economic models at subnational scales. The decentralized way in which these local needs are being met makes it difficult to develop a national scheme for integrated information management.

During the panel's July 2011 workshop, Robert Atkinson of the Information Technology and Innovation Foundation said subnational information would be particularly helpful for technology and innovation policy. Other workshop participants described subnational decompositions in several countries. Based on her extensive research on STI hot spots, Maryann Feldman of the University of North Carolina emphasized that economic growth does occur within these finer geographic units. She went on to stress that decision makers in the states and metropolitan areas need data on innovation activities at the subnational level. She suggested NCSES work with users to determine what statistics would be useful at this level and what some users have already created that could serve as useful inputs for NCSES's subnational indicators.

At the workshop, representatives of the Association of Public and Land-grant Universities (APLU) David Winwood and Robert Samors presented an overview of the Commission on Innovation, Competitiveness, and Economic Prosperity (CICEP) project. They noted that APLU would like universities and other organizations to collect dozens of measures on a wide range of topics, especially numbers, types, and dollar amounts for research activities funded by private-sector entities (e.g., consortia, trade associations, companies); similar information for federal, state, or foundation sources of funding; numbers of students participating in work-related activities, regardless of whether they earn academic credit for those activities; numbers of full-time equivalent employees engaged in sponsored research-related activities; and equity investments in university technology development activities by graduates of the institutions, as well as other types of investors.[17]

In summary, comparing the almost two dozen subnational measures requested by users at the panel's workshop with the information from APLU's questionnaire exercise and with findings in the literature on regional innovation systems (e.g., Cooke and Memedovic, 2003, p. 17; Hollanders, 2013, p. 80) reveals the following to be high-priority subnational STI indicators: (1) academic R&D expenditures; (2) federal R&D expenditures (some of which are directed to academic institutions and private firms); (3) industry expenditures on R&D, including support for academic research; (4) non-R&D innovation expenditures; (5) STI equity investments (from various sources, including venture capital); (6) sales of products new to the firm (noting distortions introduced by multiplant, multisector firms); (7) share of the population aged 26-64 with tertiary degrees or engaged in certificate training programs; (8) employment in knowledge-intensive manufacturing and services; (9) knowledge transfer and other linkages between academic institutions and industry (e.g., public-private scientific copublications per million

[17]This information was circulated to participants at an APLU workshop in October 2012. APLU staff developed a list of 11 first-tier and 23 second-tier priority metrics from a pilot test of a questionnaire administered to its membership universities.

population); and (10) infrastructure investments (e.g., broadband access). These and many other indicators were requested at both the national and subnational levels and are included in Box 2-2 presented earlier. NCSES produces indicators in many of these categories at the state level.[18] However, users are interested in an expanded set of subnational indicators at finer geospatial scales. Although expenditures on STI activities are at times determined by state legislators, venture capital investments and some economic outcomes are better observed in metropolitan areas or in smaller economic clusters.

One should not presume, however, that arriving at national aggregates based on subnational data is at all straightforward. The panel was advised during the workshop that the pursuit of more subnational STI indicators at the state and even local levels is fraught with problems of distribution and aggregation.

CAUTIONS, POSSIBILITIES, AND LIMITATIONS

Although the production of indicators across many fields has an established history, at least three major cautions regarding their use are important to note.

First, indicators can send mixed signals that require expert judgment for interpretation. For example, it is commonly held that increased innovation—which is key to advancing living standards—enhances job creation, and policy makers discuss spurring innovation as a job creation tactic. However, innovation can lead to fewer jobs if the process or managerial expertise increases efficiency. On the other hand, short-term displacement of workers in one industry or sector can be counterbalanced in the longer term by the development of new products, services, and even sectors and by increased market demand if process efficiencies drive down prices (see Pianta, 2005; Van Reenen, 1997). One way to be cautious about mixed signals is to develop STI indicators that support analysis of time scales, sectors, and geographic locations.

Second, once a given metric becomes widely used, it may change the behavior of the people and practices it attempts to measure. The worst thing a metric can do is not only deliver a bad (i.e., misleading) answer but also incentivize bad practice—that is, decisions or policies that are counterproductive (see, e.g., West and Bergstrom, 2010). It is important that indicators not send distorted signals to users.

Third, not everything that counts can be counted, and not everything that can be counted counts. Some outcome measures that reflect the importance of R&D and innovation to society are elusive. For example, social well-being is difficult to measure, yet one of the key interests of policy makers is the return on investment of public funding for S&T for the good of society.

For this study, Bronwyn Hall and Adam Jaffe prepared a commissioned paper that in part takes up the notion of a policy-driven framework for STI indicators. As mentioned in Chapter 1, Hall and Jaffe (2012, p. 39) give a balanced view of the extent to which users can rely on indicators to address key issues, making a strong case for the need for improved metrics that can be used for analytical purposes. Their observations are worth quoting at length here:

> ***Overall level of public investment in R&D***. Implicitly, the Congress and the President are continuously deciding what overall level of resources to invest in new knowledge creation through the R&D process. Ideally, this would be informed by data showing the marginal rate of return on these investments. But marginal rates of return are very difficult to measure. Economists and others have made estimates of the average rate of return to R&D investments (Hall et al., 2010). Within the model, the marginal rate of return declines with the intensity of R&D investment (R&D/GDP) other things equal, so a high average rate of return is a necessary but not sufficient condition to justify increased investment.
>
> In the absence of explicit information, R&D intensity measures do provide some implicit evidence on the rate of return. Economic models typically presume that there are diminishing returns to increased R&D expenditure, so that the rate of return to R&D will fall as R&D/GDP rises. This means that if today's U.S. R&D/GDP ratio is lower than at another point in time, we may be able to infer that the rate of return in the U.S. today is higher than it was at that point of time, assuming that nothing else has changed. The same argument applies when comparing R&D intensities across countries, although it is even more difficult to assume that other things are equal in that case. Thus if we have some reason to believe that the investment level was right at some point in time, then we might be able to infer that the implied high rate of return in the United States today justifies a higher level of investment (and vice versa if today's U.S. R&D intensity is higher than at some other time or place). However, given all the uncertainties, it would probably be better to attempt to measure the return to R&D spending in this case.
>
> ***Overall level of public investment in education and training***. The issues with respect to the optimal level of investment in education and training are analogous to those related to R&D. We would, ideally, like to have measures of the rate of return; measures of the current ratio of investment to GDP may provide indirect evidence on the rate of return, at least relative to other times or places. In addition, public policy may view having an educated public as a desirable end in itself, over and above any return it may provide in terms of innovation and technology. If so, then data on years of schooling and degrees awarded are useful policy indicators independent of their indirect implications for the economic rate of return.
>
> Education and training also take many forms and occur in

[18] NCSES's state-level indicators are available in the following areas: elementary, secondary, and tertiary education; workforce; financial R&D inputs; R&D outputs; and S&T in the state economy (including venture capital activity, Small Business Innovation Research awards, and high-technology business activity).

many different contexts. We have better data on what occurs in formal educational institutions than we have on training that occurs on the job, or is otherwise provided by firms without recourse to formal educational institutions.

Allocation of both of above by scientific/technical area or area of ultimate application. Even more than the overall determination of public investment, the government must continuously decide the allocation of public resources for R&D and education/training across scientific and technical fields, and across areas of application. Again, within the model the most relevant information for these decisions would be the marginal rates of return. And again, these are hard to measure, and measurements of average rates of return are incomplete as indicators of marginal rates. In addition, there are substantial spillovers across scientific fields (e.g., the importance of computer science for DNA analysis) so that localized rates of return may not capture the true importance of some fields.

The relevance of investment intensity measures as indirect indications of marginal rates of return is more complex in the context of allocation across fields or sectors. If the inherent technological opportunity is greater in a given sector, then its marginal returns are higher at any given level of investment. Thus it is possible, for example, that our much higher level of public investment in research in health sciences than in other fields represents an implicit belief that technological opportunity, and hence marginal returns, are higher in that area than in others. On the other hand, no other country in the world devotes such a large share of its public research investment to health sciences. Unless the variation of technological opportunity across fields is different in different countries, comparative benchmarking on sectoral allocations may provide indirect evidence on rates of return. As noted above, however, this is a particularly problematic sector due to the difficulty of measuring output properly and the fact that health improvements are not completely captured by national income accounts.

Allocation of federal R&D and training resources by types of institutions (e.g., intramural versus extramural or universities versus firms). Allocation of public resources across different kinds of institutions raises the same issue of relative rates of return as allocation across sectors. In addition, different kinds of institutions play different roles in the STI system. Hence, indicators reflecting intermediate outputs of the research process, and flows of knowledge within the system, might be informative about imbalances within the system. It would also be useful to construct and publicize more detailed statistics on the demand for S&T skills in certain areas, including starting salaries, in a timely manner.

Science and technology policy choices other than spending. Many government policy choices explicitly or implicitly affect the STI system, including R&D subsidies (and other tax policies), intellectual property rules, and mechanisms for the transmittal of funds (e.g., basic research grants, contract research, prizes, etc.). It is not clear that indicators, as we normally think of them, shed light on the relative efficacy of different policy choices of this kind. But the data collected as the basis for indicators can also be used by social scientists to study the relative effectiveness of different mechanisms. In fact, these data are essential for this purpose.

Immigration policy (as applied to scientific/technical workers). Indicators related to the number and fields of scientific and technical workers, combined with the level of investment in research, may be useful for informing the nature and extent of visa programs to allow more technically trained immigrants to work in the United States.

Indicators for use by university administrators or firm managers. Firm managers and university administrators face many of the same choices as governments: how much to spend and what to spend it on. Many of them rely to some extent on benchmarking, that is, observing the spending patterns of their immediate competitors. Therefore, the same kinds of data as described above can be useful, preferably broken down by sector and by geography.

SUMMARY

This chapter has presented a long list of indicators that users want to have, mainly to address key STI policy issues. Users requested many more indicators, but these did not rise to the level of importance of those listed in this chapter. This chapter also has offered two recommendations, focused on (1) continuation of NCSES's Grants and Fellowships Program and (2) collaboration between NCSES and other statistical agencies to develop a consistent taxonomy of science and engineering fields and occupations.

With the focus on NCSES's decision-making challenges in the future, the panel took on the task of prioritizing the measures NCSES should produce in the near term and identifying the processes it should develop to satisfy users' future demands. The results of this effort are presented in Chapters 3 and 8, where the panel respectively identifies key policy-relevant STI indicators and strategic organizational principles for continuously developing those indicators as technology, economic, and policy environments change globally.

3

Data Resources for Indicators

Measuring capacity and change in science, technology, and innovation (STI) has a long history, dating back decades in economics and management research. Since the 1950s, under congressional mandate, the U.S. National Science Foundation (NSF) has produced measures of research and development (R&D), as well as education and occupational statistics specifically for science and engineering (S&E) fields. This chapter describes the data resources used by NSF's National Center for Science and Engineering Statistics (NCSES) to develop its STI indicators. These resources include its own surveys; administrative records and statistics from U.S. and international agencies, organizations, and companies; and bibliometric analysis of publications in peer-reviewed S&E journals. The main purpose of the chapter is to identify the high-priority indicators that NCSES already produces and the types of indicators that require further development. The panel also examined the long-standing issue of balance between indicators addressing human resources and R&D and innovation.

The analysis in this chapter, as well as the companion Appendix F,[1] is based on databases that are available to the public. NCSES's surveys yield scores of data for variables that are not released to the public for privacy and confidentiality reasons or because resources are inadequate to create data series from all relevant survey questions. In other words, NCSES has already collected or has access to some of the information desired by users of STI indicators.

[1]Appendix F contains a catalog of STI data and statistics from NCSES and other sources, including OECD; the United Nations Educational, Scientific and Cultural Organization, Institute of Statistics; Eurostat; Statistics Canada, Canadian Socio-economic Information Management System; the World Intellectual Property Organization; other U.S. agencies; and private sources. The appendix also contains a heat map analysis, illustrating a tool that could be used to identify areas for which there is either a paucity or an abundance of statistics.

NCSES DATABASES

NCSES communicates its science and technology (S&T) data through various media ranging from *InfoBriefs*, to Detailed Statistical Tables (DSTs), to table generation tools. The three table generation tools—the Integrated Science and Engineering Resource Data System (WebCASPAR), the Scientists and Engineers Statistical Data System (SESTAT), and the Survey of Earned Doctorates (SED) Tabulation Engine (National Center for Science and Engineering Statistics, 2013b)—are supported by application-specific database systems. The Industrial Research and Development Information System (IRIS) is a searchable database of prepopulated tables.

WebCASPAR hosts statistical data for S&E at U.S. academic institutions (National Science Foundation, 2012e). This database is compiled from several surveys, including:

- NSF SED[2]/Doctorate Records File;
- NSF Survey of Federal Funds for Research and Development;
- NSF Survey of Federal Science and Engineering Support to Universities, Colleges, and Nonprofit Institutions;
- NSF Survey of Research and Development Expenditures at Universities and Colleges/Higher Education Research and Development Survey;
- NSF Survey of Science and Engineering Research Facilities;
- NSF-National Institutes of Health (NIH) Survey of Graduate Students and Postdoctorates in Science and Engineering; and

[2]SED data on race, ethnicity, citizenship, and gender for 2006 and beyond are available in the SED Tabulation Engine. All other SED variables are available in WebCASPAR except for baccalaureate institution. For more details on the WebCASPAR database, see https://ncsesdata.nsf.gov/webcaspar/ [April 2014].

- National Center for Education Statistics (NCES), Integrated Postsecondary Education Data System (IPEDS)
    - IPEDS Completions Surveys;
    - IPEDS Enrollment Survey;
    - IPEDS Institutional Characteristics Survey (tuition data); and
    - IPEDS Salaries, Tenure, and Fringe Benefits Survey.

SESTAT (National Science Foundation, 2013d) is a database of more than 100,000 scientists and engineers in the United States with at least a bachelor's degree. This is a comprehensive data collection on education, employment, work activities, and demographic characteristics, covering 1993 to 2010. The SESTAT database includes data from:

- the National Survey of College Graduates (NSCG),
- the National Survey of Recent College Graduates (NSRCG),
- the Survey of Doctorate Recipients (SDR), and
- an integrated data file (SESTAT).

IRIS (National Center for Science and Engineering Statistics, 2013a) is a database containing industrial R&D data published by NSF from 1953 through 2007. It comprises more than 2,500 statistical tables, which are constructed from the Survey of Industrial Research and Development (SIRD). It is, therefore, a databank of statistical tables rather than a database of microdata of firm-specific information. The data are classified by Standard Industrial Classification and North American Industrial Classification codes (as appropriate), and by firm size, character of work (basic, applied, development), and state. Employment and sales data for companies performing R&D are also included in IRIS.

The data outlined above focus on academic and industrial R&D expenditures and funding and on human capital in S&T. NCSES conducts five surveys to capture R&D support and performance figures for various sectors of the economy. The *National Patterns of Research and Development Resources* series of publications presents a national perspective on the country's R&D investment. R&D expenditure and performance data are available, as well as employment data on scientists and engineers. The *National Patterns* data are useful for international comparisons of R&D activities. The data also report total U.S. R&D expenditures by state. The data series spans 1953 through 2011 and is a derived product of NCSES's above-referenced family of five active R&D expenditure and funding surveys:

- Business Research and Development and Innovation Survey (BRDIS; for 2007 and earlier years, the industrial R&D data were collected by the SIRD);
- Higher Education Research and Development Survey (HERD; for 2009 and earlier years, academic R&D data were collected by the Survey of Research and Development Expenditures at Universities and Colleges);
- Survey of Federal Funds for Research and Development;
- Survey of Research and Development Expenditures at Federally Funded R&D Centers (FFRDCs); and
- Survey of State Government Research and Development.[3]

Two surveys are under development at NCSES: the Microbusiness Innovation Science and Technology Survey (MIST), which will capture firms with fewer than five employees,[4] and the Survey of Postdoctorates and Early Career Researchers.[5]

NCSES's data catalog clearly covers a wide range of information on the STI system, but underlying challenges influence the portfolio and timeliness of the indicators produced.

First, response rates are declining for many of NCSES's major surveys.[6] Although most of NCSES's surveys had response rates above 80 percent in the early 2000s, by 2010 the SDR, NSRCG, BRDIS, and HERD saw response rates fall into the 70-75 percent range, while the NSCG hovered around the 80 percent mark, down from 87 percent just 2 years prior.[7] Given the importance of these surveys to NCSES's indicators on innovation activities in the United States and abroad, these declining response rates represent a critical issue for the agency.

---

[3]For details on each of these surveys, see http://nsf.gov/statistics/question.cfm#ResearchandDevelopmentFundingandExpenditures [November 2012]. A sixth survey, the Survey of Research and Development Funding and Performance by Nonprofit Organizations, was conducted in 1973 and for the years 1996 and 1997 combined. The final response rate for the 1996-1997 survey was 41 percent; see http://www.nsf.gov/statistics/nsf02303/sectc.htm [January 2014]. This lower-than-expected response rate limited the analytical possibilities for the data, and NSF did not publish state-level estimates. The nonprofit data cited in *National Patterns* reports either are taken from the Survey of Federal Funds for Research and Development or are estimates derived from the data collected in the 1996-1997 survey. See National Science Foundation (2013c, p. 2), which states: "Figures for R&D performed by other nonprofit organizations with funding from within the nonprofit sector and business sources are estimated, based on parameters from the Survey of R&D Funding and Performance by Nonprofit Organizations, 1996-1997."

[4]A microbusiness is defined as a corporation, partnership, or sole proprietorship operating in the United States or Puerto Rico with fewer than five employees. By this definition, a microbusiness may have between zero and four employees. As micobusinesses with zero employees are numerous and have different characteristics from businesses with one to four employees, there is a case for treating them separately for analytical purposes.

[5]NCSES also added a "field of bachelor's degree" question to the American Community Survey in 2009. More detail on this development is provided in Chapter 6.

[6]This statement refers to survey nonresponse. The panel did not have data on item nonresponse rates for NCSES.

[7]This information was provided by NCSES. It is important to note, though, that the SED, Graduate Student Survey, Federal Funds Survey, Federal Support Survey, State Agency R&D Survey, and Academic Facilities Survey had response rates above 90 percent.

Second, costs per respondent have increased dramatically for almost all of NCSES's surveys. The same surveys that have seen declines in response rates also have seen marked increases in costs per respondent, which have doubled or almost tripled for some surveys. Obtaining responses from marginal respondents, especially when the goal is to achieve response rates above 80 percent, is increasingly expensive, and this is a factor in prolonged delays between fielding a questionnaire and publishing related data tables and statistics.

Third, 80 percent of NCSES's expenditures on surveys in 2010 were for human resources data. That year, BRDIS accounted for 10 percent of survey expenditures. The remaining 10 percent was for surveys on academic facilities, FFRDCs, and R&D expenditures by state governments; the Federal S&E Support to Universities Survey; and contributions to the General Social Survey for information on public attitudes toward and knowledge of S&E (see Figure 3-1).

NCSES explained to the panel the high share of expenditures on human resources surveys: gathering information on individuals is more difficult and time-consuming than gathering information from firms or government agencies that can rely on some administrative reporting for answers to questionnaires. In addition, because each survey has an assigned staff member to manage contracting, site visits, analysis, and publication of information related to the survey, approximately three-fourths of NCSES personnel costs is related to the management of human resources data and statistics.

In light of tightening budgets, NCSES is seeking to rationalize its survey operations, looking for opportunities to reduce its dependence on large, lengthy surveys while augmenting data and statistical acquisitions with a focus on satisfying user needs, as discussed in Chapter 2. To its credit, the agency has already cut costs on the human resources side. NCSES eliminated the NSRCG; 2010 was the final year of data collection for this survey. However, users will still be able to obtain data on recent college graduates, because NCSES is relying on the American Community Survey (ACS) for a sampling frame for other surveys and for data on new college graduates. This decision—and the implications for indicators on human capital—is examined in Chapter 6.

Despite NCSES's sizable expenditures on human resources surveys, the number of tables or data series published by the agency appears to be more balanced. Figure 3-2 shows the concentration of NCSES tables, with points closer to the center of the diagram indicating greater concentration.[8] Based on the datasets described earlier in this chapter, the figure shows that business expenditures on R&D (BERD) make up 52 percent of NCSES's databases, while the category "scientists and engineers" (including academic and workforce data) accounts for 44 percent.

Figure 3-3 similarly shows the concentration of STI datasets for the United Nations Educational, Scientific and Cultural Organization (UNESCO), OECD, Eurostat, and Statistics Canada. UNESCO, Eurostat, and Statistics Canada have a greater concentration of innovation statistics and much less focus on human resources statistics than is the case for NCSES. OECD appears to place greater emphasis on measuring activities of researchers, technology trade in R&D-intensive industries, and business expenditures on R&D. For international comparability purposes, it appears that a balance would be accomplished and users would be better served if NCSES improved its portfolio of indicators on innovation activities (including innovative output), while other countries produced comparable measures on scientists and engineers (e.g., number of earned bachelor's, master's, and doctoral degrees as indicators of potential for R&D and innovation in S&T fields). This is not to say that a proliferation of indicators is preferable to the status quo. Chapters 4-6 provide detail on the types of indicators that should be targeted for further development by NCSES and its international counterparts.

NCSES'S STI INDICATORS

The National Science Board's *Science and Engineering Indicators* (*SEI*) and companion publications are major outlets for NCSES's STI indicators. Indicators also appear in NCSES's *InfoBriefs* and interactive tools on the agency's website.

More than 300 indicators are published in *SEI*, covering a wide range of topics for several countries and regions around the world and for a variety of demographic classifications.[9]

Figure 3-4 shows the distribution of indicators in several areas of importance to NCSES's clientele. Based solely on shares of different types of indicators published in *SEI* as shown in the figure, it is clear that a rich set of indicators covers science, engineering, and health degrees; the S&E workforce; education assessment; public attitudes toward S&T; trade in R&D-intensive industries; and R&D expenditure. By contrast, there is relatively sparse coverage of direct measures of innovation, public-sector R&D, and R&D conducted by nonprofit organizations—areas of keen interest for users of indicators and therefore areas in which NCSES could improve its portfolio.

---

[8]The categories for the diagram were chosen for purposes of comparison with data from international organizations. Hence the shares do not add to 100 percent. For example, science, engineering, and health (SEH) degrees make up a portion of the data on "scientists and engineers." Some organizations report SEH degrees, while others report data on scientists and engineers (including academic and workforce data). Concentration is defined here as the percentage of tables produced on a particular subtopic relative to the total tables generated by the STI database.

[9]See National Science Foundation (2012c) for the appendix of STI indicators. Appendix F of this report contains a summary table of these indicators.

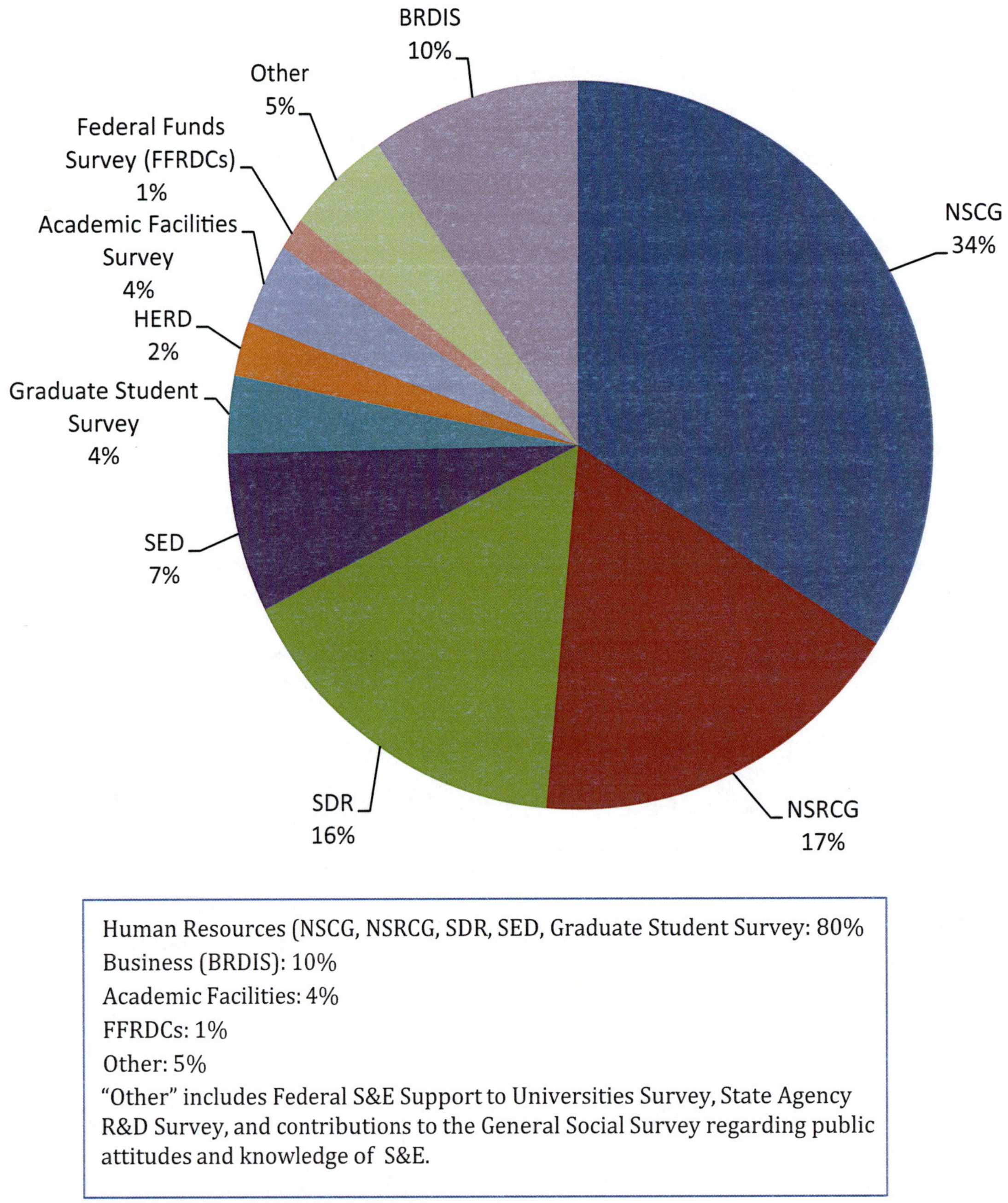

FIGURE 3-1 Percentage of NCSES survey costs, 2010.
NOTES: BRDIS = Business Research and Development and Innovation Survey; FFRDC = Federally Funded Research and Development Center; HERD = Higher Education Research and Development Survey; NCSES = National Center for Science and Engineering Statistics; NSCG = National Survey of College Graduates; NSRCG = National Survey of Recent College Graduates; R&D = research and development; SDR = Survey of Doctorate Recipients; SED = Survey of Earned Doctorates.
SOURCE: NCSES data.

HEAT MAP EXERCISE

The panel used another method—a heat map exercise—to determine the areas in which NCSES has abundant coverage of STI indicators—items that are highly correlated and potentially tell a similar story about trends in S&T activities. Perhaps this method could also be used to identify areas in which there is a paucity of information. The results of this heat map exercise for NCSES's indicators are formally presented in Appendix F and summarized here.

Initially, all of the variables were included in the heat map analysis and multidimensional scaling of the Pearson correlation matrix. The result was multiple clusters because most of the variables are tabulations of main variables, so that the

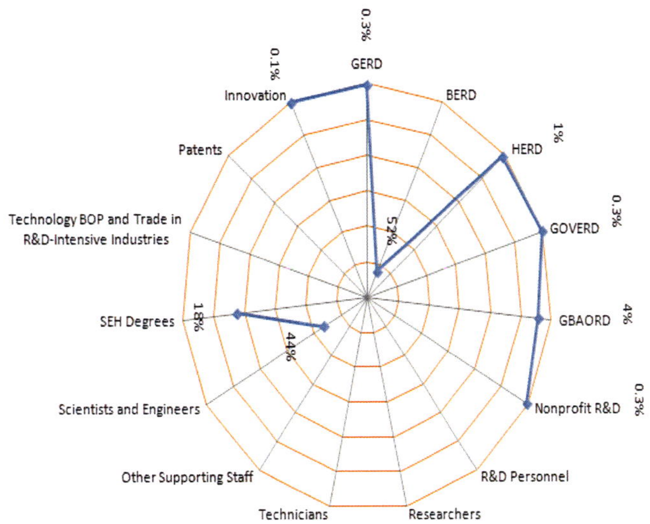

FIGURE 3-2 NCSES's concentration in STI subtopics.
NOTES: BERD = business enterprise expenditure on R&D; BOP = balance of payments; GBAORD = government budget appropriations or outlays for research and development; GERD = gross domestic expenditure on research and development; GOVERD = government intramural expenditure on R&D; HERD = higher education expenditure on research and development; NCSES = National Center for Science and Engineering Statistics; R&D = research and development; SEH = science, engineering, and health.
SOURCES: Adapted from BRDIS, see http://www.nsf.gov/statistics/industry/ [November 2012]. Federal Funds, see http://www.nsf.gov/statistics/fedfunds/ [November 2012]. R&D Expenditure at FFRDCs, see http://www.nsf.gov/statistics/ffrdc/ [November 2012]. HERD, see http://www.nsf.gov/statistics/herd/ [November 2012]. Science and Engineering State Profiles, see http://www.nsf.gov/statistics/pubseri.cfm?seri_id=18 [November 2012].

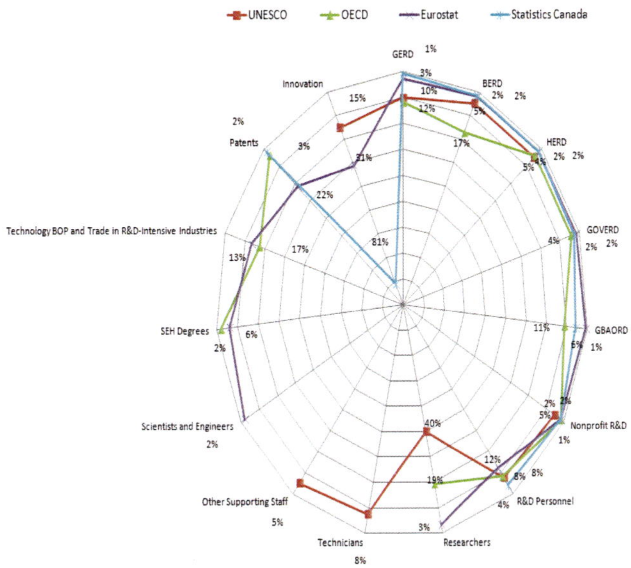

FIGURE 3-3 Subtopics of STI data produced by agencies/organizations other than NCSES.
NOTES: BERD = business enterprise expenditure on R&D; BOP = balance of payments; GBAORD = government budget appropriations or outlays for research and development; GERD = gross domestic expenditure on research and development; GOVERD = government intramural expenditure on R&D; HERD = higher education expenditure on research and development; OECD = Organisation for Economic Co-operation and Development; R&D = research and development; SEH = science, engineering, and health; UNESCO = United Nations Educational, Scientific and Cultural Organization.
SOURCES: Adapted from UNESCO, see http://www.uis.unesco.org/ScienceTechnology/Pages/default.aspx [November 2012]. OECD, see http://stats.oecd.org/Index.aspx?DataSetCode=MSTI_PUB [November 2012]. Eurostat, see http://epp.eurostat.ec.europa.eu/portal/page/portal/science_technology_innovation/data/database European Union, 1995-2013 [November 2012]. Statistics Canada, CANSIM; see http://www5.statcan.gc.ca/cansim/a33?lang=eng&spMode=master&themeID=193&RT=TABLE [November 2012].

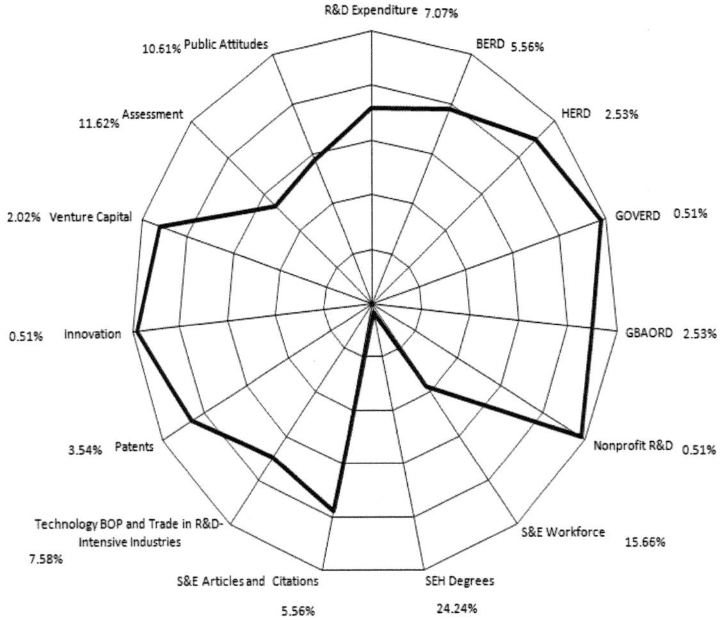

FIGURE 3-4 *Science and Engineering Indicators 2012*: Distribution of STI subtopics.
NOTES: BERD = business enterprise expenditure on R&D; BOP = balance of payments; GBAORD = government budget appropriations or outlays for research and development; GERD = gross domestic expenditure on research and development; GOVERD = government intramural expenditure on R&D; HERD = higher education expenditure on research and development; R&D = research and development; S&E = science and engineering; SEH = science, engineering, and health.
SOURCE: Adapted from *Science and Engineering Indicators, 2012*, see http://www.nsf.gov/statistics/seind12/tables.htm [January 2014].

number of variables gleaned from a single subtopic was very large. Because the aim of this analysis was to understand the redundancy in main S&T variables, the variables included in the heat map exercise were narrowed to those that address key STI topics. Therefore, the heat map exercise did not utilize all *SEI* 2012 variables, but it did capture all the major areas important to users of the statistics.

Before delving into the results of the heat map exercise, it is essential to note that the granularity and frequency that users demand of STI indicators often determine which indicators take priority in the portfolio. Certain users may want to understand levels of R&D expenditure to compare nations, while others may want the same variable for subnational comparisons (say, hot spots for innovation activities in different areas of the United States). Thus the same statistics may be produced at different geospatial or sectoral levels to satisfy different types of users. Also, highly correlated variables may have substantively different meanings and not stem from the same root logically. Therefore, the information conveyed by each of the highly correlated variables may be useful to users, offering no obvious opportunity for eliminating any specific indicator. Nevertheless, it is helpful to get a picture of the areas in which indicators have proliferated yet point to the same concept or idea, and for this purpose the heat map exercise can be instructive.

Consider the 20 variables in Table 3-1. These variables represent only a small fraction of the STI indicators produced by NCSES, but they are presumed to convey very different information about the S&E system. The R&D intensity variable is highly used by policy makers, as are measures of doctoral degree holders, trade in S&T products and intangible assets, and publications of discoveries in journals. The related heat map shown in Figure 3-5 reveals that (1) the export and value-added variables are shaded red and are therefore highly, positively correlated; (2) the education degrees and R&D intensity variables appear in the yellow area, and are not much in line with one another; (3) the variables doctoral degrees in natural science, global high-value patents, trade balance in knowledge-intensive services, and engineering journal articles as a share of total S&E journal articles are not correlated at all with one another; and (4) the variables doctorates in engineering, research articles with international coauthors, and share of citations in international literature are strongly, negatively correlated. One conclusion to be drawn here is that the highly, positively correlated variables are derived from the same root or factor, so it could be helpful to use that basis as an indicator. Again, users from the health and social services sector will want to know levels and trends for that particular sector. They are unlikely to be as satisfied with information from other knowledge-intensive sectors, even if those statistics are highly correlated with those from the health and social services sector.

Another way to view the results of the heat map exercise is to look at the multidimensional scaling of the variables.

TABLE 3-1 Variables Used in Heat Map Exercise

Indicator Label	Indicator
1. RD_%_GDP	R&D expenditures as a share of economic output (research and development [R&D] as percentage of gross domestic product [GDP] or R&D intensity)
2. Deg_NatSci	First university degrees in natural sciences
3. Deg_Eng	First university degrees in engineering
4. Doct_NatSci	Doctoral degrees in natural sciences
5. Doct_Eng	Doctoral degrees in engineering
6. S&E_Art	Science and engineering (S&E) journal articles produced
7. Eng_Share_S&E_Art	Engineering journal articles as a share of total S&E journal articles
8. Res_Art_Int_CoAuthor	Percentage of research articles with international coauthors
9. Share_Citation_Int_Lit	Share of region's/country's citations in international literature
10. Global_HighValue_Patents	Global high-value patents
11. Export_Comm_KIS	Exports of commercial knowledge-intensive services
12. HighTech_Exports	High-technology exports
13. Trade_Balance_KIS_IntAsset	Trade balance in knowledge-intensive services and intangible assets
14. VA_HighTech_Manu	Value added of high-technology manufacturing industries
15. VA_Health_SS	Global value added of health and social services
16. VA_Educ	Global value added of education services
17. VA_Whole_Retail	Global value added of wholesale and retail services
18. VA_Real_Estate	Global value added of real estate services
19. VA_Transport_Storage	Global value added of transport and storage services
20. VA_Rest_Hotel	Global value added of restaurant and hotel services

Figure 3-6 shows groupings of variables that are highly associated with one another. R&D intensity lies close to the value-added variables in the center of the diagram, suggesting relatively high correlations among those variables. One might not want to eliminate some items in the center of the multidimensional configuration, but one might want those variables gathered with less frequency or across staggered time periods. Of interest, the variables on the periphery of Figure 3-6 can provide additional information for those in the center. The former variables are grouped as follows: (1) engineering journal articles as a share of total S&E journal articles, trade balance in knowledge-intensive services and intangible assets, and global high-value patents and (2) share of region's/country's citations in international literature and percentage of research articles with international coauthors. Doctoral degrees in natural sciences and in engineering do not overlap in the space. Global value added of real estate services is negatively related to most of the variables. In addition, the variable S&E journal articles produced is far away from the variable engineering journal articles as a share of total S&E journal articles, indicating that these two variables convey independent information.

This method of mapping the portfolio of STI indicators and observing which indicators are similar at least in some dimensions of what they convey can be useful in setting priorities for NCSES's indicators program. However, there are three important caveats. First, it is important to analyze a full set of indicators, not just the selection used in this report for illustrative purposes. Second, as stated earlier, the panel believes that understanding user needs and the best indicators to address user demand takes precedence over methods that cannot weigh users' specific interests. Third, other methods, such as principal components and model-based cluster analysis, should be applied to see whether statistical artifacts arise from a given clustering method. Finally, to consider cost savings, a more comprehensive analysis would be necessary, including information on how many surveys or survey questions need to be analyzed to produce the data and whether variables can be obtained from administrative records or organizations that would produce the statistics in any event.

GAPS IN STI INDICATORS THAT NCSES SHOULD FILL

Taking the inventory of all of the indicators produced by NCSES together with the list of indicators derived using the policy-driven framework in Chapter 2, the panel identified key indicators that NCSES currently produces that satisfy user priorities and those indicators that need further development or need to be created over time.

High-Priority STI Indicators Currently Produced by NCSES

It is no surprise that NCSES has an impressive collection of data and indicators on R&D expenditures; science, technology, engineering, and mathematics (STEM) education; and the STEM workforce and talent. These are the bedrock statistics that have long been in NSF's purview. Also in NCSES's portfolio are statistics on innovation and other measures of commercial outputs and outcomes and knowledge outputs, and some information on institutions and organizations that are actors in the STI system. Hall and Jaffe (2012, p. 20) note, however, that

> There are also multiple indicators that correspond to knowledge and human capital outputs, although these measures are universally proxies that are related to the underlying concepts with substantial measurement error (e.g., degrees as a measure for the human capital of graduates; papers as a measure of new scientific knowledge; patents as a measure

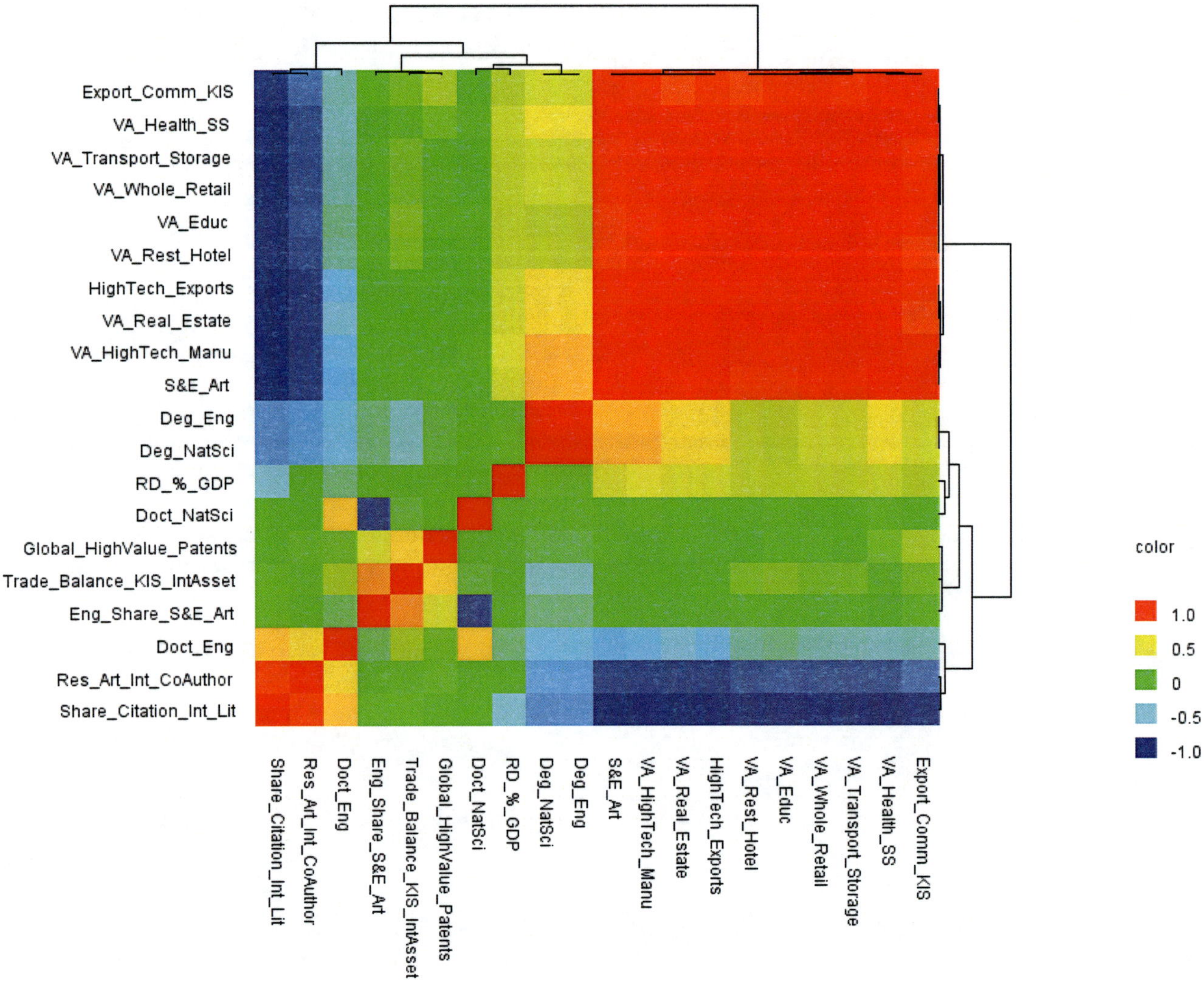

FIGURE 3-5 Heat map based on Pearson correlation matrix for 20 variables (see Table 3-1).
SOURCE: Produced by panel from the National Science Board (2012a). See Appendix F for details of heat map generation.

of new technical knowledge). The measures of innovation (as opposed to innovative activities) are much less complete, and arguably more "distant" from the underlying concepts. For example, the errors of both over-inclusion and under-inclusion in using new trademark registrations as a proxy for innovation are probably even greater than the corresponding errors in using patents as a proxy for new inventions. And a similar observation applies to using high-tech value added as an indicator for increased output that can be attributed to innovation.

Box 3-1 shows the STI indicators already produced by NCSES that users identified as priorities. Any improvements to these indicators would likely involve changes in frequency or granularity to enhance their utility. For example, some users stated that it would be helpful to have some measures available more than once per year (say, quarterly), while other indicators that fluctuate little could be reported annually or biannually or even every 3-5 years. Users were pleased with the validity of the statistics, informing the panel that having a reliable, unbiased source of data on companies, educational institutions, and other countries was an important public good. They stressed, however, that receiving these statistics in a timely manner is of paramount importance and expressed some concern about the growing gap between observation and public release of the data.

Gaps That NCSES Should Fill in the Near Term

NCSES has an opportunity in the near term to produce new or revised STI indicators in a few key areas, based on existing datasets or statistics from other agencies and organizations. The core chapters of this report—Chapters 4, 5, and 6—offer specific recommendations for improvements to NCSES's measures of innovation, knowledge capital, and

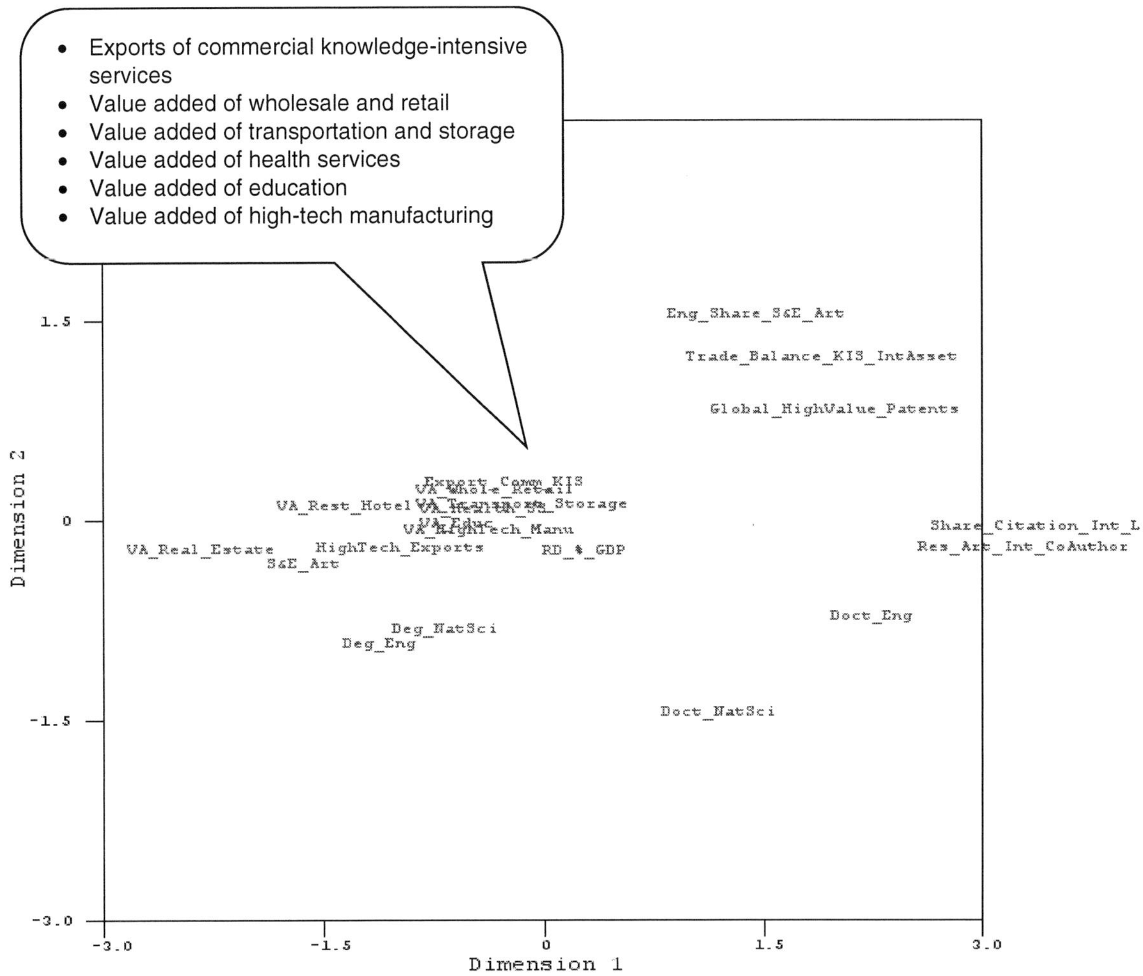

FIGURE 3-6 Multidimensional scaling of 20 measures (see Table 3-1).
SOURCE: Produced by panel from the National Science Board (2012a). See Appendix F for details of heat map generation.

human capital. NCSES should focus first on cultivating measures in areas in which the data already exist in its BRDIS and SESTAT databases and in which it has productive collaborations with other statistical agencies in the United States and abroad. Indicators that fall into this category include (but are not limited to)

- innovation breakdowns by firm size (based on number of employees) that are comparable with OECD statistics;
- expenditures on design and technical specifications, including computer systems design and design patents;
- sale of new-to-market and new-to-firm innovations as a share of turnover;
- births and deaths of businesses linked to innovation outputs (firm dynamics by geography, industry, business size, and business age);
- technology balance of trade (especially intellectual property);
- knowledge stocks and flows in specific sectors, including nanotechnology, information technology, biotechnology and agriculture research, oil and gas production, clean/green energy, space applications, weapons, health care technologies, educational technologies (massive open online courses [MOOCs]), and mining;
- advanced manufacturing outputs (information technology-based processes);
- percentage of faculty in nonteaching and nonresearch roles at universities;
- share of population aged 30-34 having completed tertiary education;
- share of youth aged 20-24 having attained at least upper-secondary-level education;
- persistence and dropout rates in education, by geographic and demographic distinctions;
- postdoctoral levels and trends in various STEM fields, by country of birth and country of highest degree;

BOX 3-1
High-Priority Indicators Currently Produced by NCSES

Research and Development (R&D)

- National R&D expenditures
  - Federal and state funds for basic research
  - Public-sector R&D (focus on advanced manufacturing, green technologies, energy-related R&D, nanotechnology, agriculture, weapons)
  - Public R&D spending as a share of gross domestic product (GDP)
  - Business R&D spending
  - Business R&D as a share of GDP
  - Industry support for R&D in universities
  - Social science R&D
- National R&D performance (by type of industry and source of funds)

Innovation

- Firms (5+ employees) introducing new or significantly improved products or processes as a share of all firms
- Firms (5+ employees) introducing new or significantly improved goods or services as a share of all firms

Commercial Outputs and Outcomes

- Medium- and high-tech manufacturing exports as a share of total product exports
- Knowledge-intensive service exports as a share of total service exports
- Value added in manufacturing
- Value added in technical services
- Trade flows of science and technology (S&T) products and services
- Information and communication technology (ICT) output and sales (intermediate and final)

Knowledge Outputs

- U.S. receipts and royalty payments from foreign affiliates
- U.S. patent applications and grants by country, technology
- U.S. trademark applications and grants by country, technology
- Patent citations

- number of postdoctoral fellows in health (specific fields); and
- labor mobility and workforce migration.

Over time, NCSES should build capacity in house and through its Grants and Fellowships Program to develop measures that are of high priority for users but require deeper knowledge of how to obtain statistically valid data or require the use of frontier methods (as described in Chapter 7).[10] Some of this information might be in the form not of quantitative measures but of in-depth case studies that NCSES could obtain from the research community and communicate through *InfoBriefs* or vignettes in the National Science Board's biennial *SEI* volume. NCSES's Grants and Fellowships Program also could benefit synergistically from cofunding opportunities with the Science of Science and Innovation Policy Program, which also resides in NSF's Social, Behavioral, and Economic Sciences Directorate. Important as well is for NCSES to develop a roadmap or strategic plan for adding new indicators or case studies, because doing so will likely require curtailing the frequency of some of its current measures. Proliferation of indicators is not the goal but rather the development of a rich portfolio of information on the global STI system desired by users.

PRIORITIZATION

Users of NCSES's data and statistics are diverse and eager to obtain information on a variety of topics that the agency has as yet been unable to produce. NCSES has access to some

[10]The solicitation for NCSES's grants competition is found at http://www.nsf.gov/funding/pgm_summ.jsp?pims_id=5265&org=SBE&from=home [January 2014].

- License and patent revenues from abroad as a share of GDP
- Triadic patent families by country
- Percentage of patent applications per billion GDP
- Percentage of patent applications related to societal challenges (e.g., climate change mitigation, health) per billion GDP

Science, Technology, Engineering, and Mathematics (STEM) Education
- Expenditures, direct and indirect costs, investments, revenues, financing for STEM education
- Enrollment data by STEM field at various levels (e.g., associate's, bachelor's, master's, doctorate) and for various types of institutions
- New degrees (e.g., associate's, bachelor's, master's, doctorate); new doctoral graduates per 1,000 population aged 25-34
- Stock of degrees (e.g., associate's, bachelor's, master's, doctorate)
- Number of high school students pursuing associate's degrees and implications for the workforce and cost of higher education
- Disciplines in which community colleges have comparative advantage
- Foreign-born STEM-educated individuals—country of birth; immigration visas, etc.
- Stay rates of foreign students
- Trends in online learning

STEM Workforce/Talent
- STEM employment
- Demographic composition of people who would enter specific occupations (e.g., clean energy, ICT, biotechnology, health services)
- Fraction of STEM degree holders that hold STEM jobs
- Earnings by degree type and occupation
- Employment in knowledge-intensive activities (manufacturing and services) as a share of total employment

Organizations/Institutions
- Public-private copublications per million population
- Number of international collaborations
- Technology transfer between academic institutions and businesses, including mechanisms
- Technology transfer (Manufacturing Extension Partnership)
- Technology transfer from national laboratories
- Research and experimentation tax credits (federal and state)

of the raw data required to satisfy user demands, and it has opportunities to provide more high-utility policy-relevant indicators in particular with the use of various methodologies. However, satisfying user demands and anticipating future demands on its databases and analytical products will require developing a strategic plan, whose execution will in turn require careful husbanding of existing resources and possibly new financial and human capital resources as well.

RECOMMENDATION 3-1: In the near term, the National Center for Science and Engineering Statistics (NCSES) should work to produce new and revised science, technology, and innovation indicators in a few key areas, using existing data from the Business Research and Development and Innovation Survey and the Scientists and Engineers Statistical Data System or from productive collaborations with other statistical agencies in the United States and abroad. Over time, NCSES should build capacity in house and through its Grants and Fellowships Program to develop measures that are high priority for users but that require deeper knowledge to obtain statistically valid data or to use frontier methods appropriately. NCSES should also develop a strategic plan for adding new indicators or case studies because doing so may require curtailing the frequency of some of its current measures.

SUMMARY

Four main themes arise from the discussion in this chapter. First, NCSES has an array of education and workforce data from which to produce indicators of STI talent. These indicators address many of the questions raised in Chapter 2 regarding STI talent. However, there is still room to improve coverage of human capital, especially regarding the mobility of S&E workers within the United States and between countries, and there is a paucity of data on training of S&E workers at the postdoctoral level and in the workplace. NCSES and statistical organizations around the world should collaborate on building capacity in these areas.

Second, in contrast to human capital indicators, international organizations such as OECD, Eurostat, UNESCO, and Statistics Canada have comparatively more developed innovation indicators than NCSES. Improving international comparability could be a mutually beneficial collaborative effort among these organizations.

Third, as NCSES determines which new indicators to produce to satisfy user needs, it must make difficult decisions about which indicators to discontinue producing or at least to publish less frequently. This chapter offers some guidance on empirical methods that could be used to help make those decisions. However, the priorities and diverse needs of users should remain primary considerations as these decisions are made.

Fourth, improving the ability to address pressing policy-relevant issues in the future will require more than raw data or statistical measures. Acquiring analytical knowledge about STI worldwide will require statistics derived from empirical research; experimental exercises that allow counterfactual analysis to reveal impacts of expenditures on R&D and innovation; and case studies that convey narratives regarding collaborative activities, networks, and other characteristics of tacit knowledge that are key drivers of the international system of innovation. NCSES is primed to take on these challenges over time, with some near-term opportunities to satisfy user demands.

4

Measuring Innovation

This chapter considers the measurement of innovation—in particular, the outputs of the country's innovation system.[1] It presents definitions of innovation, explains the importance of measuring innovation and the relevance of innovation measures for policy decisions, and examines the role of innovation surveys and their limitations. The chapter then reviews improvements to innovation surveys, in particular the Business Research and Development and Innovation Survey (BRDIS), including improving international comparability, gathering deeper information on innovations, extending BRDIS to include such items as organizational and marketing as well as "unmarketed" innovations and to track a broader array of inputs to innovation, improving the presentation of information, and improving linkages between the BRDIS data and other datasets. Finally, the chapter turns to the use of nontraditional methodologies to track innovation, such as the use of "business practice" data.

In accordance with the framework used in this study (Chapter 2), the panel's recommendations for the National Center for Science and Engineering Statistics (NCSES) regarding the measurement of innovation are driven by key policy questions and by specific questions raised by users of science, technology, and innovation (STI) indicators, as well as by the need for internationally comparable measures. The panel acknowledges that the innovation measures presented in this chapter answer policy questions only in part. It is essential that such measures be used in concert with in-depth analysis of economic and institutional environments and understanding of behavioral responses to changes in those environments. An innovation may reach forward into varied markets and backward to human capital and at times research and development (R&D) inputs. The discussion in this chapter must therefore be viewed in the context of an indicators program that is analytically strong.

DEFINITIONS

Schumpeter (1934, p. 66) provided a definition of "innovation" early in the 20th century. He defined product innovation as "the introduction of a new good . . . or a new quality of a good" and process innovation as "the introduction of a new method of production . . . or a new way of handling a commodity commercially." This definition influenced early attempts to measure the activity of innovation through case studies and surveys. Other definitions offered more recently are presented in Box 4-1.

Knowledge gained from a decade of experience with experimental surveys in Canada, Germany, the Nordic countries, and the United States was codified by OECD in the first Oslo Manual (OECD, 1992), which dealt with innovation in manufacturing products and processes. Following more measurement experience, the scope of the manual was extended to the nonagricultural economy (OECD-Eurostat, 1997). This broadened scope gave rise to a better understanding of the measurement of innovation in service industries and expansion of the definition of innovation to include not only product and process innovations but also organizational change and the development of new or the extension of existing markets. These additional components of the definition were also considered by Schumpeter.

For measurement purposes, the Oslo Manual (OECD-Eurostat, 2005, p. 46) defines innovation as follows:

[1]Edquist (2005, p. 182) defines the "system of innovation" as "all important economic, social, political, organizational, institutional, and other factors that influence the development, diffusion, and use of innovations." Lundvall (1992, p. 2) states that a system of innovation is "constituted by elements and relationships which interact in the production, diffusion and use of new, and economically useful, knowledge." Lundvall (2007, p. 106) extends this definition by focusing on the learning processes, stating: "The analysis of innovation systems may be seen as an analysis of how knowledge evolves through processes of learning and innovation." Other noted studies on national innovation systems include Freeman (1995); Furman et al. (2002); Lundvall (1988); Martin and Johnston (1999); Mowery (1992); Muller and Zenker (2001); and Nelson (1993).

BOX 4-1
Definitions of Innovation

The following definitions of innovation vary, but the common thread is the extraction of economic value from novel activities (Innovation Vital Signs Project 2007):

Innovation is "the commercial or industrial application of something new—a new product, process or method of production; a new market or sources of supply; a new form of commercial business or financial organization."

Schumpeter 1983

Innovation is the "intersection of invention and insight, leading to the creation of social and economic value."

Council on Competitiveness 2005

Innovation covers a wide range of activities to improve firm performance, including the implementation of a new or significantly improved product, service, distribution process, manufacturing process, marketing method or organization method.

European Commission 2004

Innovation—the blend of invention, insight and entrepreneurship that launches growth industries, generates new value and creates high value jobs.

Business Council of New York State 2006

The design, invention, development and/or implementation of new or altered products, services, processes, systems, organization models for the purpose of creating new value for customers and financial returns for the firm.

Advisory Committee on Measuring Innovation in 21st Century Economy, Department of Commerce 2008

An innovation is the implementation of a new or significantly improved product (good or service), or process, a new marketing method, or a new organizational method in business practices, workplace organization or external relations. Innovation activities are all scientific, technological, organizational, financial and commercial steps which actually, or are intended to, lead to the implementation of innovations.

OECD-Eurostat 2005

Innovation success is the degree to which value is created for customers through enterprises that transform new knowledge and technologies into profitable products and services for national and global markets. A high rate of innovation in turn contributes to more market creation, economic growth, job creation, wealth and a higher standard of living.

Innovation Vital Signs Project 2007

SOURCE: Aizcorbe et al. (2009).

> . . . the implementation of a new or significantly improved product (good or service), or process, a new marketing method, or a new organizational method in business practices, workplace organization or external relations.

Innovations can be distinguished from inventions by the criterion that innovations are implemented in the marketplace. By definition, therefore, innovation is a new product or idea that has been commercialized. This does not mean that an innovation is necessarily widely distributed or diffused in a market. It does mean that neither a new product that is not marketed nor a new process that does not help get a product to market is considered an innovation. Note that later in this chapter, the panel considers the possibility of collecting information on "unmarketed" innovations to address important policy questions.[2]

By contrast, the Oslo Manual (OECD-Eurostat, 2005, p. 18) identifies such inputs as R&D, capital expenditures, and training as "innovation activity":

> . . . all scientific, technological, organisational, financial and commercial steps which actually, or are intended to, lead to the implementation of innovations. Some innovation activities are themselves innovative; others are not novel activities but are necessary for the implementation of innovations. Innovation activities also include R&D that is not directly related to the development of a specific innovation.

[2]The European Community Innovation Survey (CIS) has been gathering such information, which has been shown to be policy relevant. For CIS 2012, the relevant question is 4.1 (Eurostat, 2012).

WHY MEASURE INNOVATION?

The underlying analytical framework used in this study posits that new and improved products and processes, new organizational methods, and new marketing concepts and strategies are key measures of the output of a country's innovation system. Innovation affects both economic performance measures, such as productivity growth, profits, and job creation, and noneconomic variables, such as life expectancy and environmental response. Meanwhile, the rate of innovation responds to inputs such as R&D spending; the availability of science, technology, engineering, and mathematics (STEM) labor; regulatory policies; and other variables.[3]

From this perspective, the panel strongly believes that NCSES needs to improve its ability to measure and track innovation. Improved measures of innovation are necessary to assess the impact of federal, state, and local innovation policies, such as the amount and direction of federal R&D funding, support for STEM education at the graduate level, and regulation of new products and services. In addition, having good measures of innovation output facilitates comparison of the United States with other countries in a key area that promotes economic growth. NCSES's mandate in the America Creating Opportunities to Meaningfully Promote Excellence in Technology, Education, and Science (America COMPETES) Act (U.S. House of Representatives, 2010) includes the curation and dissemination of data on "United States competitiveness in science, engineering, technology, and research and development." Innovation is an important element for such comparisons. Without improved direct indicators of innovation outputs, policy analysis will continue to rely on imperfect indicators of innovation, such as number of patents granted; inputs to innovative activities, such as R&D spending and number of STEM workers; and broad performance measures for the economy, such as productivity.

A voluminous literature addresses the historical link between patents and innovation, and the topic remains controversial today. However, there is little doubt that the more than doubling of patent grants over the past 20 years—from 96,511 in 1991 to 224,505 in 2011—did not reflect an equally sharp rise in the volume of innovation. As recent litigation has shown, multiple patents have become a bargaining chip rather than an accurate measure of innovation output. Moreover, a recent NCSES *InfoBrief* suggests that patents are less important to companies than other forms of intellectual property protection (Jankowski, 2012). Based on data from the 2008 BRDIS, firms across virtually every industry are much more likely to cite trademarks, copyrights, and trade secrets than either design or utility patents as important for intellectual property rights. Also, the propensity to patent is highly industry dependent.

Similarly, input measures such as R&D spending cannot adequately substitute for measures of innovation output, especially if policy makers are concerned about government funds being used efficiently. Moreover, recent evidence shows that most firms that report innovations do not report R&D spending. For example, Table 4-1, drawn from the 2011 BRDIS results, shows that 79 percent of all firms with new or significantly improved products or processes did not report R&D spending. In other words, focusing only on firms that report R&D spending raises the possibility of missing much of the innovation taking place in the economy.[4]

Many innovations developed and introduced by start-up companies may not be associated with formal R&D, or R&D may not be recorded as such because start-ups often have no revenues against which to record expenses. More broadly, advances in information and communication technologies and the importance of innovations such as Google's search service, Amazon's e-commerce website, and Apple's iTunes store signal how the nature of the innovation process has changed, increasingly depending on investments in innovative assets other than R&D (e.g., data, organizational know-how) and with take-up primarily in the service sector. Indeed, in fiscal year 2006, the year before the iPhone was introduced, Apple reported spending less than 3 percent of its revenues on R&D—relatively low for a technology company.[5]

Finally, broad measures of economic performance, such as productivity, do contain information about the country's innovation system, especially over longer periods such as decades. However, productivity growth is a weak signal in the short run because it is affected by external factors such as the business cycle and by external factors such as the availability of finance.

The panel acknowledges that developing useful measures of innovation output is far more difficult than measuring inputs or patents, for example. The space of economically significant innovations is both diverse and expanding. Inno-

---

[3]One prominent model is presented by Crépon and colleagues (1998). This three-equation model has the following structure: (1) RD = f (X_r); (2) Innovation = g(RD, X_i); and (3) Productivity/other economic output = h(Innovation, Xp), where X's are vectors of other factors relevant to the output (see also Mairesse and Mohnen, 2010; OECD, 2009).

[4]Sectors in which more than 50 percent of firms indicated that they were engaged in either product or process innovation between 2009 and 2011 were *manufacturing*—petroleum and coal products; chemicals (particularly basic chemicals; soap cleaning compound, and toilet preparation; paint, coating, adhesive, and other chemicals); machinery (particularly agricultural implement; engine, turbine, and power transmission equipment); computer and electronic products (particularly communications equipment; semiconductor and other electronic components; navigational, measuring, electromedical, and control instruments; other measuring and controlling devices); and transportation equipment (particularly guided missile, space vehicle, and related parts); or *nonmanufacturing*—information (particularly software publishers). See Appendixes H-K for specific details.

[5]Apple's R&D/sales ratio has historically been relatively low compared with companies such as Google, Microsoft, Intel, and Samsung. For example, Apple's R&D/sales ratio was at 2.27 percent in fiscal year 2012, approximately one-third of Samsung's R&D/sales ratio. However, Apple may have benefited from R&D spending by suppliers such as Samsung and Intel.

TABLE 4-1 Research and Development (R&D) Firms Are More Likely to Innovate, but Most Innovating Firms Do Not Do R&D

	% of Firms with New or Significantly Improved Products or Processes			% of Total Firms in Scope*
R&D Status of Firm	Doing R&D	Not Doing R&D	Total	
Firms Doing R&D	64	N.A.	21	5
Firms Not Doing R&D	N.A.	12	79	95

NOTE: A firm that does R&D may produce many innovations, while a firm without R&D may produce only one innovation. However, BRDIS does not report on the number of innovations per firm.

*Approximately 1.2 million firms on a weighted basis are in scope, that is, reported to the survey.

SOURCE: Calculated by the panel from the 2011 BRDIS, Table 49; see Appendix K. For more information on innovation statistics from the 2011 cycle of BRDIS, see *Business R&D and Innovation: 2011*, which will be published in 2014. For previous cycles, the full set of detailed statistical tables is available in *Business R&D and Innovation: 2008-2010* (NSF 13-332) at http://www.nsf.gov/statistics/nsf13332/start.cfm [April 2014]. Innovation statistics from the 2008 BRDIS are found in Tables 33-36, the 2009 BRDIS innovation statistics are found in Tables 82-85, and the 2010 BRDIS innovation statistics are found in Tables 126-129. Relative standard errors are available from NCSES staff upon request.

vation can include everything from the successful development and approval of gene therapy, to the development of the special high-strength glass used for smartphone screens, to the steady improvement of unmanned aerial vehicles used by the military in Afghanistan, to logistical warehouse management systems. Other significant contemporary innovations include Google's AdSense/AdWords technology for targeted online advertising and Apple's invention of the App Store concept, which enables small programming firms to sell and get paid for lightweight mobile applications. Although some may disagree, moreover, the creation of credit default swaps in the 1990s was almost certainly an economically significant innovation of a type that should be considered by NCSES.[6]

Nevertheless, it is clear that as a national goal, policies that encourage bringing more innovations to market are useful if they generate economic growth and jobs and improve the nation's competitiveness. As discussed below, measures of innovation outputs are needed, and providing such measures is clearly within the purview of NCSES to support such policies.

RECOMMENDATION 4-1: The National Center for Science and Engineering Statistics should develop additional indicators for measuring innovation outcomes that would complement existing data on patents, inputs to innovation activities, and broader measures of economic performance.

POLICY RELEVANCE OF INNOVATION MEASURES

Information should not be collected in a vacuum. It is good business practice for NCSES to focus on developing new indicators that are useful for informing policy decisions, especially given the current fiscal environment. This section identifies some innovation-relevant policy questions that such indicators could help address. This is not to say that the indicators suggested here are relevant only for government policy makers; many of them should be useful to university administrators and business managers as well.

As noted above, the value of a measure of innovation is that it offers the possibility of a direct link between innovation outcomes and relevant policy variables, many of which are innovation inputs. In particular, one or more innovation measure can make it possible to test the impact on innovation of such factors as R&D spending, the availability of STEM labor, regulation, market power, globalization of research, immigration policy, and tax policy (see Box 4-2). Note that these factors can have either a positive or a negative effect on innovation, and it is essential for policy makers to know which way the arrow points and under what conditions.

Firm-level analysis also links investment in innovation to improved outcomes. Using the third Community Innovation Survey (CIS3) for 16 European countries, Jaumotte and Pain (2005) found that the proportion of firms engaged in innovation spending was closely correlated with the proportion of successful innovators in a country. Their results suggest that a 1 percentage point increase in aggregate innovation spending was associated with an increase of 0.85 percentage points in the probability of being a successful innovator and an increase of 0.7 percentage points in the share of new products in turnover.[7]

An analysis of innovative companies could shed light on such key policy questions as the appropriate amount and direction of federal R&D funding, limits on or encouragement of immigration of skilled workers, support for STEM

[6]According to one account, the concept of credit default swaps was developed in 1994 to help Exxon fund its huge potential liability for the Exxon Valdez disaster (see Tett, 2009).

[7]In this study, innovation spending included not only intramural R&D but also extramural R&D, other acquisitions of knowledge (such as rights to use patents, licenses, and software), investment in capital goods, expenditures on training, and spending necessary for the innovation to be placed on the market (for example, marketing or design expenditures). Non-R&D expenditures were sometimes more important than those on R&D, and many of the countries with the highest proportion of successful innovators also had the highest propensity to engage in non-R&D innovation spending—showing the importance of taking a broad view of innovation inputs.

BOX 4-2
Selected Innovation-Related Policy Questions

- What kinds of innovations create jobs and what kinds of jobs?
- What specific innovations have made the most important contributions to economic growth and over what time period?
- Where should the government spend its R&D dollars?
- Can government procurement policy be changed to encourage innovation?
- What new industries, occupations, and skill sets are being created by recent innovations, and are current educational institutions equipping enough people with the skills needed to take maximum advantage of these changes?
- How important is formal R&D to the innovative process, as opposed to organizational changes and continued on-the-job tinkering by employees and their managers?
- Does the globalization of R&D increase or decrease the ability to turn research into marketed innovations?
  - Are U.S. firms conducting more or less of their research activities offshore and does it even matter where this activity takes place?
  - Is the United States innovating more or less than other countries?
- Are small companies more likely to innovate than big companies? In which types of firms—small or large, young or mature—do the most important innovations occur, and is this pattern stable?
- How do service firms innovate and create new knowledge?
- What are the sources of funding for innovation activities, including R&D?
- Are potential innovations not reaching market because of too much regulation, taxes that are too high, or too much market power in product markets?
- Are potential innovations being stalled because of insufficient funding, not enough skilled workers, or too many obstacles to skilled immigrants?
- Are innovations not being brought to market or implemented because of misguided patent policies?

education at the graduate level, regulation of new products and services, rules for university technology transfers, and the appropriate role of R&D tax credits. Perhaps more important, innovation measures should help identify bottlenecks in the innovation system—cases in which most of the necessary inputs are present, but innovation apparently falls short of expectations. These are potentially high-return areas for policy, where relatively small policy changes could potentially lead to much better results.

The most striking potential bottleneck is in biomedical research, with the pharmaceutical industry and the National Institutes of Health having struggled over the past decade with how to turn increases in scientific knowledge into marketable innovations that improve human health outcomes. In this context, innovation measures could be helpful in two key ways. First, they would quantify the perception that commercialization of biomedical research is slower than expected. Second, they could help researchers offer insight into different possible explanations for the gap between knowledge and the market, providing useful guidance for policy makers.

On the broadest level, measures of innovation can help answer important policy-relevant questions such as those listed in Box 4-2. The list of such questions is easily much longer. However, current STI indicators are not fully capable of answering many of these questions.

THE ROLE OF INNOVATION SURVEYS

Currently, the most widely used tool for measuring innovation is the innovation survey. Innovation surveys ask firms about types of innovation in which they are engaged, reasons for innovating (or not), collaboration and linkages among firms or public research organizations, and flows of knowledge; they also collect quantitative data on sales of product innovations and spending on a range of innovation activities beyond R&D. They are designed to collect direct information on the elements of the knowledge-to-production system without asking about particular innovations or details thereof. They provide fairly detailed guidance to respondents about what is or is not an innovation, and allow for comparison of levels of reported innovation across organizations/countries and for analysis of the determinants of reported innovation and their link to economic output (Stone et al., 2008).

The Business Research and Development Innovation Survey (BRDIS)

In 2008, the National Science Foundation (NSF) launched the BRDIS, which collects key information on innovation as well as a wide range of other variables. Although the 2008 BRDIS was a pilot survey, it did yield some data on the incidence of product and process innovation among firms by sector (including services), size class, and whether respondents reported R&D activity (Borousch, 2010).[8] Questions on innovation were augmented in the 2009 and 2010 versions of the survey, allowing for comparison of innovation statistics across several countries.[9] Box 4-3 has selected innovation questions from the 2011 version of BRDIS (U.S.

[8]The data are based on the 2008 BRDIS, which was launched in January 2009. See National Science Foundation (2013a) for the BRDIS questionnaire.

[9]It is widely known that the innovation statistics from BRDIS and the CIS lack comparability; see an explanation in Hall (2011). Also see the discussion of the disparities between the U.S. and European innovation statistics later in this chapter.

BOX 4-3
Innovation Questions on the 2011 Business Research and Development and Innovation Survey (BRDIS)

Product (good or service) Innovation

A product innovation is the market introduction of a **new** or **significantly** improved good or service with respect to its capabilities, user friendliness, components, or sub-systems.

- Product innovations (new or improved) must be new to your company, but they do not need to be new to your market.
- Product innovations could have been originally developed by your company or by other companies.

1-11 During the three years 2009 to 2011, did your company introduce:

a. New or significantly improved goods (Exclude the simple resale of new goods purchased from other companies and changes of a solely aesthetic nature)? . . . . . ☐ Yes ☐ No

b. New or significantly improved services? . . . . . ☐ Yes ☐ No

1-12 If you answered "yes" to either 1-11, line a, or 1-11, line b, were any of your product innovations during the three years 2009 to 2011:

a. New to your market?. . . . . ☐ Yes ☐ No

Your company introduced a new or significantly improved good or service to your market before your competitors. (It may have been available in other markets.)

b. New only to your company?. . . . . ☐ Yes ☐ No

Your company introduced a new or significantly improved good or service that was already available from your competitors in your market.

1-13 Using the definitions above, please give the percentage of your total sales in 2011 from:

a. New or significantly improved goods and services introduced during 2009 to 2011 that were **new to your market** . . . . . %

b. New or significantly improved goods and services introduced during 2009 to 2011 that were **new only to your company**. . . . . %

c. Goods and services that were **unchanged or only marginally modified** during 2009 to 2011 (include the resale of new goods or services purchased from other companies) . . . . . %

d. **Total sales in 2011** . . . . . 1 0 0 %

Process Innovation

A process innovation is the implementation of a **new** or **significantly** improved production process, distribution method, or support activity for your goods or services.

- Process innovations must be new to your company, but they do not need to be new to your market.
- The innovation could have been originally developed by your company or by other companies.
- Exclude purely organizational innovations.

1-14 **During the three years 2009 to 2011, did your company introduce:**

a. New or significantly improved methods of manufacturing or producing goods or services? . . . . ☐ Yes ☐ No

b. New or significantly improved logistics, delivery or distribution methods for your inputs, goods, or services? . . . . ☐ Yes ☐ No

c. New or significantly improved supporting activities for your processes, such as maintenance systems or operations for purchasing, accounting, or computing? . . . . ☐ Yes ☐ No

SOURCE: Available: http://www.nsf.gov/statistics/ [January 2014].

Department of Commerce, 2012), which contained sections on: the 2011 survey was in the field as of this writing (see Box 4-3 for selected innovation questions from the 2011 survey). The 2011 version of the BRDIS (U.S. Department of Commerce, 2012) contained sections on:

- innovation, reported by the nature of the innovation (new product or new process), and the percentage of sales coming from newly introduced products;
- R&D spending, reported by business code, country, state, funder, and performer;
- characteristics of R&D spending, reported by type (basic versus applied versus development) and area of focus (e.g., biotechnology, nanotechnology);
- R&D employees, by type (scientists and engineers versus technicians and technologists); and
- use of intellectual property, including patents, and trademarks.

NCSES's efforts to revise BRDIS are aimed at gathering more information on innovation activities, going beyond simple "yes/no" questions on whether a firm introduced or significantly improved goods, services, or processes. These efforts also are aimed at enhancing comparability with key questions in the CIS and ensuring that the questions on innovation are answered by firms that do not conduct R&D. Comparability of the BRDIS and CIS data also depends on surveying similar populations of firms and on the techniques used to derive estimates from the data.

As befits a cutting-edge survey, BRDIS is still evolving. It is important to recognize that such concepts as invention, innovation, and technology diffusion are on a continuum, and there is still debate regarding their respective space on that continuum. As NCSES develops surveys, new datasets, and new indicators of innovation activities, it will be important to attempt to establish rigorous standards for defining these terms. Such standards will have implications for innovation surveys internationally and for the comparability of data produced by those surveys. NCSES's role in the working group of OECD's National Experts on Science and Technology Indicators (NESTI) gives the agency a good opportunity to contribute to the development of more precise standardized definitions and their implementation through surveys such as the CIS and BRDIS when the Oslo Manual on innovation (OECD-Eurostat, 2005) is next revised.

In addition to the lack of standard definitions, other factors can limit the comparability of U.S. and European data on the innovativeness of firms (see the discussion of international comparability later in this chapter in the section on improvements to BRDIS). These factors include the use of different survey frames, of different size cutoffs (BRDIS uses businesses with 5 or more employees, while other countries use businesses with 10 or 20 more employees), the impact of a lengthy R&D and innovation survey, sampling errors, and weighting issues. NCSES and OECD are actively collecting evidence to assess the factors that may drive biases in international comparisons. (See Table 4-2 for a list of websites for innovation surveys from regions around the world.)

Complete cognitive testing of the questions used on innovation surveys is ongoing in the United States and Europe. Nevertheless, the data are useful for preliminary snapshots. OECD-NESTI has been conducting a cognitive testing research project since 2011, with participation by representatives from the United States and other countries. The project is examining how businesses define and measure innovation. This effort is motivated by the sizable gap between estimates in the United States and Europe on the incidence of innovation among firms, with 2008 BRDIS data being used as the baseline for U.S. innovation statistics. The 2009 and 2010 BRDIS instruments included some systematic variations aimed at determining whether the way innovation questions were being asked substantially influenced the incidence answers provided. During a discussion with panel members, NCSES staff indicated that the conclusion to date is that there is no major difference in the statistics due to the phrasing of the questions. The weak result on innovation changed little when the questions were moved to the front of the 2009 BRDIS (see Table 4-3). The relatively low reported incidence of innovation in the United States compared with European nations remains an issue for further research by NCSES and its partners in OECD-NESTI.[10] In addition, the size dependence of the propensity to innovate may contribute to the weak results, suggesting that the Table 4-3 should be produced for at least three employment cutoffs (see Recommendation 4-2 later in this chapter).

Other U.S. Public and Private Innovation Surveys

BRDIS is not the only source of innovation-related data in the United States. Other agencies collecting information on innovation include the Bureau of Labor Statistics (BLS) in the U.S. Department of Labor, the Census Bureau, the Bureau of Economic Analysis (BEA), the National Institute of Standards and Technology (NIST) in the U.S. Department of Commerce, and the Economic Research Service in the U.S. Department of Agriculture. Indeed, if the subject is broadened to STI, then at least six agencies collect these data, including the National Center for Education Statistics, which collects data on STEM education.

The surveys administered by these agencies contain a wide range of useful global, national, state, and local indicators. However, most rely on policy indicators or indicators of inputs to innovation activity; they contain very few measures

[10]This is a reminder that innovation is difficult to measure, and international comparability is difficult to achieve. Not all business strategy is "innovation." Improved business value and other success indicators can come from various quarters. Can companies distinguish innovation from, say, "continuous improvement" or other noninnovation business enhancers? How can innovation and its impact on a business be measured? These are basic questions that the cognitive testing project is considering.

TABLE 4-2 Websites for Innovation Surveys from Regions Around the World

Region	Countries	Website
European Union (EU) 27	Austria, Belgium, Bulgaria, Cyprus, Czech Republic, Denmark, Estonia, Finland, France, Germany, Greece, Hungary, Ireland, Italy, Latvia, Lithuania, Luxembourg, Malta, Netherlands, Poland, Portugal, Romania, Slovenia, Slovakia, Spain, Sweden, UK	http://ec.europa.eu/enterprise/policies/innovation/policy/innovation-scoreboard/index_en.htm
Rest of Europe	Norway, Switzerland	http://www.ssb.no/innov_en/ http://www.kof.ethz.ch/en/surveys/
Latin America	Argentina, Brazil, Chile, Columbia, Cuba, Ecuador, Mexico, Panama, Peru, Trinidad and Tobago, Uruguay	http://www.ricyt.org
Asia	Japan, Korea, Malaysia	http://anstep.net/indicators.htm http://www.kistep.re.kr/en/c3/sub3.jsp http://www.kistep.re.kr/eng/publication/survey.jsp
Pacific	Australia, New Zealand	http://www.ausinnovation.org/articles/new-thinking-new-directions-building-innovation-capability-in-australia.html http://www.msi.govt.nz/get-connected/science-and-innovation-in-nz/
Other	Canada, South Africa	http://www.statcan.gc.ca/concepts/index-eng.htm http://www.hsrc.ac.za/en/departments/cestii/innovation-survey

SOURCE: Adapted from Hall and Jaffe (2012).

TABLE 4-3 Percentage of Firms Reporting Innovation, BRDIS 2008-2011

Type of Innovation	2006-2008 (%)	2007-2009 (%)	2008-2010 (%)	2009-2011 (%)
Goods	4.6	5.8	5.5	5.4
Services	6.9	9.1	7.8	6.8
Manufacturing/production methods	3.6	4.4	4.1	3.8
Logistics/delivery/distribution methods	3.1	4.1	3.7	3.2
Support activities	6.9	9.3	8.2	7.2

NOTE: In the 2008, 2009, 2010, and 2011 BRDIS, companies were asked to identify innovations introduced in 2006-2008, 2007-2009, 2008-2010, and 2009-2011, respectively.
SOURCE: Tabulations from the 2008-2011 BRDIS.

of innovative outputs. This gap illustrates the importance of BRDIS, and it also suggests that these agencies, to the extent possible, should be working together (see Recommendation 8-2 in Chapter 8).

Limitations of Innovation Surveys

The panel strongly supports the increased use of innovation surveys as a way to measure the output of the innovation system. Later in this chapter, new, nontraditional methodologies for measuring innovation are discussed. In this section, however, the panel underscores the inherent limitations of any survey of innovation inputs and outputs.

First, surveys are time-consuming and expensive for the companies asked to complete them, as well as for those that must compile, validate, and publish the results. The quality of the survey depends not only on the number of responses, but also on the amount of time and level of personal attention paid to them by respondents, which surely vary across companies. Surveys of individuals, such as those relating to the career paths of scientists and engineers discussed later in this report, are becoming increasingly problematic because of declining response rates.[11] NCSES also experiences

[11]See the National Research Council (2013a) report on survey nonresponse. The report finds that "for many household surveys in the United States, response rates have been steadily declining for at least the past two decades. A similar decline in survey response can be observed in all wealthy

delays in obtaining data on firms, as is clear from the delays in publishing statistics from BRDIS.

A second limitation of innovation surveys is that the nature of the economy is rapidly changing, largely as a result of innovation itself. As a consequence, new and interesting subjects (such as outsourcing of R&D or the rise of "big data")[12] on which survey questions and answers would be desirable are constantly coming to the attention of policy makers and researchers. Yet because of the many lags built into the survey process—from survey design, to survey deployment, to compilation and validation of results, sometimes across multiple agencies—results may be stale or of limited use once they are published. Timing also is important because rates of dissemination of new knowledge and products are important variables of interest. Furthermore, in the current budgetary environment, it will be difficult if not impossible for NCSES to mount new surveys in the future without cutting back on those it already undertakes.

Third, large, multinational companies that report R&D expenditures report where they conduct the R&D (and thus spend money on salaries, equipment, and the like). They rarely if at all, however, report to the public and government statistical agencies the impacts of their R&D on their operations, and specifically the extent to which the location of the R&D leads (or does not lead) to the creation of jobs and income in the United States.

Fourth, the traditional classification of economic transactions into goods and services can miss important innovations that do not fall easily into either of these categories. A good example is the rise of data as a separate economic category or product. Innovations in constructing and analyzing databases are becoming increasingly important in an ever-expanding range of industries—not only Internet search, but also retailing and manufacturing and now even health care. These innovations are important, but their economic value is difficult to measure and categorize.

Fifth, a more general challenge is that some innovations—the precise share is unknown—do not result in increased output, at least as conventionally measured. Accordingly, "production functions" that relate R&D to increased output may not be capturing well the real economic impact (or spillover effects) of some innovations. For example, the activity of someone in Germany downloading a free application written in the United States will not show up in either the U.S. trade or output statistics. Similarly, an increase in Internet search activity will not show up as an increase in consumer expenditures (Mandel, 2012).

Sixth, although economists have long viewed innovation as unfailingly positive, extending the so-called "production-possibility" frontier, recent events suggest that this is not always the case. Various (but not all) financial innovations, for example, had perverse effects that contributed to the financial crisis of 2007-2008, which in turn led to the Great Recession (see, e.g., Litan, 2010). Likewise, while innovations and communication that have facilitated the growth of trade can make some countries better off, some countries or portions of their population (unskilled workers in particular) may be made worse off in a globalized economy (see, e.g., Samuelson, 1962).

Finally, the virtue of innovation surveys is that they provide an overall picture of firms' innovation activity. However, experience suggests that the subjective nature of survey responses poses some difficulties for interpretation. In particular, the percentage of self-reported innovative firms varies across countries in unexpected ways. According to the 2009 BRDIS, for example, only 6.4 percent of U.S. firms had introduced new or significantly improved goods in the previous 3 years, while only 10.3 percent reported new or significantly improved services. By contrast, the percentage of German firms reporting themselves as innovative in the period 2005-2007 was far higher, at 79.9 percent.[13] In part, this large and surprising disparity may reflect differences in survey methodology and questions. Even within the European Union (EU), however, unexpected differences are seen, with Germany reporting far higher rates of innovation than the Netherlands (44.9 percent) and the United Kingdom (45.6 percent).

The panel does not have recommendations for NCSES for responding to each of these challenges. Rather, the challenges are described here to give a sense of how changes in the economy lead to difficulties in measuring innovation and its impact. The panel does have some thoughts about general processes NCSES can follow to stay abreast of these changes and meet these challenges, which are outlined later in this and subsequent chapters of the report.

IMPROVEMENTS TO BRDIS

This section examines improvements that could be made to BRDIS in five areas: (1) international comparability; (2) deeper information on innovations; (3) extensions to cover organizational and marketing innovations, unmarketed innovations, and a broader array of inputs to innovation; (4) improvements to the presentation of information; and (5) better linkages between BRDIS data and other datasets.

countries" (National Research Council, 2013a, p. 1). The report also documents declining response rates for a variety of surveys, including panel and longitudinal surveys (pp. 14-30).

[12]The term "big data" technically refers to a collection of data so large and complex that it is difficult to store in one place and difficult to manage. Recently, however—and in this report—the term has come to refer to data that are gathered from sources other than surveys, such as administrative records, websites, genomics, and geographic sensors, to name a few. Statistically, the term and its characteristics are still being codified.

[13]The U.S. statistic was calculated from the NSF summary sheet on BRDIS. The statistics for German firms are found in Eurostat (2010).

International Comparability

One impediment to understanding and assessing innovation in the United States is the lack of comparability between U.S. STI indicators and those developed by other countries.

Comparability of BRDIS and CIS data requires surveying similar populations of firms in comparable size categories and using similar techniques to derive estimates of innovation from the raw data collected. Also needed for comparability are statistics using the same set of industries typically used in statistics for other countries. An example is the core set of industries used by Eurostat for comparison of innovation statistics among EU countries, including mining; manufacturing; and selected service industries, such as wholesale trade, transport services, financial services, information technology (IT) services, R&D services, and business services. These data could be used to compile a simple indicator of the share of product-process innovative firms, defined as firms that have implemented a product or process innovation.

Deeper Information on Innovations

Even with more comparable statistics on innovation, it still will not be clear to users that firms are representing the same or similar things when they report product or process innovations. The BRDIS questions do not give enough information to provide a full understanding of what the resulting data and statistics mean. For example, users have no independent measure of whether the innovations of firms that innovate but do not conduct R&D are more or less important than those of firms that conduct R&D. Users would have more confidence in and understanding of the BRDIS innovation measures if they knew that knowledge input measures were correlated with actual performance. Users would have even more confidence in the measures if they knew what firms meant by innovation—how closely company reporting on the survey matched NCSES's definition of innovation.

Without greater detail on specific innovations, moreover, the surveys paint an exceedingly broad picture of innovation. Knowing, for example, that 60 percent of a country's firms have introduced some type of innovation does not help in understanding why and how innovation happened, what impacts it has on the economy, and how it can be encouraged. Indicators should provide not only a level but also insight into how that level was achieved. Microdata from innovation surveys connected with other data on a firm might help achieve this goal, but this approach has as yet not been exploited. Most innovative firms introduce both product and process innovations, as well as organizational or marketing innovations (discussed below), and the impacts of the innovations are likely to depend on many other business decisions (see OECD, 2010; Stone et al., 2008).

In a recent study conducted in Australia (Arundel et al., 2010), respondents were asked to describe their most important innovation, and experts then classified these examples according to whether they met the requirements for an innovation. The study results provide valuable information on how respondents perceive innovations and on what innovation indicators represent. They also indicate which types of innovations are deemed most important for business performance; in many cases, examples of new organizational methods and new marketing concepts and strategies were cited.

NCSES might want to consider including an open question in BRDIS on the most important innovation or innovations produced by the firm over the previous 3 years. Alternatively, NCSES might want to consider commissioning a study similar to that of Arundel and colleagues (2010) on a subset of responses to BRDIS to determine what firms are measuring as innovation. It would be useful to have firms of different sizes, different sectors, and different geographic locations represented in such a study.

Extensions to BRDIS

There are three important ways in which BRDIS could be extended to make it more useful for policy purposes. First, the innovation questions could be broadened to include organizational and marketing innovations. Second, it might be appropriate for policy makers if the survey included the category of "unmarketed" innovations, reflecting the reality that an innovation may be ready to be marketed but be held back by such factors as regulation, financial constraints, or the market power of incumbents. Third, the survey could track a broader array of inputs to innovation, including non-R&D investments such as design, market research, and databases.

Organizational and Marketing Innovations

The communications sector, broadly defined to include communications-related hardware and software as well as telecommunications providers, has clearly been one of the most innovative sectors of the economy in recent years. Within this sector, smartphones are a prominent example of product innovation, while increases in mobile broadband speed exemplify process innovation. Other recent innovations in the communications sector, however, do not fit so neatly into the product and process categories. These innovations include social media, such as Twitter, Facebook, and LinkedIn; search, exemplified by Google and Microsoft; web marketing and advertising, exemplified by Google's AdSense/AdWords technology (see Box 4-4); and the availability of mobile applications through Apple's App Store.

To identify such successful and important innovations as AdSense/AdWords, NCSES may need to broaden the scope of the innovation questions on BRDIS to include organizational and marketing innovations as identified by the Oslo Manual (see Box 4-5). In addition, NCSES might consider introducing a question about significant new data algorithms as a category of innovation. Such a question would require

BOX 4-4
Anatomy of a Marketing Innovation

Google's AdSense/AdWords advertising technology is one of the most successful marketing innovations ever achieved. This technology, originally implemented piecemeal from 2000 to 2003, allows Google to place advertisements automatically on its own webpages or those of partner sites, depending on the webpages' content. At the same time, an advertiser can bid for the right to be included in these placements. A retailer selling umbrellas, for example, can bid for the right to place an advertisement on any webpage using the term "rain."

The AdSense/AdWords technology was clearly an innovation and proved remarkably difficult for competitors such as Yahoo! and Microsoft to match. Moreover, the technology allows Google to monetize successfully such product innovations as Gmail.

SOURCE: Edelman et al. (2007).

significant new conceptual development, but would help capture technological developments in some of the most dynamic sectors of the economy.

Unmarketed Innovations

A second potential expansion of the BRDIS innovation questions could be extremely useful to policy makers. As noted earlier, according to the current definition of innovation, a product must be bought to market, or implemented, to be considered an innovation. A new product, for example, does not count as an innovation unless it can be purchased or used by a consumer or a business.

From the perspective of the macroeconomic impact of innovation on growth and productivity, such a requirement makes sense. An innovation cannot affect economic performance unless it can be marketed or implemented. It is also important to understand, however, why some industries and some countries appear to produce innovations at a rapid pace, while others are less productive in turning innovative activities into marketable innovations. Part of the explanation is that promising innovations can be stalled by a variety of factors.

First, a tightening of approval regulations for scientific, political, or legal reasons can result in fewer innovations coming to market, even if the pace of technological advance stays the same.[14]

Second, the nature of payment systems can have a similar dampening effect on marketable innovation, especially in the case of health-related innovations that require long and expensive testing and are dependent on the willingness of Medicare and insurance companies to reimburse. For example, pharmaceutical companies avoided bringing vaccines to market because the reimbursement for them was relatively low, and the potential exposure to lawsuits was relatively high (Offit, 2005).

Third, in some industries, the market power of incumbents may be an important deterrent to the commercialization of

BOX 4-5
Four Forms of Innovation

The Oslo Manual (OECD-Eurostat, 2005, pp. 48-52) identifies four forms of innovation. The 2012 Community Innovation Survey (Eurostat, 2012) employs the following definitions of these forms of innovation:

Product: A product innovation is the market introduction of a new or significantly improved good or service with respect to its capabilities, user friendliness, components or sub-systems. Product innovations (new or improved) must be new to your enterprise, but they do not need to be new to your market. Product innovations could have been originally developed by your enterprise or by other enterprises or institutions. A good is usually a tangible object such as a smartphone, furniture, or packaged software, but downloadable software, music and film are also goods. A service is usually intangible, such as retailing, insurance, educational courses, air travel, consulting, etc.

Process: Process innovations must be new to your enterprise, but they do not need to be new to your market. The innovation could have been originally developed by your enterprise or by other enterprises, and excludes purely organizational innovations

Organizational: An organizational innovation is a new organizational method in your enterprise's business practices (including knowledge management), workplace organization or external relations that has not been previously used by your enterprise. It must be the result of strategic decisions taken by management, and excludes mergers or acquisitions, even if for the first time.

Marketing: A marketing innovation is the implementation of a new marketing concept or strategy that differs significantly from your enterprise's existing marketing methods and which has not been used before. It requires significant changes in product design or packaging, product placement, product promotion or pricing, and excludes seasonal, regular and other routine changes in marketing methods.

SOURCE: The 2012 Community Innovation Survey (Eurostat, 2012).

[14]The panel notes that some regulations achieve societal goals that increase social well-being and that advancing innovation clearly is not the only means of advancing a society's goals.

potential innovations. For example, if companies lack access to important sales channels that are controlled by incumbents, then they may not find it worthwhile to commercialize a new product even if that product would be economically viable in a more competitive market.

Finally, in some cases, a potential innovation may not be introduced to market because of the need for other, complementary innovations. A classic example is Chemcor, an ultrahard glass invented in the 1960s by Corning. Because of the cost of production, Chemcor did not find a place in the market until 2006. Renamed Gorilla Glass, it became the product of choice for cell phone screens.

The concept of unmarketed innovations is implicit in an existing question on the CIS:

> During the three years 2010 to 2012, did your enterprise have any innovation activities that did not result in a product or process innovation because the activities were
>
> - abandoned or suspended before completion
> - still ongoing at the end of 2012

The panel believes NCSES should consider adding a similar question to BRDIS. NCSES might also consider asking respondents to rank the main reasons why the outputs of their innovation activities have not yet been marketed or implemented, including not yet ready, lack of funding, lack of sufficient skilled labor, the need to meet regulatory requirements, or blocked from market by competitors. The panel acknowledges that such questions may be difficult for respondents to answer, although such a question on factors hampering product and process innovation has been asked in the CIS for some years (Eurostat, 2012). If such questions were added to BRDIS, then the insights they would yield into bottlenecks for innovation could be quite useful.

A Broader Array of Inputs to Innovation

The panel is aware of the budget constraints faced by NCSES. With these constraints in mind, the panel notes that BRDIS could collect other types of data that would be helpful for policy makers. Hall and Jaffe (2012), whose paper was commissioned for this study, note that it would be helpful to have more information on the amount spent on different kinds of innovation activities. For example, BRDIS currently asks respondents whether they "acquired more than 50 percent ownership in another company for the primary purpose of acquiring their IP [intellectual property]." However, it would also be useful to know how much the company spent on the acquisition of external knowledge, marketing and design of the improved goods or services, and other innovation activities. These questions are drawn from question 5.1 of the CIS, which asks specifically about:

- "acquisition of advanced machinery, equipment, software and buildings to be used for new or significantly improved products or processes (excluding expenditures on equipment for R&D);
- acquisition of existing know-how, copyrighted works, patented and non-patented inventions, etc. from other enterprises or organisations for the development of new or significantly improved products and processes;
- in-house or contracted out training for your personnel specifically for the development and/or introduction of new or significantly improved products and processes;
- in-house or contracted out activities for the market introduction of your new or significantly improved goods or services, including market research and launch advertising;
- in-house or contracted out activities to design or alter the shape or appearance of goods or services; and
- other in-house or contracted out activities to implement new or significantly improved products and processes such as feasibility studies, testing, tooling up, industrial engineering, etc." (Eurostat, 2012, p. 6).[15]

While having reliable data on these innovation-related expenditures would be useful, it may be difficult for companies to report these expenditures if they do not already record them in their accounts. Interviews and testing with companies would help in discerning which types of expenditures can be reported reliably. Expenditures on training and design may be among the most feasible to measure and are important for non-R&D innovation activities. Data on innovation-related expenditures may also provide useful input for the development of statistics on knowledge-based capital and "intangible assets" (discussed in Chapter 5).

Improvements to the Presentation of Information

NCSES has long been focused on providing information on levels of R&D funding by performing and funding sector. The agency's core tables on R&D report dollar amounts of R&D, both nominal and real, and R&D as a share of gross domestic product (GDP). In addition, BRDIS collects a wealth of other policy-relevant information that has the potential to provide guidance for policy makers. In February 2012, for example, NCSES published an *InfoBrief* on business use of intellectual property.

At the same time, NCSES could greatly improve the usefulness of its surveys by quickly publishing more "cross-tab" tables on key policy issues, or by making it easier for researchers and policy makers to access quickly the underly-

[15]There are some differences between the 2010 and 2012 CIS for this and other questions. The 2012 version includes copyrighted works under acquisition of existing knowledge. It also clarifies that the activities may be in-house or contracted.

ing data necessary to construct such tables.[16] A crosstab table shows the interrelationship of two variables. For example, BRDIS collects information on the amount of research spending outside the United States by U.S.-based companies. That information is important in itself, and NCSES has published the breakdown of foreign versus domestic R&D spending for different industries. From the perspective of policy, however, it is also useful to know whether companies that conduct a higher share of their R&D overseas relative to other companies are more or less likely to report innovations. That is, is globalization of R&D linked with a higher or lower propensity to innovate? Such a table, if published by NCSES, could help guide policy makers, stimulate research, and inform public debate on U.S. innovation policies.

RECOMMENDATION 4-2: The National Center for Science and Engineering Statistics should build on its Business Research and Development and Innovation Survey (BRDIS) to improve its suite of innovation indicators in the following ways:

- **tabulate the results from BRDIS using the same cutoffs for firm size (as well as comparable industry sectors) that are used by OECD countries in order to facilitate international comparisons;**
- **fund research exploring precisely what companies mean when they report an innovation or report no innovation on BRDIS—such research would help inform current policy debates;**
- **broaden the innovations tracked by BRDIS to encompass organizational and marketing innovations, as well as new data algorithms;**
- **consider adding a section to BRDIS on unmarketed innovations, giving respondents the opportunity to cite the main reason these innovations have not yet been marketed or implemented;**
- **as funds permit, extend BRDIS to gather information on innovation-related expenditures in such areas as training and design; and**
- **publish more results from BRDIS that link innovation to business characteristics, including the amount of research and development spending by U.S.-based companies outside of the United States. Production and distribution of such cross-tabulations should be timely, and they should address contemporary policy questions.**

The globalization of research is only one example of an area in which presentation of a crosstab table could be helpful for policy makers. Box 4-6 identifies examples of policy-relevant questions for which a crosstab table could be useful, especially if published in a timely manner. With access to the raw data, it is relatively easy for NCSES to construct high-interest crosstab tables and publish them quickly. Such a process is important to the agency's mission of providing statistics that are relevant to policy makers and the public.

> **BOX 4-6**
> **Examples of Questions That Could Be Informed by Policy-Relevant Crosstab Tables**
>
> Compared with other companies:
>
> - Are companies that perform research overseas more likely to report innovations?
> - Are companies that perform research in California more likely to report innovations?
> - Are companies that collaborate with other companies more likely to report innovations?
> - Are companies that partner with academic institutions more likely to report innovations?
> - Are companies that have a high percentage of science, technology, engineering, and mathematics personnel more likely to report innovations?
> - Are companies that acquire intellectual property more likely to report innovations?
> - Are companies in regulated industries more likely to report unmarketed innovations?
> - Are companies engaged in health care research more likely to report unmarketed innovations?
> - Are companies created to commercialize academic research more likely to report unmarketed innovations?

The panel also found that NCSES's limited access to some datasets constrains the timely development of innovation indicators. For instance, some data are available at the U.S. Census Bureau's research data centers before NCSES has on-site access to those data at its Survey Sponsor Data Center (SSDC). This is the case for the BRDIS data, which the Census Bureau collects on behalf of NCSES. Ready access to these data is imperative if NCSES is to satisfy the demands of users for timely innovation statistics. Housing the BRDIS data in the SSDC would be one key way to improve the timeliness of the statistical analysis of the data and the publication of related R&D and innovation indicators.

RECOMMENDATION 4-3: The Survey Sponsor Data Center at the National Science Foundation should house the Business Research and Development and Innovation Survey data, improving access to the data for National Center for Science and Engineering Statistics staff who develop the research and development statistics.

[16] Timeliness and relevance are both data quality measures (see Recommendation 8-1 in Chapter 8).

Linkages Between BRDIS Data and Other Datasets

The data gathered in BRDIS could be used to begin developing statistics on high-growth firms and "gazelles." The Manual on Business Demography Statistics (OECD-Eurostat, 2008, Chapter 8, p. 61) defines high-growth enterprises as "all enterprises with average annualised growth greater than 20% per annum, over a three year period. . . . A size threshold has been suggested as 10 employees at the beginning of the growth period." Gazelles are the subset of high-growth enterprises that are up to 5 years old (OECD-Eurostat, 2008, Chapter 8, p. 63). These thresholds are arbitrary and based only on convention. NCSES could conduct its own sensitivity analysis to fine-tune the definitions of high-growth firms and gazelles.[17]

During the panel's July 2011 workshop, several speakers[18] mentioned the importance of tracking trends in the sustainability of jobs in these types of firms during economic downturns (even if total employment is small). It would also be useful to have firm data by age classes to determine over time whether high-growth firms or gazelles in particular have a higher incidence of innovation activity relative to other firms. In his presentation at the July 2011 workshop, Hollanders showed that high-growth firms are significantly more innovative than other firms in his dataset. The connection between high-growth firms and innovation is complex, and these data would help researchers better understand it. Statistics on high-growth firms and gazelles could also be used to answer the question of whether these types of firms drive economic and job growth. A simple table could compare the economic characteristics of high-growth and other firms that are and are not innovative, ideally over time.

At the panel's September 2011 meeting in Washington, DC, representatives from BLS, the U.S. Census Bureau, and BEA mentioned that linking certain datasets among them would yield reasonable numbers on gazelles. A table with these numbers could be added to the *Science and Engineering Indicators* or become the foundation of an *InfoBrief*. The following indicators could be produced using BRDIS and other data on high-growth firms and gazelles: number of high-growth enterprises as a percentage of the total population of active enterprises with at least n-number of employees, and number of gazelles as a percentage of all active enterprises with at least n-number of employees that were born 4 or 5 years ago. These indicators would be comparable to those produced in several other countries, thus increasing users' understanding of the comparative position of the United States on an aspect of the country's innovation capacity.

NCSES has a unique set of data in BRDIS, which, if combined with other datasets, could be instrumental in answering these and other important questions. Integrating data on firm dynamics (and the related employment effects) would take time and resources. During his presentation at the workshop, Haltiwanger described three Census Bureau datasets that, together with BRDIS data, would allow NCSES to develop indicators of business dynamics:

- Longitudinal Business Database—tracks all establishments and firms with at least one employee, including start-ups, from 1976 to the present;
- Integrated Longitudinal Business Database—tracks all nonemployer firms and integrated-with-employer firms from 1994 to the present; and
- Longitudinal Employer-Household Dynamics—tracks longitudinally all employer-employee matches and transitions (hires, separations, job creation, and job destruction) from 1990 to the present.

Questions from the Census Bureau's 2007 and 2012 Economic Census, Company Organization Survey, and Management and Organizational Practices Survey can also yield useful information on R&D and other innovation activities for establishments. In addition, infrastructure datasets can track relationships between start-up and young high-growth firms and large, mature firms, and can be linked further to patent and citation data. Important as well is to link data on firm dynamics to those on innovation outputs, such as patent and citation data.

Haltiwanger proposed that indicators track firm dynamics by geography, industry, business size, and business age. Hollanders noted that European countries and other OECD members are continuing to fine-tune their measures of firm dynamics. NCSES's indicators on this dimension could further the international comparability of its STI indicators. Building the foundations for indicators of firm dynamics using BRDIS and other datasets would give NCSES a productive platform for developing several STI indicators that are policy relevant.

Clearly, developing publishable statistics on high-growth firms and gazelles is a multistage task requiring data acquisition and linkage in addition to use of the data available from BRDIS. A good first step would be for NCSES to explore linking its BRDIS data with data on firm dynamics from BLS.

RECOMMENDATION 4-4: The National Center for Science and Engineering Statistics (NCSES) should begin a project to match its Business Research and Development and Innovation Survey data to data from ongoing surveys at the U.S. Census Bureau and the Bureau of Labor Statistics. It should use

[17]Petersen and Ahmad (2007) present a technique for conducting this type of analysis in OECD (2007).

[18]Howard Alper, University of Ottawa; Robert Atkinson, Information Technology and Innovation Foundation; John Haltiwanger, University of Maryland; Hugo Hollanders, United Nations University's Maastricht Economic and Social Research Institute on Innovation and Technology (UNU-MERIT); and Brian MacAulay, National Endowment for Science, Technology, and the Arts.

the resulting data linkages to develop measures of activities by high-growth firms, births and deaths of businesses linked to innovation outputs, and other indicators of firm dynamics, all of which should be tabulated by geographic and industry sector and by business size and business age to facilitate comparative analyses. NCSES should conduct a sensitivity analysis to fine-tune meaningful age categories for high-growth firms.

USE OF NONTRADITIONAL METHODOLOGIES

Traditionally, NCSES and its predecessors have used surveys to trace the inputs and outputs of the innovation system. As noted earlier, however, executing a survey is an expensive and time-consuming process, requiring writing and testing questions, identifying the universe of potential respondents, collecting the data, cleaning and validating the data, analyzing the data, and then finally publishing the results. Thus, for example, NCSES did not publish the topline R&D spending results from the 2009 BRDIS until March 2012.

Another issue with government surveys is that those being surveyed may not respond, and when they do, they are almost always guaranteed confidentiality. As a result, some or even many results must be withheld to avoid disclosing, even indirectly, the responses of individual companies. In the basic table showing R&D spending by industry and source of funding from the 2010 and 2011 BRDIS (Table 2 in *InfoBrief* 13-335), for example, 37 and 36 (respectively) of the 154 cells are suppressed. The significance of this problem grows when more detailed industries or subnational areas are considered. Take, for example, the interesting question of whether the location of R&D conducted by the IT industry is becoming more or less concentrated in Silicon Valley. If BRDIS were to ask companies to apportion their research spending by metro area, the data for most metro areas across the country would likely have to be suppressed because information on a specific firm might be identified in cases involving a small number of firms in a given sector or region. It might even be difficult to obtain useful data at the state level for the same reason.

Until fairly recently, there was no good alternative to surveys for collecting data on innovation inputs and outputs. Increasingly, however, businesses and individuals are generating detailed electronic data in the normal course of their economic activity. Such data are available either in a firm's administrative records, publicly on the web, or from third parties that collect them in the normal course of economic activity, and can be obtained in digital form from a given firm. Such nontraditional data are referred to here as "business practice" data. Examples of these data include several datasets derived from the Internet: (1) reports on innovations, (2) help-wanted ads, (3) journal articles on new products and services, (4) altmetric measures of scientific and engineering activities,[19] and (5) trademark databases.[20] This section examines the advantages and disadvantages of business practice data from the perspective of NCSES. Further discussion of the use of these types of data for developing STI indicators appears in Chapter 7 of this report.

Conceptual Background

The use of business practice data as a supplement to traditional surveys is under active consideration in statistical agencies. Robert Groves, then director of the Census Bureau, recently delineated the key issues entailed in using these data (see Box 4-7), noting: "Combined, the 'big data' and the benchmark survey data can produce better statistics."

The advantages of business practice data include

- Timeliness—Collecting and analyzing survey data is a lengthy process. Because most business practice data today are digital, they can be available for analysis in near real time.
- Detail—Survey data offer limited detail by industry and geographic location. Business practice data can be used to track innovation activity by detailed industry or subnational area.
- Flexibility—Survey methods require making some assumptions about the nature of innovation many years in advance. The ability to adjust the measuring tool easily is helpful, particularly when new categories of innovation emerge.

At the same time, business practice data have disadvantages that make them an imperfect substitute for conventional surveys. These disadvantages include

- "Institutional drift"—Business practice data are generated by normal business activity, but because patterns of business activity change over time, interpretations of business practice data are not necessarily stable.
- Difficulty with cross-industry/cross-country comparisons—Different industries and countries may have very different business practices.

---

[19]"Altmetrics are alternative measures that can supplement citation counts and journal impact factors as measures of the impact of scholarly communications. Such measures are generally derived from online activity such as mentions, downloads, tweets, blog posts, Facebook "likes," bookmarking and other similar evidence of attention" (Travis, 2013).

[20]Note that not all business practice data fall into the category commonly referred to as "big data." For example, administrative records are included in the panel's definition of business practice data but are not considered "big data." Also note that the term "big data" had not been formally codified by statistical agencies at the time this report was being written. Lastly, "big data" is not a panacea (see Boyd and Crawford [2011] for important caveats).

BOX 4-7
"And Now, for Something a Little Different"
Excerpt from a Blog Post by
Robert Groves, Director, U.S. Census Bureau

If we had access to customer purchase transactions volume, we might construct models blending our benchmark sample survey data with the continuous transaction data, to produce more timely and more disaggregated estimates. The strength of the transaction data will be their timeliness and the large number of transactions they reflect; their weakness will be that they do not include many transactions conducted in ways other than those the data reflect (e.g., cash might be omitted). The strength of our benchmark survey will be its statistical coverage of the entire population of business units; its weakness is its lack of timeliness and its relatively small sample size of firms. Combined, the "big data" and the benchmark survey data can produce better statistics.

Sometimes the link between our sample surveys and the big data will be time, other times it will be space. "Big data" will be useful for constructing small area estimates. For example, internet sites listing asking prices for houses may be accompanied with exact location of the units. Their strength is that they offer millions of records of prospective sales; their weakness is that they don't cover all areas of the country, not all sales are covered, and asking prices are not sale prices. Our sample survey on residential sales offers statistical coverage of all sales, but its sample size is too small to provide statistics on all areas. Combining the two data series might offer more spatial detail.

At other times, the link between the big data and our sample survey data may be measures that are highly correlated to our key statistics. For example, we might have access to traffic volume data continuously streaming based on traffic cameras, with location codes to permit fine spatial detail. Our sample survey reports of commuting times from home to place of work might be enhanced by statistically combining them with the traffic count data from available cameras. The strength of the traffic camera counts would be very fine grain detail on time; the weakness would be coverage of all roads and counts of commercial traffic as well as private cars.

SOURCE: Available: http://directorsblog.blogs.census.gov/2012/06/27/and-now-for-something-a-little-different/ [January 2014].

As a result, using business practice data requires innovative statistical techniques to standardize measures across time, industries, and location. In addition, key indicators must be maintained consistently over time to provide a benchmark.

RECOMMENDATION 4-5: The National Center for Science and Engineering Statistics should make greater use of business practice data to track research and development spending and innovation-related jobs at a more detailed geographic and occupational level than is possible with government survey data.

BOX 4-8
Examples of Web-Based Data on Innovation

Lists of Top Innovations

- http://www.fastcompany.com/1738506/the-10-most-innovative-companies-in-health-care
- http://www.rdmag.com/Awards/RD-100-Awards/2011/06/R-D-100-2011-Winners-Overview/
- Small Business Administration—http://www.sba.gov/content/sba-announces-winners-2011-tibbetts-awards
- Technology Review—http://www.technologyreview.com/tr50/
- http://my.clevelandclinic.org/media_relations/library/2011/2011-10-6-cleveland-clinic-unveils-top-10-medical-innovations-for-2012.aspx

Innovation Data Reported by Companies

- New England BioLabs products—http://www.neb.com/nebecomm/newprod.asp
- GE products—http://www.ge.com/products_services/directory/by_product.html
- Corning—http://www.corning.com/displaytechnologies/en/index.aspx

Site That Collects Innovation Data

- ProductDb: http://productdb.org

Examples of Business Practice Data[21]

Understanding of innovation would be facilitated if NCSES regularly (annually or biannually) published a list of significant innovations brought to market. Such a list would be useful for assessing trends in innovation and how they relate to growth and jobs. Also helpful would be to have a list of major potential innovations under development, based on publicly reported data. Such lists have occasionally been manually constructed in the past (see Box 4-8). In 1982, for example, the Small Business Administration sponsored a project that entailed gathering information from 46 trade magazines and identifying and coding 8,074 innovations—4,476 in manufacturing industries. This information included

- model name, trade name, or trademark;
- name and description of the innovation;

[21]Nonsurvey methods for extracting data for the development of STI indicators are discussed in greater detail in Chapter 7.

- year of introduction;
- innovation type (product or process);
- innovation significance (new class, new type, significant improvement, updated model); and
- origin of technology and source of funding.

Today, more efficient techniques for creating lists of innovation are available, based on web scraping. One can also look at the propagation of technologies by examining the publication of product manuals. The panel considers these techniques in greater detail in Chapter 7.

SUMMARY

In this chapter, the panel has offered five recommendations whose implementation would improve NCSES's indicators program with respect to measuring innovation. These recommendations address four topics: (1) new and improved measures of innovation inputs and outcomes based on existing data, with clear implications for economic performance; (2) analysis, based on existing data, that is more comparable across countries than is currently the case; (3) improved data resources and accessibility; and (4) augmented survey questions that would ensure better reporting on innovation activities by firms. The panel believes NCSES should focus first on activities that use existing data while further developing capabilities for linking its data with those of other agencies and using frontier tools to produce STI indicators in a more timely fashion.

5

Measuring the Three K's: Knowledge Generation, Knowledge Networks, and Knowledge Flows

Knowledge generation can occur formally through directed research and experimental development in academic institutions, firms, and public and nonprofit institutions. Knowledge generation can also occur informally in a working environment through the activities and interactions of actors in an organization or the general economy. People are the critical input for knowledge generation, whether as individual researchers; in research teams; or even in collectives such as organizational subunits, entire organizations, or nation-states.[1] Therefore, indicators of knowledge generation focus on attributes of human capital inputs and related outputs. Knowledge can be acquired by using codified (written) sources such as publications or patents, or in tacit form by hiring people with the needed knowledge or participating in networks where the knowledge is stored (Chapter 6 focuses on knowledge embodied in people). Knowledge can be both an intermediate input and a final output and can depreciate over time.[2]

Knowledge networks link actors, organizations, and technologies in the global economy, revealing new discoveries and transferring knowhow on the development of new techniques, processes, and at times breakthroughs that can be commercialized (Chapter 4 focuses on innovation). Knowledge networks include research collaborations, coinventorships, coauthorships, and strategic alliances.[3] *Knowledge flows* transmit across knowledge networks and point to comparative advantage, presence in other markets, and access to foreign technologies. To use acquired knowledge, recipients must have absorptive capacity.[4]

Knowledge generation, diffusion, and use, as well as conduits for knowledge flows, are all key elements for economic growth (Romer, 1990). Therefore, it is critically important for the National Center for Science and Engineering Statistics (NCSES) to produce indicators of these varied dimensions of knowledge at the national, international, and subnational levels.

Quite a few data elements, such as research and development (R&D), patents, bibliometrics, and trade in technology, capture knowledge generation, networks, and flows (referred to as "the three K's"). NCSES has been collecting these data for several decades in order to publish indicators on these topics, drawing on both its own and other data sources, such as the Bureau of Economic Analysis for data on global multinational R&D activities. International R&D is well covered by NCSES's Business Research and Development and Innovation Survey (BRDIS). While NCSES has good measures of knowledge creation, however, a number of complex issues remain unaddressed, and challenges for measurement remain in the area of knowledge flows.

Therefore, the purpose of this chapter is to discuss the dynamics and outcomes of scientific R&D. To illustrate specific uses of science, technology, and innovation (STI) indicators in this context, the focus is on the policy questions that can be addressed using indicators on the three K's; however, it should be noted that these indicators have several other uses. Box 5-1 highlights key policy questions relating to the generation and transfer of knowledge.[5] While raw data on R&D expenditures and patent citations are useful for understanding whether the United States is falling behind other countries in R&D expenditures and outcomes, more sophisticated statistics are required to address other

[1]See Phelps and colleagues (2012, p. 7) for a description of repositories of knowledge. Romer (1990, p. S84) makes the following distinction between knowledge as an intermediate and final output: ". . . knowledge enters into production in two distinct ways. A new design enables the production of a new good that can be used to produce output. A new design also increases the total stock of knowledge and thereby increases the productivity of human capital in the research sector."

[2]See Huang and Diewert (2011) for methods of measuring knowledge depreciation.

[3]For an extensive definition of knowledge networks, see Phelps et al. (2012, p. 61, endnote 1).

[4]OECD (2013a) gives definitions of knowledge flows and classifications of indicators of knowledge flows in science, technology, and innovation sectors.

[5]See Appendix B for the full list of policy questions.

BOX 5-1
Policy Questions Related to Knowledge Generation, Networks, and Flows

- What new technologies or fields are emerging from current research?
- Is the United States promoting platforms in information and communication technology, biotechnology, and other technologies to enable innovation in applications?
- Is the United States falling behind other countries in R&D expenditures and outcomes?
- How much are U.S. companies spending to be present in emerging markets? How much R&D are they conducting in these nations?
- Is the United States losing or gaining advantage by buying and selling its R&D abroad?
- Is the United States benefiting from research conducted in other countries?

issues pertaining to the competitiveness of U.S. companies and the benefits of buying and selling R&D internationally. The focus of this chapter is on the latter set of indicators.

A recent OECD (2012c) study titled *Knowledge Networks and Markets in the Life Sciences* describes key aspects of the three K's in which indicators require further development. The following findings are particularly in accord with those presented in this chapter:

- Individuals, firms, and countries are not uniformly linked to knowledge networks.
- Evidence gaps persist with respect to capturing differences between knowledge production and use (as in the case of R&D), capturing partnerships and their financial dimension, monitoring the combined outward and inward dimensions of knowledge flows, and going beyond intellectual property indicators as measures of knowledge outputs.
- Measurement standards need to be adapted if improvements are to be achieved in the interoperability of STI data sources across different domains, such as R&D, patents, other forms of registered intellectual property, scientific publications, innovation survey data, and administrative sources. Solutions need to be developed that address the impact of knowledge flows on the interpretation, relevance, and international comparability of existing STI indicators.

NCSES is poised to make important contributions to the improvement of indicators on the three K's. Collaborative efforts with other agencies in the United States and abroad should be fruitful for this endeavor.

CODIFIED DEFINITIONS

The internationally accepted definition of "research and experimental development"—more commonly referred to as R&D—comes from OECD (2002, p. 30): "creative work undertaken on a systematic basis in order to increase the stock of knowledge, including knowledge of man, culture and society, and the use of this stock of knowledge to devise new applications."[6] In BRDIS, NCSES expands on this definition, providing the following guidance (U.S. Department of Commerce, 2011, p. 3):

> R&D is planned, creative work aimed at discovering new knowledge or developing new or significantly improved goods and services. This includes (a) activities aimed at acquiring new knowledge or understanding without specific immediate commercial applications or uses (basic research); (b) activities aimed at solving a specific problem or meeting a specific commercial objective (applied research); and (c) systematic use of research and practical experience to produce new or significantly improved goods, services, or processes (development).
>
> The term "research and development" does NOT include expenditures for:
>
> - costs for routine product testing, quality control, and technical services unless they are an integral part of an R&D project;
> - market research;
> - efficiency surveys or management studies;
> - literary, artistic, or historical projects, such as films, music, or books and other publications; and
> - prospecting or exploration for natural resources.

The term "science and technology" (S&T) covers a wide range of activities, including R&D, but is rarely defined in the literature, perhaps because its breadth leads to its being used in different ways in different contexts. The United Nations Educational, Scientific and Cultural Organization (UNESCO) (1984, p. 17) provides a definition of the term that is used for this chapter:

> For statistical purposes, Scientific and Technological Activities (STA) can be defined as all systematic activities which are closely concerned with the generation, advancement, dissemination, and application of scientific and technical knowledge in all fields of science and technology, that is the natural sciences, engineering and technology, the medical and the agricultural sciences (NS), as well as the social sciences and humanities (SSH).[7]

[6]This is the definition used in all OECD, European Union, African Union, and Latin American countries. All elaborate on this definition in their survey instruments as the United States has done to incorporate definitions for basic, applied, and experimental development.

[7]Also included in the definition of S&T are "scientific and technological services" and "scientific and technological education and training," the

Because S&T includes but is not limited to R&D,[8] the focus of this chapter is on indicators of foreign direct investment in R&D and trade in knowledge-intensive services. Measurement of intangible assets also is touched upon, although the panel does not view the development of such measures as more appropriate for NCSES than for the Bureau of Economic Analysis.

MEASURING SCIENCE AND TECHNOLOGY: MAJOR GAPS IN INTERNATIONAL COMPARABILITY

Comparability is a universal challenge for statistics and for indicators based on those statistics. The comparability of data can be affected by the survey techniques used to collect the data and the conversion of the data into statistics through the use of weighting schemes and aggregation techniques. These problems are amplified when statistics are used to create indicators, as the indicators may be a combination of statistics (e.g., an average, a sum, or a ratio) with different comparability problems. In addition to the international or geographic comparison of indicators that describe an aspect of a system (e.g., R&D as a percentage of gross domestic product [GDP]), there are problems with intertemporal and intersectoral comparisons. Users of indicators need to recognize that all statistics and indicators have a margin of error beyond which they should not be pushed. The problem is growing as response rates to official surveys continue to decline.

International comparisons entail fundamental issues such as language (e.g., the Japanese term for "innovation" is actually closer to what most Americans think of as "technology"), and NCSES is to be congratulated for supporting a project with OECD and the European Union (EU) on the cognitive testing of survey questions in multiple languages. Differences in institutions (e.g., the accounting for the European Union Framework program across EU member states) pose problems, as do cultural differences (e.g., the Nordic world has access to "cradle to grave" linked microdata on individuals) and differences in governance structures (e.g., the importance of subnational R&D programs in some countries). These differences can limit comparability and increase the margin of error that should be applied to international comparisons of statistics and indicators.

In the area of S&T indicators, a number of key comparability problems are well known. OECD compiles S&T statistics, monitors the methodology used to produce them, and publishes international comparisons and has documented the problems summarized below.

Research and Development[9]

Each country depends for its R&D data on the coverage of national R&D surveys across sectors and industries. In addition, firms and organizations of different sizes are measured, and national classifications for firm sizes differ. Countries also do not necessarily use the same sampling and estimation methods. Because R&D typically involves a few large organizations in a few industries, R&D surveys use various techniques to maintain up-to-date registers of known performers. Analysts have developed ways to avoid double counting of R&D by performers and by companies that contract with those firms or fund R&D activities of third parties. These techniques are not standardized across nations.

R&D expenditure data for the United States are somewhat underestimated for a number of reasons:

- R&D performed in the government sector covers only federal government activities. State and local government establishments are excluded from the national figures.[10]
- In the higher education sector, R&D in the humanities is excluded, as are capital expenditures.[11]
- R&D expenditures in the private nonprofit sector include only current expenditures. Depreciation is reported in place of gross capital expenditures in the business enterprise sector.

Allocation of R&D by sector poses another challenge to the comparability of data across nations. Using an industry-based definition, the distinction between market and public services is an approximate one. In OECD countries, private education and health services are available to varying degrees, while some transport and postal services remain in the public realm. Allocating R&D by industry presents a challenge as well. Some countries adopt a "principal activity" approach, whereby a firm's R&D expenditures are assigned to that firm's principal industrial activity code. Other countries collect information on R&D by "product field," so the R&D is assigned to the industries of final use, allowing reporting companies to break expenditures down across product fields when more than one applies. Many countries follow a combination of these approaches, as product breakdowns often are not required in short-form surveys.

definitions of which are found in United Nations Educational, Scientific and Cultural Organization (1978).

[8]The OECD *Frascati Manual* (OECD, 2002, p. 19) notes that "R&D (defined similarly by UNESCO and the OECD) is thus to be distinguished from both STET [scientific and technological education and training] and STS [scientific and technological services]." The *Frascati* definition of R&D includes basic research, applied research, and experimental development, as is clear from NCSES's presentation of the definition in the BRDIS for use by its respondents.

[9]This description draws heavily on OECD (2009, 2011) and Main Science and Technology Indicators (MSTI) (OECD, 2012b).

[10]NCSES reports state R&D figures separately.

[11]In general, OECD's reporting of R&D covers R&D both in the natural sciences (including agricultural and medical sciences) and engineering and in the social sciences and humanities. A large number of countries collect data on R&D activities in the business enterprise sector for the natural sciences and engineering only. NCSES does report data on social science R&D.

The *Frascati Manual* (OECD, 2002) recommends following a main activity approach when classifying statistical units, but recommends subdividing the R&D by units or product fields for firms carrying out significant R&D for several kinds of activities. This applies to all industry groups and, at a minimum, to the R&D industry (International Standard Industrial Classification [ISIC] Rev. 3, Division 73, or North American Industry Classification System [NAICS] 5417 in North America), although not all countries follow this method.

Comparability problems are also caused by the need to preserve the confidentiality of survey respondents (see Chapter 4). National statistical practice will prevent publication of the value of a variable if it is based on too few responses. This not only results in suppression of a particular cell in a table, but also requires additional suppression if there are subtotals that could be used to infer the suppressed information. The result is reduced comparability, which can be overcome only by microdata analysis under controlled conditions.

In principle, R&D institutes serving enterprises are classified according to the industry they serve. When this is not done, the percentage of business enterprise expenditure on R&D (BERD) performed by what is most likely a service industry is overestimated compared with estimates for other countries.

Finally, R&D performers recently have been asked in surveys to break down their R&D activities across sites in different national territories or regions. Estimating R&D intensity by region or other subnational unit presents additional challenges. The existence of multinationals headquartered in a given country that conduct R&D and trade in R&D services worldwide makes it difficult to pinpoint where the R&D is funded and performed and where it has impact. For example, the R&D could be funded by a head office in Rome, performed in a research institute in Israel, and have an impact on consumers of the resulting product in the United States.

Government Budget Appropriations or Outlays for R&D (GBAORD)[12]

GBAORD data are assembled by national authorities using statistics collected from budgets. This process entails identifying all the budget items involving R&D and measuring or estimating their R&D content. The series generally cover the federal or central government only. GBAORD is a good reflection of government priorities based on socioeconomic objectives. These statistics often are used for cross-country comparisons, particularly to address such questions as: Is the United States falling behind other countries in R&D expenditures and outcomes? While it is not necessarily the case that high government expenditures foreshadow international preeminence in S&T, it is important to understand whether such expenditures indeed lead to better employment, health, and security outcomes.

However, comparability problems arise because some countries do not include in their GBAORD estimates funding for general support of universities (e.g., the United States) or R&D funded as part of military procurement (e.g., Japan, Israel). Moreover, it currently is not possible for all countries to report, on the basis of budget data, which sectors are responsible for performing the R&D funded by government.

Business Enterprise Expenditures on R&D[13]

BERD statistics convey business R&D expenditures. OECD breaks down business R&D expenditure data into 60 manufacturing and service sectors for OECD countries and selected nonmember economies. The reported data are expressed in national currencies (as well as in purchasing power parity U.S. dollars), at both current and constant prices.

When assessing changes in BERD over time, it is necessary to take account of changes in methods and breaks in series, notably in terms of the extension of survey coverage, particularly in the service sector, and the privatization of publicly owned firms. Identifying new and occasional R&D performers is also a challenge, and OECD countries take different approaches to this challenge in their BERD surveys. In addition, not all activities related to foreign affiliates' R&D are recorded in company transactions. There are intracompany transfers (e.g., intracompany mobility of researchers) with no monetary counterparts that lead to R&D efforts that do not appear in the statistics as R&D spending by foreign affiliates. The increasing internationalization of R&D and other economic activities also makes it difficult to accurately identify inflows of R&D funds to companies and their precise nature (as discussed later in this chapter). For example, there is a growing need to measure international R&D transactions properly and to deal with the problem of nonpriced transfer of R&D within multinational enterprises. All of these issues require expert data manipulation and statistical analysis, thereby presenting challenges to the international comparability of indicators derived from these statistics.

Technology Receipts and Payments[14]

Technology receipts and payments, including those for R&D services, show a country's ability to sell technology abroad and its use of foreign technologies, respectively. Further qualitative and quantitative information is needed to analyze a country's deficit or surplus because a deficit (surplus) on the technology balance does not necessarily indicate the lack (presence) of competitiveness.

[12]This section is based on OECD (2011) and OECD (2012b).

[13]This section is based on OECD (2011).

[14]This section is based on OECD (2011).

Measurement errors may lead to underestimation or overestimation of technology transfers. Licensing contracts provide payment channels other than technology payments, and payment/receipt flows may be only part of the total price paid and received. Alternatively, national tax and control regulations on technology receipts and payments may bias data on technology flows, notably for international transfers of multinationals. If royalties are less taxable than profits, then they may be preferred to other transfer channels and exceed the value of technology transferred. On the other hand, if limitations are imposed on royalty remittances, then some portion of repatriated profits will represent remuneration of technology transfer.

Summary

Each of the above reasons for international incomparability of some S&T measures goes beyond what NCSES can deal with on its own. An OECD Working Party, the National Experts on Science and Technology Indicators (NESTI), has been in place for 50 years to discuss these issues and support collaboration to resolve them. Nonetheless, there are some areas in which NCSES has opportunities to adjust definitions and improve methodologies to obtain more accurate STI indicators. For example, finer-grained size classes for firms would allow a better understanding of the relationship between firm size and innovation (as discussed in Chapter 4). In addition, improved measures of business enterprise R&D would shed some light on the question of whether the United States is increasingly depending on knowledge generated in other countries. And better measuring of technology receipts and payments would show which countries are net buyers or sellers of knowledge-intensive services. Recommendations for how NCSES could go about improving these measures appear later in this chapter.

TRADITIONAL INDICATORS OF THE THREE K'S

Patent[15] data and bibliometrics (data on publication counts and citations) can be used to measure new knowledge, knowledge networks, and knowledge flows.

Patents

Patent administrative records—including citations, claims, technical classifications, families,[16] and countries where the patents are effective—contain a wealth of information about invention. They also contain detail on inventors and applicants and on the regulatory and administrative processes of the patenting system.[17] Patent information is useful for determining when a new product or process was developed and its linkages to prior inventions and to research that was the foundation for the invention. Observing where patents are registered can also yield clues to how new knowledge is diffused from nation to nation.

Patent data often are used to develop indicators of knowledge generation, flows, and linkages. OECD's (2008) *Compendium of Patent Statistics 2008* gives several examples:

- Patent-based statistics can be derived that reflect the inventive performance of countries, regions, and firms.
- The inventors' addresses can be used to monitor linkages, including the internationalization of and international collaboration in S&T activities.
- Knowledge networks can be determined by observing cooperation in research and diffusion of technology across industries or countries in specific technological areas.
- The market strategy of businesses can be inferred from information contained in the patent file.

At the same time, information derived from patent records must be used with caution (OECD, 2006):

- The value distribution of patents is skewed as many patents have no industrial application (and hence are of little value to society), whereas a few are of substantial value.
- Many inventions are not patented because they are not patentable, or inventors may protect them using other methods, such as secrecy or lead time.
- The propensity to patent differs across countries and industries.
- Differences in patent regulations make it difficult to compare counts across countries.
- Changes in patent law over the years make it difficult to analyze trends over time.

The panel emphasizes the first point on the above list: patents may be used strategically in some sectors of an economy to deter competition. Andrew Updegrove of

[15]"Patents are an exclusive right issued by authorised bodies to inventors to make use of and exploit their inventions for a limited period of time (generally 20 years). Patents are granted to firms, individuals or other entities as long as the invention is novel, non-obvious and industrially applicable. The patent holder has the legal authority to exclude others from commercially exploiting the invention (for a limited time period). In return for the ownership rights, the applicant must disclose information relating to the invention for which protection is sought" (Khan and Dernis, 2006, p. 6).

[16]"A patent family is the same invention disclosed by a common inventor(s) and patented in more than one country" (United States Patent and Trademark Office, http://www.uspto.gov/main/glossary/#p [June 2013]). The European Patent Office has the following definition: "A patent family is a set of either patent applications or publications taken in multiple countries to protect a single invention by a common inventor(s) and then patented in more than one country. A first application is made in one country—the priority—and is then extended to other offices" (http://www.epo.org/searching/essentials/patent-families.html [June 2013]).

[17]As administrative records, patent applications and grants are a rich microdata source that do not rely on surveys and do not generate the respondent burden associated with traditional statistical surveys.

Gesmer Updegrove LLP captured this sentiment by saying, "Patents don't give value; they cause friction" (Updegrove, 2012). Therefore, the notion that substantial patent activity is an indicator of major leaps in S&T capabilities or innovation is not necessarily the case. In some instances, patenting could have a negative impact on knowledge creation and innovation. Thus observed patent activity as an indicator of knowledge generation or innovation should be determined sector by sector.

In his presentation to the panel in February 2012, Stuart Graham, chief economist at the United States Patent and Trademark Office (USPTO), outlined USPTO's Economic Data Agenda. In the near term, the agency will improve its databases, particularly the Patent Assignment, Trademark Casefile, and Trademark Assignment datasets. Over time, USPTO is also "considering providing a forum that would facilitate the posting of additional matched datasets, papers and findings" and working with other agencies to create "matched datasets to other economically-relevant information." For NCSES's activities on STI indicators, particularly those related to producing better measures of knowledge generation, flows, and networks, continued collaboration with USPTO should be beneficial. NCSES already relies on USPTO data for basic measures of patenting activity. However, linking basic research outputs to patents and trademarks (including the human capital and demographic markers that are indicated on the records) and ultimately to outcomes that have significant societal impacts would be of great benefit to users of NCSES indicators. In addition, these linked files would be helpful to researchers who work with the datasets of USPTO, NCSES, and others to understand relationships and rates of return in the STI system.

The panel makes no explicit recommendation here for NCSES to do more than continue to explore wider use of patent indicators and to engage in international cooperation on the development of indicators based on patent records to address user needs. There is no standard method for calculating indicators from patent data, and as noted earlier, analysis of these data without reservation can lead to incorrect inferences and misleading policy decisions. It is important to improve data quality and analytical techniques in this area—an active role for NCSES in collaboration with other agencies and organizations worldwide. As NCSES continues to disseminate patent data as part of its STI indicators program, it would be valuable to users to have clear cautions regarding the use and misuse of these statistics for decision-making purposes.

Bibliometrics

Publication is a major vehicle for disseminating and validating research results. Bibliometric data on publication counts and citations thus are a valuable source for measuring scientific performance, tracking the development of new technologies and research areas, and mapping linkages among researchers. Publication counts are based on science and engineering (S&E) articles, notes, and reviews published in a set of the world's most influential scientific and technical journals (Ruegg and Feller, 2003, p. 31).

A number of characteristics can be used for categorization of publications and indicator development. Fields are determined by the classification of each journal. Publications are attributed to countries by the author's institutional affiliation at the time of publication. Indicators of coauthorship appear to be affected by two factors. The first is language, although this has become less of an issue as English has become the language most commonly used internationally by researchers. The second is geographic location, although the effect of information and communication technologies on knowledge flows has undoubtedly lessened its effect. The quality of publications can be measured both by the quality of the journal and by how often it is cited in other publications. Citations can also be used to measure knowledge flows and linkages between different research areas. Coauthorship provides an additional measure of linkages and often is used as an indicator of collaboration patterns.

NCSES currently publishes a number of indicators based on bibliometric data. These include counts of S&E articles, shares of articles with domestic or international coauthors, counts and shares of citations and top-cited articles, and citation rates. These indicators can be used primarily to measure the output of scientific research. For example, counts of articles and citations and shares of world totals show how the United States is faring compared with other countries or regions. These indicators can also be used to measure the extent of collaboration and linkage. An example is the network maps used in the report *Knowledge, Networks and Nations: Global Scientific Collaboration in the 21st Century*, by the UK Royal Society (The Royal Society, 2011). These network maps are based on authorship of articles and show patterns of collaboration between countries. They are based on numbers of jointly authored research papers, with linkages being displayed when the collaboration between two countries amounts to 5-50 percent of the overall publication output of one of the partners. The OECD (2010) report *Measuring Innovation: A New Perspective* uses citation data to measure the interrelatedness of different research areas.[18]

Bibliometric data potentially can be used to create a number of additional indicators to provide further detail on linkages across research areas or by geographic location. This information can be particularly valuable for mapping the development of new research areas, such as green technologies, or the spread of general-purpose technologies.

There are some limitations to the use of bibliometric analysis for the production of S&T indicators, particularly when used to measure causal relationships, such as socioeconomic impacts of funding basic science. It is also difficult to isolate how much research networks have changed because

[18]This report references the citation technique used in Saka et al. (2010).

of a given research funding award granted or the existence of a new collaborative agreement. Impact factors and Hirsh's (h) index, commonly used by bibliometricians, do not allow for comparisons with counterfactual analysis. Furthermore, measures must be normalized to be helpful for comparing research outputs, or they are no better than "nose-prints"—metaphorically, signs of high window-shopping activity, with no true indication that a substantive purchase has occurred. There are ways for numbers of patents and articles to be inflated by their producers without substantive advances in S&T having been achieved. Bornmann and Marx (2013) state that ". . . mere citation figures have little meaning without normalization for subject category and publication year. . . . We need new citation impact indicators that normalize for any factors other than quality that influence citation rates and that take into account the skewed distributions of citations across papers." Bornmann and Marx describe techniques using percentiles to create normalized indicators, an improvement on impact factors and Hirsh's (h) index.[19] To its credit, the National Science Board (for which NCSES produces the *Science and Engineering Indicators* [*SEI*] biennial volumes) is mentioned by Bornmann and Marx as one of the federal agencies that uses percentile ranks of publications. Although this is good practice, it is important to note that these indicators are not appropriate for impact assessment, for which counterfactual evidence is necessary.

RECOMMENDATION 5-1: The National Center for Science and Engineering Statistics should expand its current set of bibliometric indicators to develop additional measures of knowledge flows and networking patterns. Data on both coauthorship and citations should be exploited to a greater extent than is currently the case.

BUSINESS R&D SERVICES AND INTANGIBLE ASSETS

Although NCSES publishes a rich set of data on R&D expenditures and performance, measures of spillover effects still are needed to aid in determining the effects of scientific investment on socioeconomic outcomes. Policy makers would benefit from such measures in addressing such questions as: What is the effect of federal spending on R&D on innovation and economic health, and over what time frame? What is the international balance of trade in R&D services? How much R&D do U.S. multinational companies conduct outside the United States, and how much R&D do foreign multinational companies carry out in the United States? How much are U.S. companies spending to be present in emerging markets? How much R&D are they conducting in these nations?

This section addresses the question of how R&D data can best be exploited, focusing in particular on the measurement of trade in R&D services. BRDIS contains a rich dataset on R&D that is only partially exploited in present indicators. Given the size and complexity of BRDIS, however, a trade-off is entailed in terms of the time and resources needed to process these data. BRDIS can be exploited by researchers within and outside government, subject to appropriate restrictions to protect respondents, but only if a researcher database is provided with sufficient metadata[20] to define all the variables and the degree of imputation for each.

At the same time, the panel acknowledges that further exploitation of BRDIS would require additional resources and might also involve a trade-off in terms of the timeliness of the release of key R&D indicators. The time required to process and release R&D statistics increased significantly following the introduction of BRDIS, which is a longer and more complex survey than its predecessor, the Survey of Industrial Research and Development. The panel views timeliness as an important factor in determining the value of R&D and other indicators and encourages NCSES to place high priority on reducing the time lag in the release of BRDIS data.

Trade in R&D Services[21]

One important aspect of R&D is R&D services, which are services for the performance of R&D provided by one organization for another. R&D services are for the most part provided by companies and organizations involved in biotechnology; contract research (including physical, engineering, and life sciences firms); and professional, scientific, and technical areas (including social sciences and humanities). These are companies or organizations categorized under NAICS code 5417 (scientific R&D services). Specifying NAICS codes for R&D services (as does BRDIS) is important, since firms in almost any industry can buy or sell R&D services. For example, Boeing can buy services to fill a gap in its R&D program for wing design; Walmart can sell its knowledge, based on R&D, on supply chains; and extraction firms can buy or sell R&D services related to extraction.

Currently, R&D services are captured through the use of a number of indicators published in the *SEI*. These include R&D by sector and location of performance, funding of R&D

[19]"The percentile of a publication is its relative position within the reference set—the higher the percentile rank, the more citations it has received compared with publications in the same subject category and publication year" (Bornmann and Marx, 2013, p. 2).

[20]Metadata describe the data and how they were constructed.

[21]"Services are the result of a production activity that changes the conditions of the consuming units, or facilitates the exchange of products or financial assets. These types of service may be described as change-effecting services and margin services respectively. Change-effecting services are outputs produced to order and typically consist of changes in the conditions of the consuming units realized by the activities of producers at the demand of the consumers. Change-effecting service are not separate entities over which ownership rights can be established. They cannot be traded separately from their production. By the time their production is completed, they must have been provided to the consumers" (European Commission, 2009, Chapter 6, paragraph 17).

by companies and others, R&D performed abroad by U.S.-owned companies, R&D performed in the United States by foreign multinationals (foreign direct investment in R&D), and exports and imports of R&D and testing services. For the *SEI*, data on R&D performance and funding are taken from BRDIS, while the Bureau of Economic Analysis (BEA) provides the data on foreign direct investment in R&D and on international trade in R&D testing services.

NCSES is expanding its data-linking activities to match BRDIS microdata with BEA survey microdata on U.S. foreign direct investment. The agency also has undertaken fruitful interagency collaboration with BEA to integrate R&D into the system of national accounts.

The panel deliberated on globalization and its impact on the research enterprise in the United States. An immediate policy question was how much R&D, measured in terms of expenditures, was outsourced to countries such as Brazil, China, or India, and whether R&D was performed by foreign affiliates or purchased from other companies. A related question was how much knowledge produced by U.S. R&D is being purchased by other countries, and which countries are leading purchasers. These are important but also complex questions that present a number of difficult challenges for data collection.

The panel thus commissioned a paper on this subject by Sue Okubo (2012). The paper reviews the current work of BEA in this area and compares it with recent NCSES work on BRDIS.[22] Several observations follow from this comparison.

One key observation in Okubo's paper is the difference between the classifications used by BEA and NCSES and the fact that BEA measures trade in R&D and testing services, whereas NCSES in BRDIS measures R&D services only. While BEA and NCSES are cooperating on survey activity, the panel emphasizes the importance of this cooperation's leading to comparability of the data produced by these and other agencies (see Recommendation 5-2 later in this section).

The surveys on international transactions administered by BEA and the R&D surveys[23] carried out by NCSES follow different guidance: BEA follows the sixth edition of the International Monetary Fund's (IMF) (2011) *Balance of Payments and International Investment Position Manual*, while NCSES follows the *Frascati Manual* (OECD, 2002a). However, the two approaches are not far apart. The IMF manual includes some R&D and intellectual property elements that are consistent with the *Frascati Manual*. Therefore, the geographic and ownership scope of BEA's international transaction surveys and that of the BRDIS are conceptually close. For example, BEA's international transaction surveys encompass any company with activities in the United States, regardless of ownership. The surveys cover transactions of U.S.-located units of foreign multinational enterprises with entities outside the United States, including transactions with their own foreign parents, and affiliated and unaffiliated trade. Similarly, for the United States, the surveys cover affiliated and unaffiliated trade and transactions by purely domestic companies (no relationship with any multinational enterprise). BRDIS also covers any company with activities in the United States, regardless of ownership, and foreign affiliates of U.S. multinational enterprises.

On the other hand, BRDIS treats foreign parent companies differently from the way they are treated in both BEA's trade surveys and BEA's surveys of foreign direct investment. Other differences exist between BRDIS and BEA data on the international balance of payments in R&D trade: BEA's testing services, which are part of the research, development, and testing measure, may include R&D and non-R&D components, and R&D is treated by NCSES basically as a cost measure, while transactions are treated more like market values. Moris (2009, p. 184) suggests a matrix for use in parsing the data from BEA's trade surveys and R&D surveys (including BRDIS).

A second key observation in Okubo's paper relates to the results of the BEA surveys with respect to the sale of R&D and testing services abroad. For 2010, the largest buyers of U.S. R&D and testing services were Bermuda,[24] Ireland, Japan, the Netherlands, and Switzerland, accounting for 6.6 percent of the total trade of $30.9 billion. Such a distribution of trade statistics is rare, as is illustrated by trade in professional, business, and technical (PBT) services. In 2010, the largest buyers of U.S. PBT services were Germany, Ireland, Japan, Switzerland, and the United Kingdom, accounting for 37 percent of total trade; the largest sellers of PBT services to the United States—the countries to which these services were outsourced—were Germany, India, Japan, the Netherlands, Switzerland, and the United Kingdom, which accounted for 40 percent of total U.S. payments for these services (Okubo, 2012). The dominance of the leading countries in the sale and purchase of PBT services is seen in other trade figures, but not in the sale and purchase of R&D and testing services. This difference in the concentration of R&D and testing services merits further analysis.

In summary, the questions that beg to be answered are: Under what circumstances does the R&D activity of multi-

[22]In September 2012, NCSES inaugurated a website with two new publications on the International Investment and R&D Data Link project. The site will also house future publications on the BRDIS link (National Science Foundation, 2013b). It should be noted that BEA plans to incorporate R&D as investment in the core economic accounts in 2014.

[23]The NCSES surveys referred to include BRDIS and its predecessor, the Survey of Industrial Research and Development.

[24]If one were to start with R&D performers only and then look at their R&D exports and imports, legitimate non-R&D performers that only import their R&D from overseas would be eliminated from the analysis. This exercise would require access to the microdata, which are not publicly available. However, NCSES could conduct this analysis and publish the statistics and rankings. There is no escape from accounting and transfer price issues, such as allocated costs that are not related to actual R&D trade. R&D performance data for multinational enterprises are not immune to this issue. Conditioning on performance for trade flows can eliminate unwanted R&D and training data.

national corporations enhance U.S. economic performance, including leadership and strength in S&T? What effect do tax laws have on the location of R&D services? Clearly, the R&D activity of multinational corporations has grown, but the data available with which to analyze and track this activity have limitations. BRDIS includes data on domestic and foreign activities of firms and can provide a more detailed picture of R&D activities than has previously been possible or been fully exploited. Specifically, BRDIS offers more information on R&D service production and flows of R&D services in the United States and in U.S. firms abroad than has heretofore been published. Understanding outsourcing and trade in R&D services is particularly important because the developed economies are dominated by service industries. BRDIS data also can support measures of payments and receipts for R&D services abroad, by leading countries, which is critically important for policy purposes.

RECOMMENDATION 5-2: The National Center for Science and Engineering Statistics (NCSES) should make greater use of data from its Business Research and Development and Innovation Survey to provide indicators of payments and receipts for research and development services purchased from and sold to other countries. For this purpose, NCSES should continue collaboration with the U.S. Bureau of Economic Analysis on the linked dataset.

The panel believes NCSES can provide these estimates and, if necessary, include appropriate questions on BRDIS in 2013 and subsequent years. The 2008, 2009, and 2010 BRDIS did not allow NCSES to collect all of the elements described above, but the 2011 and 2012 questionnaires are more comprehensive in this dimension, collecting data on R&D production, funding, and transactions. Data would be available with which to produce statistics on payments and receipts for R&D services involving U.S. company affiliates at home and abroad and on how those data differ, if at all, from the BEA measures. Similar information on foreign company affiliates from other sources could be used for parallel comparisons.[25] NCSES could consider developing two series—payments and receipts for R&D services—for three to five leading countries. The resulting statistics would show what knowledge creation is being outsourced and which countries are buying U.S. knowledge. This information would enable users to track trends over time and have a better understanding of knowledge flows and the formation of R&D networks.

Over time, this exercise would provide answers to a range of questions: Is the United States losing or gaining advantage by buying and selling its R&D abroad? Is the United States benefiting from research conducted in other countries? What is the United States learning from other countries, and what are other countries learning from the United States? In what technological areas are other countries accelerating development using knowledge sourced in the United States? What is the role of multinational enterprises in transferring R&D capacity from country to country? The data could also be used in regression analysis to answer another important question: What impact does the international flow of R&D have on U.S. economic performance? Users of the data on international flows of R&D services are likely to be interested in seeing how emerging economies are advancing in R&D capacity, in what fields U.S. companies are sourcing or outsourcing R&D and whether it is increasingly being sourced or outsourced in specific countries, and which countries 5-10 years from now may be the hub of new scientific knowledge—possibly countries in Latin America, the Middle East, or sub-Saharan Africa.

Intangible Assets

Until recently, the important role of knowledge-based capital (KBC) was rarely recognized, one exception being Nakamura's (1999) research on intangibles[26] and the "New Economy." This situation has changed primarily as a result of the pioneering research of Corrado and colleagues (2005) on intangibles. In their 2006 paper, these authors point out that most knowledge-based investment is excluded from measured GDP and from most productivity and economic growth models. The authors recognize three broad categories of KBC: computerized information (software and databases); innovative property (patents, copyrights, designs, trademarks); and economic competencies (including brand equity, firm-specific human capital, networks joining people and institutions, and organizational know-how that increases enterprise efficiency). Another important form of KBC is human capital that is not firm specific, such as most human capital that is created through education.[27] The World Bank (1997) estimates that for most countries, intangibles, including human capital more broadly defined, represent the majority of a country's wealth.[28] By all accounts, failing to recognize KBC in any analysis of economic growth or the potential for innovation is a significant omission.

For this reason, a major development in the measurement of KBC occurred when the status of R&D was changed in the 2008 System of National Accounts (SNA) from an expense to an (intangible) capital investment. Efforts are still ongoing both in the United States (see, e.g., U.S. Bureau of Economic Analysis, 2010) and internationally to integrate R&D fully into national accounts. This work requires not only high-

[25]See, for example, Eurostat 2010 statistics (Eurostat, 2013). Also see statistics for Germany (Deutsche Bank Research, 2011) and on the Indian engineering R&D offshoring market (NASSCOM and Booz & Company, 2010). These two reports cite private company estimates, as well as published Eurostat statistics.

[26] Part of the broad category of KBC; see, e.g., OECD (2012a).

[27]Human capital is discussed in Chapter 6.

[28]World Bank intangibles include human capital, the country's infrastructure, social capital, and the returns from net foreign financial assets.

quality data on R&D, but also methods for estimating the depreciation of R&D capital, appropriate R&D deflators, and the estimation of price changes. Although the integration of R&D into the SNA is mainly the responsibility of BEA, NCSES has an important role through its long-standing expertise in the collection of R&D data.

The estimates of Corrado, Hulten, and Sichel for the United States give a sense of the relative importance of various components of KBC as defined above (Corrado et al., 2006). Almost 35 percent of their measured KBC either is currently in GDP (computer software) or is in GDP beginning with estimates for 2013 (mainly scientific R&D). Some data on nonscientific R&D (e.g., social science R&D) are now collected through National Science Foundation (NSF) surveys. Total nonscientific R&D is estimated by Corrado, Hulten, and Sichel to be in excess of 20 percent of total R&D. The largest portion of the unmeasured component, economic competencies, accounts for somewhat less than 40 percent of spending on business intangibles.

More than 70 percent of spending on economic competencies is for firm-specific resources. This spending includes employer-provided worker training and management time devoted to increasing firm productivity. Examples given for management time are time for strategic planning, adaptation, and reorganization. Corrado, Hulten, and Sichel used management consulting industry revenues, trends in compensation, and numbers of individuals in executive occupations to estimate spending in the management time category. Sixty percent of advertising expenditures is allocated to business spending on brand equity intangibles.[29]

A number of researchers have estimated KBC for individual countries following the lead of Corrado, Hulten, and Sichel. These individual countries include Australia (Barnes, 2010; Barnes and McClure, 2009), Canada (Baldwin et al., 2008), China (Hulten and Hao, 2012), Finland (Jalava et al., 2007), France and Germany (Delbecque and Bounfour, 2011), Japan (Fukao et al., 2007, 2009, 2012; Miyagawa and Hisa, 2012), the Netherlands (van Rooijen-Horsten et al., 2008), and the United Kingdom (Gil and Haskel, 2008; Marrano et al., 2009). Corrado and colleagues (2012) recently completed KBC estimates for the 27 EU countries and the United States. In addition, the methodology for estimating individual components of KGC has been refined, most notably by Gil and Haskell (2008).

A discussion paper by Corrado and colleagues (2012) provides the broadest view of the importance of KBC as it covers the largest number of countries.[30] In their estimates, the United States stands out for two reasons as compared with regional EU country averages: it has the largest share of intangible investment in GDP (11 percent), and it is the only country/region for which intangible investment is a larger share of GDP than tangible investment. In all country/regional comparisons, however, the rate of growth in intangible investment exceeds that in intangible investment. The authors report three main results. First, capital deepening is the dominant source of economic growth once intangibles are recognized. Second, deepening of intangible capital accounts for one-fifth to one-third of the growth of labor productivity. Finally, the contribution of intangible capital in some large European countries (e.g., Germany, Italy, and Spain) is lower than that in the United Kingdom and the United States. However, there are significant country differences in the distribution of intangibles by broad types: computerized information, innovative property, and economic competencies.

Aizcorbe and colleagues (2009) review various definitions of innovation; propose how measures of innovation like that addressed by Corrado, Hulten, and Sichel could be integrated into a satellite account; and outline future BEA plans. They note that whether advertising and marketing expenditures should be treated as investment is being debated. They question whether cumulating all firms' advertising expenditures should be registered as increasing aggregate output. In addition, they comment on the difficulty of measuring spending on organizational change. As Corrado, Hulten, and Sichel also recognize, they describe how developing deflators and depreciation rates for most intangibles can be difficult. Their paper calls for cultivation of sources for spending on the development and implementation of new business models, the creation of new artistic originals (see below), the design of new products, and intermediate inputs to innovation. Finally, they hope to work toward better price and depreciation estimates and, in cooperation with the Census Bureau and NSF, the publication of firm and establishment innovation statistics.

Since Corrado, Hulten, and Sichel published their first paper on intangibles in 2005, U.S. government agencies have moved forward to measure and recognize intangibles more comprehensively. As mentioned above, efforts are under way to capitalize R&D and fully integrate it into the SNA. Investment in artistic originals is incorporated into U.S. GDP in 2013 (Aizcorbe et al., 2009).[31] BEA-defined artistic originals include theatrical movies, original songs and recordings, original books, long-lived television programming, and miscellaneous artwork (Soloveichik, 2010a,b,c,d, 2011a,b). For many years, mineral exploration, a relatively small component, has been recognized as investment in U.S. GDP.

Many reports and monographs and at least one book have been produced on KBC. Many of them have been published since 2005. An interim project report from OECD (2012a)

[29] More information on how business spending in intangibles was estimated is available in Corrado et al. (2005).

[30] The years covered vary in Corrado et al. (2012): the earliest beginning year is 1995, and the latest is 2009. Regions include Scandinavian (Denmark, Finland, and Sweden), Anglosaxon (Ireland and the United Kingdom), Continental (Austria, Belgium, France, Germany, Luxembourg, and the Netherlands), and Mediterranean (Greece, Italy, Portugal, and Spain).

[31] See Chapter 7 of this report for more detail on how Aizcorbe and colleagues at BEA are using administrative records and web-based data in the agency's project to capitalize intangible assets for inclusion in the SNA.

echoes the Corrado and colleagues (2012) conclusion that intangibles have been estimated to account for a substantial share of labor productivity: 20-25 percent across Europe and 27 percent in the United States. In addition, the OECD report notes that there are substantial spillovers from and repeated use of KBC, and that global competitiveness may increasingly be determined by KBC. After offering answers to the question of why business is investing in KBC, the OECD report focuses on policy questions. The policy challenges discussed with respect to KBC are in the areas of taxation, competition, intellectual property rights, personal data, and corporate reporting. Other publications focus on KBC more from an accounting or business perspective. Lev (2001) uses movements in stock market prices to estimate the impact and importance of intangibles. A long report by Stone and colleagues (2008), written from the business/accounting perspective, includes a long list of references. Among its contributions are a summary of efforts to measure firm- and aggregate-level innovation and a taxonomy of possible types of measures—indicator indices, monetary, and accounting. Many authors recognize the complexity of measuring and estimating the contribution of KBC to economic growth.

The potential definition of KBC is far broader then that employed by Corrado, Hulten, and Sichel. Aside from including all formal education, not just employer-provided training, Stone and colleagues (2008) cite two major categories—relational capital and open innovation. Relational capital refers to relationships with external stakeholders, including customers and suppliers. Its value can encompass the complementarity of user needs, such as customers and advertisers using Google for similar purposes. Companies that use open innovation post R&D and commercialization challenges on web-based forums or "marketplaces" that are accessible to communities of scientists, engineers, and entrepreneurs. A component of the Corrado, Hulten, and Sichel definition that is featured less prominently in related research, including that of Stone and colleagues (2008), is general networking. Stone and colleagues comment that general networking is particularly useful for businesses operating in emerging economies. Facebook provides a form of social capital/networking that by extension has information and business value. Each of these expansions or extensions of the Corrado, Hulten, and Sichel definition of intangibles presents substantial measurement challenges.

As stated by Stone and colleagues (2008, p. II-4), "Intangible assets are not innovations, but they may lead to innovations." And as stated by Ben Bernanke in the concluding sentence of a 2011 speech, "We will be more likely to promote innovative activity if we are able to measure it more effectively and document its role in economic growth." The open question, however, is which KBC leads to economic growth and to what degree, and is this part of the challenge of making a direct and quantifiable connection between innovative activity and economic growth? Certainly some components of KBC have been studied extensively to document their role; scientific R&D is the prime example. Other components of KBC have been less well studied; organizational know-how is an example. The importance of KBC as an STI indicator depends on the drawing of connections. However, it is critical to recognize both KBC and tangible capital as factors that may be important indicators of future growth. Although the panel believes work on intangible assets may generate useful STI indicators, it believes NCSES should not seek to produce these statistics on its own, but support and draw on the work of other agencies, particularly BEA, in this area. However, NCSES still has an important role to play through its collection of high-quality R&D data, and it may also be able to contribute with other data sources. This might be the case, for example, if NCSES were to begin collecting data on innovation-related expenditures, as outlined in Chapter 4.

RECOMMENDATION 5-3: The National Center for Science and Engineering Statistics (NCSES) should continue to report statistics on knowledge-based capital and intangible assets obtained from other agencies as part of its data repository function. In addition, NCSES should seek to use data from the Business Research and Development and Innovation Survey on research and development and potentially also on innovation-related expenditures as valuable inputs to ongoing work in this area.

Indicators of General-Purpose Technologies

"General-purpose technology" (Lipsey et al., 2005) is a term used to describe technologies with the potential to transform the economy and activities across a broad range of sectors and industries (Jovanovic and Rousseau, 2005). Earlier examples are steam, electricity, and internal combustion, while more recent examples include information and communication technologies, biotechnology, nanotechnology, and green technologies. Given their potential importance for innovation and growth, tracking the development of these technologies and their diffusion and application is important to inform policy. In this area, there is one particular policy question that users of STI indicators are eager to have answered: Is the United States promoting platforms in information and communication technology, biotechnology, and other technologies to enable innovation in applications?

Bresnahan and Trajtenberg (1995) outline three characteristics of general-purpose technologies: their pervasiveness across sectors, their development and improvement over time, and their ability to spur innovation in their own and other sectors. These characteristics are useful for guiding the measurement of general-purpose technologies. Tracking knowledge generation in these technologies, their diffusion to other sectors, and the linkages among them is important for understanding innovation and other sources of growth in the economy.

Measuring general-purpose technologies poses two main difficulties. The first is that not all of these technologies can be properly identified as belonging to a particular sector, because they are spread across different industry classifications. The second difficulty arises in identifying the use of these technologies in other sectors. Clearly, the extent of these difficulties varies according to each such technology. Information and communication technology is by far the best covered in statistics in terms of both industry classification and identification of investments in other sectors.

A number of the data sources discussed in this chapter can be used to generate indicators of general-purpose technologies. For example, patents and trademarks can be used to measure the use of such technologies for knowledge creation in sectors other than those in which they were developed, and both patent and bibliometric data can be used to measure the linkages among general-purpose technology sectors. R&D data provide an indicator of knowledge generation in sectors that develop general-purpose technologies, as do broader measures of investment in these technologies. In addition, the BRDIS contains data on the percentage of R&D in energy applications, environmental protection applications, software, medical clinical trials, biotechnology, and nanotechnology. These data can potentially be used to investigate the extent of R&D in these technologies across sectors (thus giving a picture of how "general-purpose" these technologies are).

NCSES currently publishes a number of statistics on general-purpose technologies—particularly for information and communication technology, but increasingly also for green technologies. The panel encourages NCSES to continue this work and also to build on current indicators in this area. In particular, NCSES should examine possibilities for better coverage of the diffusion and uptake of general-purpose technologies in sectors other than those in which they were developed, using both BRDIS and other data sources.

RECOMMENDATION 5-4: The National Center for Science and Engineering Statistics (NCSES) should develop a suite of indicators that can be used to track the development and diffusion of general-purpose technologies, including information and communication technologies, biotechnology, nanotechnology, and green technologies. NCSES should attempt to make greater use of data from the Business Research and Development and Innovation Survey for this purpose while also exploring the use of other sources, such as patent and bibliometric data.

Subnational Issues in Measuring New Knowledge and Knowledge Networks

Compared with the measurement of innovation, the measurement of knowledge production is more clearly connected to geographic location. A number of initiatives by successive administrations have emphasized the ability to locate federal research grants on S&T down to very detailed levels within neighborhoods. Of course, some of this detail is spurious. Establishment data may link to some postal address while the actual economic activity is being carried out over a territory of some size, depending on the industry. Moreover, much of the value derived from these targeted investments comes from the trade of goods and services, which is dispersed geographically.

Still, disaggregating is certainly possible to levels much finer than the states. For example, universities are well-behaved geographic phenomena in that they remain in one place, and their relation to various nested administrative hierarchies is straightforward. Their laboratories and research facilities are similar to those of other establishments; in fact, some of them resemble industrial facilities with loading docks, employees, and so on. The movement of goods and people in the university establishment can be accounted for in the production of scientific work.

Some success appears to have been achieved in gathering data on some of the basic science output tied to spatial units. Geographic identifiers appear in many contexts, including author lists of publications and patent applications. With some care, and some level of error, these outputs can be linked to a location. But difficulties are entailed in measuring the impacts of research investments, particularly with spatial disaggregation. Particularly challenging to measure is the geographic instantiation of a knowledge network and the flows of knowledge from place to place.

As a reference for understanding the national system of R&D, it may be worthwhile to examine the results of a major study conducted in Canada in 2011 (Jenkins et al., 2011). A six-member expert panel carried out a full review of the country's federal R&D support programs. While one important theme concerned Canada's balance between tax credits and direct R&D support, the authors' comprehensive study of the whole system of R&D support programs bears examination for application to the United States. The Canadian panel surveyed more than 60 institutes and programs engaged in supporting business innovation. The distribution was highly skewed, with a few relatively large entities and many small ones. Because each was created under a distinct charter, there is little coherence in the criteria used to evaluate effectiveness, a common problem worldwide. The tendency, as in other countries, is to concentrate on generating more investment in R&D rather than on providing mechanisms for industry to obtain the assistance needed to overcome current problems in operations. Certain gaps also became evident from this comprehensive analysis, leading the Canadian panel to offer recommendations for short-term measures to improve the effectiveness of the country's innovation system. The Canadian panel notes that the responsibility for fostering innovation cuts across many functions of government and therefore requires a system-wide perspective and whole-of-government priority. That panel's recommendations include making encouragement of innovation in the Canadian economy a stated objective of federal procurement policies and programs, restructuring procurement procedures to allow more latitude for innovative solutions to emerge, and reorienting existing federal laboratories to serve sectoral needs.[32]

The important message in the present context is that certain aspects of the innovation system emerge from a comprehensive view of the whole. Canada invested the efforts of a distinguished panel in such a process, with clear results for managing its system. That panel's analysis also raised the question of how to compare existing R&D support programs. Although NCSES, as a statistical office, does not conduct evaluation, it should be in a position to provide information on government programs that would support the evaluation done by others.

SUMMARY

In this chapter, the panel has offered four recommendations regarding the development of indicators of knowledge generation, knowledge networks, and knowledge flows. The focus is on techniques that should be used to develop indicators that users want for specific market sectors and that improve the international comparability of the data. The panel also suggests that the production of certain measures is not in NCSES's purview, and these measures should instead be acquired from other agencies. In the near term, NCSES should give priority to using tools that are readily available at the agency and continuing existing collaborations with other agencies while developing new techniques and cultivating new linkages over time.

[32]The report lists the following sectors (p. 3-13): Goods Industries (agriculture, forestry, fishing and hunting; manufacturing; construction; utilities; and oil and gas and mining); Services Industries (transportation and warehousing; information and cultural industries; wholesale trade; retail trade; finance and insurance, real estate and rental and leasing; professional, scientific, and technical services; and other services); and Unclassified Industries.

6

Measuring Human Capital

The National Center for Science and Engineering Statistics (NCSES) produces a rich set of human capital indicators, ranging from elementary school education; to postdoctoral training; to employment in science, technology, engineering, and mathematics (STEM) occupations. These measures convey the magnitude, composition, and quality of human capital; funding of education; deployment of human capital in industry, government, and academe; and human capital creation within industry (see Hall and Jaffe, 2012). NCSES's academic surveys provide information on academic funding for science and engineering (S&E) research, federal spending among fields of study, and spending on academic infrastructure.[1] The education surveys provide the data needed to measure the pipeline and pathways into higher education in STEM fields. Measured by online downloads (unadjusted for length of views), the most widely viewed statistics in the National Science Board's *Science and Engineering Indicators* (*SEI*) relate to education and the workforce, making these statistics one of NCSES's most important products and making NCSES an international leader in S&E education statistics.

This chapter summarizes the human capital issues that current and potential users of NCSES statistics say are of high value to them and then details NCSES's statistical resources for developing data that meet user needs. The discussion includes opportunities for obtaining data from currently untapped sources to produce statistics that accord more closely with user needs and the panel's recommendations for new directions in human capital indicators.

WHAT USERS WANT

The America Creating Opportunities to Meaningfully Promote Excellence in Technology, Education, and Science (America COMPETES) Reauthorization Act of 2010 calls for NCSES to provide information on STEM education, reflecting the desire of Congress for such information. NCSES has many elements in its datasets with which to satisfy this requirement.

To gain a better sense of what other current or potential users of NCSES human capital indicators would like to have, the panel interviewed users of science, technology, and innovation (STI) indicators for this study. During these interviews, panel members asked one key question about the policy relevance of indicators: What are the most pressing policy issues that your agency/organization/institution encounters for which it finds STI indicators useful to have? This question yielded several fruitful responses that appeared to revolve around the central theme of this report—using STI indicators to capture change (see Box 6-1).

NCSES'S EXISTING HUMAN CAPITAL INDICATORS

NCSES produces a broad range of data on human capital, many of which are included in the *SEI*. These include data on enrollments and degrees by demographic classification, including citizenship and place of birth, as well as postdoctoral fellowships. The *SEI* report contains information on students by type of financial support in graduate school, including support from the federal government, by field of study. These data include stay rates and intent to stay in the United States. Data also are available in the *SEI* on tertiary degrees conferred in other countries.

NCSES's Scientists and Engineers Statistical Data System (SESTAT) is a comprehensive database on education, employment, work activities, and demographic characteristics. SESTAT collects information from three biennial sample surveys of individuals: the National Survey of College Graduates (NSCG), the National Survey of Recent College Graduates, and the Survey of Doctorate Recipients (SDR). The SDR has galvanized an international large-scale data collection initiative—the Careers of Doctorate Holders survey—by OECD; the United Nations Educational,

[1] The National Science Foundation (NSF) includes social sciences and psychology in its definition of S&E (see Regets, 2010).

BOX 6-1
Policy Questions Related to Education and the Workforce

The Changing Economy and the Impact on Supply and Demand for Education and the Workforce

- Is the United States producing the skills and competences needed to meet current and anticipated demand for science, technology, engineering, and mathematics (STEM) workers?
- How does the workforce respond to changes in the demand for skills?
- How mobile are science and engineering workers between employers, between public- and private-sector jobs, and between academic and non-academic jobs?
- How many people, possessing what kinds of skills, are needed to achieve a robust science, technology, and innovation (STI) system?
- What fields other than STEM are important for advancing STI?

Changes in the Demography of the U.S. Population and in the Structure and Technology of Education

- Is the population of science and engineering researchers aging, and if so, at what rate?
- How many science and engineering doctorate holders take nontraditional pathways into the STEM workforce?
- Does this vary by race/ethnicity, gender, or the existence of a disability?
- How important are community colleges in developing human resources for STEM talent?
- How will the rapid growth of STEM courses on the Internet, sponsored by major universities, contribute to the size and competence of the nation's STEM workforce?

Changes in the Stocks and Flows of Foreign Students in the United States and Around the World

- What are the career paths of foreign-born STEM-educated or foreign-trained individuals?
- What is the mobility of STEM labor between countries?
- How much do foreign students benefit from federal funding of graduate training?
- What are the stay rates for foreign students?
- If they stay, then what field or occupation do students enter? How long does it take a STEM student to acquire a study or work visa?
- Which degrees are students most commonly sponsored to acquire?

International Comparisons of Scientific Talent Stocks and Flows

- Where does the United States rank among nations on elements of advancement of scientific knowledge?
- In which fields is the United States a net exporter of knowledge?

Scientific and Cultural Organization's (UNESCO) Institute for Statistics; and Eurostat (see Auriol, 2010). This survey is expected to provide an opportunity for international comparisons of doctorate recipients.

The Survey of Earned Doctorates (SED) annually surveys individuals who have newly received a Ph.D. The SED is used in the latest *SEI* report to produce:

- Table 2-1—fraction of doctorate holders who earned a credit from a community college, broken out by ethnicity;
- Tables 2-4 and 2-5 and Figures 2-4 and 2-5—sources of support for graduate students, broken out in various ways (similar information is available in the National Science Foundation [NSF]/National Institutes of Health [NIH] Survey of Graduate Students and Postdoctorates in Science and Engineering);
- Figures 2-19 and 2-20—total number of Ph.D.'s earned, broken out by field and demographics (similar data are available from the Integrated Postsecondary Education Data System [IPEDS]);[2]
- Table 2-10—median time to degree; and
- Tables 2-12 and 2-13 and Figure 2-25—number of U.S. Ph.D.'s, by country of origin.

It appears that many of the most interesting data on doctorates can be obtained from other sources, so, as discussed in greater detail later in this chapter, the SED appears to be a good candidate for less frequent administration, assuming its use as a frame for the SDR can be sorted out.[3]

[2]IPEDS is an institutional database housed at the National Center for Education Statistics. It contains information on higher education institutions, including community colleges.

[3]Further discussion of the potential rationalization of surveys is presented later in this chapter.

In 2009, following recommendations by the National Research Council (2008), the U.S. Census Bureau added a "field of bachelor's degree" question to the American Community Survey (ACS).[4] The Census Bureau codes the open-ended ACS responses into degree field categories that are based on the Classification of Instructional Program (CIP) codes. For the 2010 and 2013 NSCG, the degree field information from the ACS was used to create a degree field stratification variable, differentiating S&E from non-S&E degree fields. This variable was combined with degree level, occupation, and key demographic stratification variables to select the NSCG sample. NCSES and the Census Bureau are currently evaluating the ACS degree field responses to determine the consistency between that information and the degree field information on the NSCG. The findings from this study will help determine whether NCSES changes the way it uses the ACS degree field responses for NSCG stratification purposes.

NCSES draws heavily on major sources for elementary, secondary, high school, and some postsecondary statistics for its human capital indicators. These sources include the U.S. Department of Education, National Center for Education Statistics (including IPEDS); the U.S. Department of Commerce, Census Bureau; OECD; the UNESCO Institute for Statistics; the Programme for International Student Assessment; the Center for the Study of Education Policy, Illinois State University; the Higher Education Research Institute, University of California, Los Angeles; U.S. Citizenship and Immigration Services; the U.S. Department of Homeland Security; and the American Association of Engineering Societies. Statistics from statistical bureaus of foreign countries (e.g., China, Germany, Japan, South Korea, United Kingdom) also are reported. Statistical results from peer-reviewed articles can illuminate contemporary issues, such as trends in international higher education (Becker, 2010) and international migration of high-skilled workers (Davis and Hart, 2010; Defoort, 2008; Shachar, 2006). Major data sources for workforce statistics are the Department of Labor, Bureau of Labor Statistics (Occupational Employment Statistics), and the Department of Commerce, Census Bureau (ACS and Current Population Survey).

There is fervent interest in the United States and abroad regarding the contributions of foreign-born scientists and engineers to knowledge creation, entrepreneurship, innovation, and economic growth. Analyzing data from NSF's NSCG, Stephan and Levin (2001, p. 59) conclude that "immigrants have been a source of strength and vitality for U.S. science" and that they made "exceptional contributions to the physical sciences." Hunt and Gauthier-Loiselle (2010), also using NSF's NSCG, as well as state-level data, found that college-graduate immigrants have a higher incidence of patenting compared with their native cohort, primarily because the former immigrants have a higher propensity to obtain S&E degrees. Borjas (2005, p. 56), using NSF's SED and SDR data to examine labor-market outcomes related to immigration of high-skilled immigrant workers, found that "an immigrant-induced 10-percent increase in the supply of doctorates in a particular field at a particular time reduces the earnings of that cohort of doctoral recipients by 3 percent."[5] Franzoni and colleagues (2012) conducted a survey to study the mobility of scientists in four different fields, for 16 countries.[6] India was found to have the highest share of scientists working outside of the country, while the United States was the first or second most likely destination for scientists working outside their home country. Furthermore, using NSF's data on stay rates, Kerr (2008, p. 536) found that "frontier expatriates do play an important role in technology transfer"; however "ties between U.S. ethnic research and entrepreneurial communities and their home countries" are important for the transfer of tacit knowledge that is critical for innovation.

Using the Public Use Microdata Sample (PUMS) of the 1990 U.S. Census, Dun & Bradstreet data on high-technology firms, and in-depth interviews to study the impact of immigration on entrepreneurship, Saxenian (2002) found a high concentration of S&E workers from India and Taiwan in Silicon Valley. Indian and Taiwanese engineers facilitated codevelopment of technologies and technology sharing between firms in Silicon Valley and in their respective native regions. Some researchers are beginning to explore the concept of a "reverse brain-drain" or "brain sharing," whereby migration of foreign-born, U.S.-educated talent back to their home countries either is beneficial for collaborative research and demand for U.S. high-technology products or is viewed as creating competitive high-technology research platforms abroad.

Mobility of students and workers is an indicator of knowledge flows and knowledge networks, and it shows where there exist new sources of science and technology (S&T) talent, high potential for creative ideas and collaboration,

---

[4]The National Research Council's Panel to Assess the Benefits of the American Community Survey for the NSF Science Resources Statistics Division made the following recommendation: "The National Science Foundation should use current data from the American Community Survey to evaluate the degree to which the American Community Survey with the field-of-degree question would allow for the production of mandated indicator reports in the future" (National Research Council, 2008, p. 7).

[5]Card (2009, p. 18), using Current Population Survey data, found evidence that "comparing high immigration cities like Miami and Los Angeles to low immigration cities like Philadelphia or Detroit, the relative wages of workers in the lowest skill group are about 3-4% lower, while relative wages for those in the highest skill group are 3-4% higher."

[6]Franzoni and colleagues (2012) note that when they conducted their survey, NSF had only recently begun to use SDR data to publish statistics on scientists and engineers who were educated in the United States and then migrated to another country. The GlobSci database includes the following fields: biology, chemistry, materials and Earth, and environmental sciences. The countries included in the study were Australia, Belgium, Brazil, Canada, Denmark, France, Germany, India, Italy, Japan, the Netherlands, Spain, Sweden, Switzerland, the United Kingdom, and the United States. Notably, data on China were not available.

new markets for high-technology goods and services, and potentially future competition for the development of high-technology products. Having data by field of degree, by occupation, for specific metropolitan areas, and for specific countries makes it possible to conduct a rich analysis of migration flows of high-skilled students and workers (see Hunter, 2013; Wadhwa, 2009).

The NSF/NIH Survey of Graduate Students and Postdoctorates in Science and Engineering is the only survey used by SESTAT that provides coverage of recipients of foreign-earned degrees, but is limited to those at the postdoctorate level. The Department of Homeland Security's database of Labor Conditions Applications (LCA) (for H-1B visa employers) provides a benchmark for the location of newly hired foreign doctorate recipients. The NSCG could use the ACS to identify doctorates granted outside the United States to workers in the United States. The ACS information (age, degree level, and year of immigration to the United States) could guide the NSCG sampling strata, allowing unique sampling rates for those likely to have earned doctoral degrees abroad. There is also an international component of the SDR—the International Survey of Doctorate Recipients—that captures U.S. doctorate recipients outside the United States. NCSES currently publishes a range of statistics on published papers, including countries, countries' shares of cited papers, and international collaborations.

NCSES also uses its Business Research and Development and Innovation Survey (BRDIS) for employment statistics. It publishes *InfoBriefs* on research and development (R&D) employment intensity, domestic and foreign R&D employment, and company-performed R&D expenditures per R&D employee (see, e.g., Moris and Kannankutty, 2010). However, these data do not account for the entire STEM workforce. Headcounts and related statistics are available for the United States and worldwide for employment, R&D employment, R&D employment by occupation and gender, and highest degree earned. For the United States, there are also counts of H-1B and L-1 visa holders.[7] Full-time-equivalent (FTE) counts are available for S&E workers, as is the number of these FTEs that are funded by the federal government in the United States (see Burrelli, 2010).[8]

NCSES could publish statistics other than those released in the 2010 *InfoBrief* on employment statistics. In particular, NCSES could use BRDIS to count a defined group of non-U.S. citizens holding H-1B or L-1 visas and employed in the United States as R&D scientists and engineers for businesses. This statistic would not cover all H-1B or L-1 visa holders working for these firms, because some may be working in non-R&D activities. Because about one-third of H1-B visa holders are in computer-related occupations,[9] and only a modest proportion are included as scientists on the Immigration and Naturalization Service's occupational list (see U.S. Department of Homeland Security, 2011b), the BRDIS number depends on how many computer-related workers a firm classifies as R&D scientists and engineers. Still, this would be a valuable new measure as it would count the stock of H-1B or L-1 visa holders working in a well-defined activity, while most of the standard data relate to flows based on less precise measures of what workers do. It would also be useful to have more up-to-date measures, because the demand for H-1B or L-1 visas varies from year to year with the economic outlook in high-tech sectors.

THE SURVEY PROBLEM

As noted in earlier chapters, the survey approach to data collection is growing increasingly expensive, while response rates are dropping (a trend that applies to all surveys). As budgets become tighter, the higher costs will make it increasingly difficult to continue maintaining the current set of surveys. A short-term approach for dealing with these challenges is to implement small, incremental changes that respond to immediate budgetary shortfalls. An alternative approach, however, is to reconsider the whole system of data collection over the longer term to identify opportunities for cost savings and quality-enhancing improvements.

The SESTAT surveys provide a prime example. In lieu of four independent surveys, the sampling frames are interlinked, as illustrated in Figure 6-1. Each sampling frame ties back to a more complete census (or census-like) source. For the NSCG, this used to be the long-form sample of the U.S. census, now replaced by the ACS. The NSCG samples about 0.25 percent of its population. The National Survey of Recent College Graduates is based on a two-stage sample design that selects institutions, then samples students from those institutions. It samples 0.4-0.8 percent of its target population. The SDR ties back to the SED, which is actually a census of all doctoral graduates in a given period. The SDR sampled about 4 percent of its target population in 2008.

These sampling frames were not designed around the requirements of consistent geographic coverage or maintenance of a panel over time. If the sampling rate is set below

[7]The H-1B and L-1 visas are for foreign workers in specialty occupations in fields that require highly specialized knowledge and intracountry transferees, respectively. For the fields covered, see U.S. Department of Homeland Security (2011a).

[8]All of the statistics reported in this brief are special tabulations from the Student and Exchange Visitor Information System database, maintained by Immigration and Customs Enforcement of the U.S. Department of Homeland Security.

[9]See National Science Board (2012a, p. 3-50) for indicators of foreign-born workers in the United States. The U.S. Department of Homeland Security (2012, p. 13) shows that approximately 51 percent of H1-B visa petitions approved were in computer-related occupations in fiscal year 2011. The U.S. Department of Labor reports positions certified for the H-1B program. STEM-related occupations accounted for approximately 49 percent of positions certified for H-1B visas for fiscal year 2011 (see U.S. Department of Labor, 2012b, p. 60). The computer-related occupations included in these statistics are computer science; computer science, engineering; and computer science, mathematics. Those three occupations made up 42.5 percent of the positions certified for H-1B visas in fiscal year 2011.

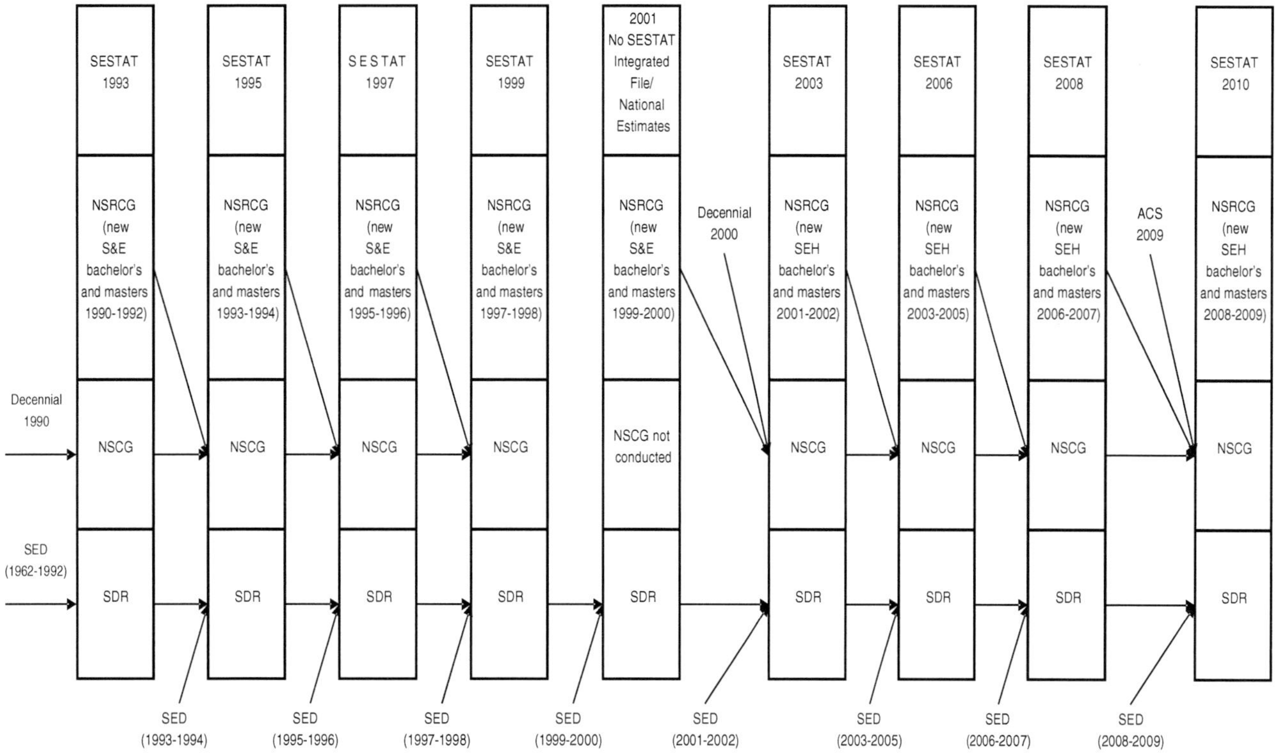

FIGURE 6-1 SESTAT surveys in the 1990s, 2000s, and 2010s.
NOTE: ACS = American Community Survey; NSCG = National Survey of College Graduates; NSRCG = National Survey of Recent College Graduates; S&E = science and engineering; SDR = Survey of Doctorate Recipients; SED = Survey of Earned Doctorates; SEH = science, engineering, and health; SESTAT = Scientists and Engineers Statistical Data System.
SOURCE: NCSES SESTAT surveys in the 1990s, 2000s, and 2010s.

1 percent in a state with a low number of institutions and graduates, then making any reasonable set of measurements will be difficult without suppression to protect the confidentiality of respondents. This is particularly true if the sampling of institutions is not designed spatially. A design that allowed for consistent geographic coverage would make the sampling frame another level deep—geographic distribution, institutions in those regions, and students who attend those institutions. This is not the case for these SESTAT surveys.

The timing of these surveys has been affected by external events. For example, the NSCG used to be administered in odd years but has switched to even years, maintaining a 2-year spacing, although for some surveys with a 5-year gap. A close tie to the decennial census is not crucial because the ACS provides data annually.

In 2008, a Committee on National Statistics panel of the National Research Council recommended[10] that NCSES reconsider the design of SESTAT following the implementation of the ACS. NCSES made the decision to terminate the National Survey of Recent College Graduates[11] because the frequency of the ACS makes it unnecessary to refresh the data from the decennial census with data on recent college graduates. Starting in 2012, NCSES switched to a new sample frame that provides coverage similar to that of the National Survey of Recent College Graduates (with the nuance of young rather than recent college graduates) at a greatly reduced cost. The switch from the National Survey of Recent College Graduates to the ACS has the added benefit of providing better estimates of foreign degree holders working in the United States (see the section on mobility later in this chapter).[12] The decision to terminate the National Survey of Recent College Graduates is a reasonable first step toward rationalizing the SESTAT surveys, but it may not be enough to remedy the rising costs and declining response rates for all the surveys. Box 6-2 gives other suggestions for gaining efficiencies in obtaining data on human capital that complement existing NCSES survey strategies.

[10]"Recommendation 7-5: The National Science Foundation should use the opportunity afforded by the introduction of the American Community Survey as a sampling frame to reconsider the design of the Scientists and Engineers Statistical Data System (SESTAT) Program and the content of its component surveys" (National Research Council, 2008 p. 75).

[11]NCSES, "Reconsidering the SESTAT Design," internal document, January 2012.

[12]Therefore, the last year of data collection using the NSRCG was 2010.

BOX 6-2
Data Options to Consider Based on Existing Surveys

- Make up-front investments in improvements to survey processes and other infrastructure, with the goal of saving money down the road:
  - Shift to web/automated phone surveys.
  - Explore techniques that narrow response windows.
  - Explore piloting nonresponse methods being developed by the U.S. Census Bureau.
- Invest in increasing the value of existing datasets through annotation and processing:
  - Unify taxonomies from past iterations of the Survey of Doctorate Recipients, Survey of Earned Doctorates, and General Social Survey using crosswalks.
  - Reduce dimensionality using cluster analysis or other tools.
- Lay the groundwork for data/text-mining techniques going forward:
  - Support outside work on semantic annotation of data on research and researchers (e.g., ORCID).
  - Improve semantic annotation of existing National Science Foundation (NSF) data:
    - Standardized citation format for NSF support (similar to SciENCV at the National Institutes of Health [NIH])
    - Department of the Interior for SESTAT datasets.
  - Explore the use of administrative records on research funding from other agencies, human resource data from universities.
  - Conduct a modest ongoing prize competition for new metrics to develop ideas, experience, future directions, multimodal acquisition (see Chapter 7 for more on this option).
  - Pilot test real-time estimation during data collection (see Chapter 7 for more on the Census Bureau's development of this idea).

POTENTIAL FOR NEW DATA SOURCES

Currently, NCSES publications do not include statistics on how many U.S. employees in S&E occupations were trained in specific foreign countries. Continued collaboration between NCSES and the Department of Homeland Security is expected to yield better indicators of STEM education and the STEM workforce.

Underutilization of existing data on STEM workers is a persistent problem. Although worker mobility is undermeasured in traditional STI employment statistics, some longitudinal studies capture data on the movement of workers with STEM degrees within and outside of traditional S&E jobs. During this study, staff from a range of agencies emphasized to the panel that they have a great deal of underutilized data, particularly regarding human capital. At the panel's September 2011 meetings, for example, Erika McEntarfer of the Census Bureau described a potential project on which her division is working that entails using data from the Longitudinal Employer-Household Dynamics Program to link workers longitudinally across jobs. Integrating these data (along with similar data on firm dynamics) into the statistics offered by NCSES would create a useful set of indicators. Trends in how macroeconomic fluctuations affect workers and knowledge flows in S&E occupations are just one potential output from these data. More descriptive data on innovators also would broaden understanding of the skill sets that lead to STI advances. Obtaining this information would require case studies, which could enhance understanding of statistics based on counting stocks and flows of individuals and knowledge capital.

At the July 2011 workshop, some presenters suggested that measures of STI talent should include the pool of students initially trained in community colleges as a pathway to bachelor's and higher degrees.[13] Because many highly skilled jobs in certain engineering fields require a master's degree and not a doctorate, NCSES has an opportunity to expand its human capital indicators by providing more information on master's-level STI talent.

More detailed data on immigrants, women, minorities, and people with disabilities in the S&E workforce would assist in answering such questions about talent. Without good counts of these individuals, the nation's full STI workforce capacity cannot be known. Also important is to consider what fields other than physical and biological sciences, technology, engineering, and mathematics are important for advances in STI. A careful assessment of network sciences, as well as behavioral and social sciences, would be valuable because these fields are partly responsible for the acceleration of innovation. For example, social networks are used in "collaboratories" to further scientific inquiry without the need for bricks-and-mortar facilities. These contributions to the scientific enterprise come from the behavioral and social sciences. For advances in basic science, it is arguably true that the physical and biological sciences are the primary reservoir of talent. The broader scope of innovation, however, including managerial and organizational elements, includes the social, behavioral, and managerial sciences as critical contributors to outputs and outcomes. Hill (2007) argues that the skill set for advancing innovation is changing:

> In the post-scientific society, the creation of wealth and jobs based on innovation and new ideas will tend to draw less on

[13] In July 2011, NCSES published an *InfoBrief* on this subject, "Community Colleges: Playing an Important Role in the Education of Science, Engineering, and Health Graduates" (Mooney and Foley, 2011). The data for that report were taken from NCSES's National Survey of Recent College Graduates.

the natural sciences and engineering and more on the organizational and social sciences, on the arts, on new business processes, and on meeting consumer needs based on niche production of specialized products and services in which interesting design and appeal to individual tastes matter more than low cost or radical new technologies.

REVISED AND NEW HUMAN CAPITAL INDICATORS

Revised and new human capital indicators are needed in the areas of labor force mobility, the supply of STEM skills and talent, the demand for STEM skills and talent, and the growth of online STEM education.

Labor Force Mobility

While current NCSES indicators provide extensive information on the STEM workforce, their educational backgrounds, and development of technical skills over time, information on the mobility of STEM workers is comparatively sparse. Given the rapidly changing nature of the economy, the panel believes it is important to better understand how the STEM workforce responds to these changes, including movement between academia and the private sector, mobility across industries, changes in occupation, mobility across regions of the United States, and international mobility.

Tracking career paths and mobility is relevant for a number of reasons. The first is related to the contribution of holders of doctorates to innovation in business. Questions to be addressed include what types of mobility patterns are seen outside of academia, in what sectors/industries and what types of jobs, and what motivates the choice to seek employment in the private sector. The second reason relates to the flip side of this issue: Are doctorate holders seeking employment in the private sector because they want to, or because of poor working conditions or lack of employment within academia? What types of skills are in demand for doctorate holders? How well do the skills and competences acquired through a doctoral education match skills needed in later employment?

One way to develop such measures is to start with the jobs people currently hold and then ask about changes. Individuals could be tracked as they moved between the following sectors: educational institution; private, for-profit; private, nonprofit; federal and state/local government; and self-employed. Data from the 2010 SDR questions A9 through A14 could be used to track movement from the principal employer to another sector.[14]

Data sources that follow individuals over time offer the best opportunity to develop statistics on STEM labor mobility. The panel's recommendations in this area involve drawing on existing dynamic databases developed by other agencies and exploiting the sampling procedure of NCSES's SDR, which provides the opportunity to follow doctorate holders over time.

Survey of Doctorate Recipients

The SDR is a longitudinal study of individuals who received a doctoral degree from a U.S. institution in a science, engineering, or health field.[15] The survey follows a sample of these individuals throughout their career, from the year of their degree award through age 75. The panel is refreshed in each survey cycle with a sample of new earners of doctoral degrees in these fields. The longitudinal nature of the survey makes it possible to examine mobility patterns over time for holders of these degrees. However, this dynamic feature of the SDR data typically has not been exploited to create indicators of the labor mobility of doctorate holders, nor have the data been linked to create a dynamic dataset. Instead, existing indicators are based only on single survey cycles.

The SDR is used to track overall changes in statistics over time, such as changes in the number of doctorate holders within academia or in shares of doctorate holders employed in the business sector. Given the longitudinal structure of the SDR sample, however, it should also be possible to use the data from this survey to create measures on the mobility of doctorate holders across different employers, occupations, or sectors. To this end, longitudinal weights would need to be constructed that would take account of changes in the sample and target population over time. One option for tracking researcher mobility in the SDR that would not require additional survey questions would be to introduce a partial panel structure to the survey in which a group of individuals would be included in the sample over a series of survey cycles.

A longitudinal database created from the SDR could be used to generate a number of useful indicators that would provide insight into the mobility of U.S. holders of doctorates in science, engineering, and health. Among these indicators would be mobility rates between the private sector and academia and from temporary to permanent positions within academia. A specific group that has received increased attention is postdoctorates. A concern is that the inability to secure a permanent position may lead many of these individuals to abandon a research career and seek other forms of employment. Longitudinal data on doctorate holders would provide a greatly enhanced ability to track this group (early-career doctorate holders are discussed in more detail below). Furthermore, these data could be analyzed to identify patterns in job mobility across occupations and across industries within the private sector.

---

[14]See http://nsf.gov/statistics/srvydoctoratework/surveys/srvydoctoratework_nat2010.pdf [July 2013] for details on the survey questions.

[15]Science, engineering, and health fields include biological, agricultural, and environmental life sciences; computer and information sciences; mathematics and statistics; physical sciences; psychology; social sciences; engineering; and health (see National Science Foundation, 2012d).

RECOMMENDATION 6-1: The National Center for Science and Engineering Statistics (NCSES) should do more to exploit existing longitudinal data. Specifically, NCSES should exploit the longitudinal panel structure of the Survey of Doctorate Recipients (SDR) in the following ways:

- **create indicators of researcher mobility over time, by constructing longitudinal weights for the SDR that take account of changes in the sample and target population over time—these weights should be constructed both for subsequent survey cycles and for existing data;**
- **create a dynamic database for researcher use in which data from the SDR over time would be linked at the level of the individual; and**
- **enhance coverage of recent doctorate recipients to better track their initial employment and career path in the first years after they receive their Ph.D., which could potentially be accomplished by including an additional module in the SDR or by exploiting that survey's longitudinal capacities or both.**

Longitudinal Employer-Household Dynamics (LEHD) and Baccalaureate and Beyond Longitudinal Study (B&B)

LEHD is a longitudinal, employer-employee database within the U.S. Census Bureau that is developed through the integration of a variety of data sources, including federal and state administrative data on employers and employees and Census Bureau censuses and surveys. The data, which are quarterly, follow both individuals and firms over time and thus can be used to track job flows across firms and industries, as well as labor mobility across industries and occupations. These data would thus appear to be well suited for the development of indicators of the mobility of the STEM (or science, engineering, and health) workforce.[16]

The B&B study, which is conducted by the National Center for Education Statistics (NCES),[17] follows a cohort of bachelor's degree recipients during the 10 years after they have finished college. For example, the B&B:93/03 study followed approximately 11,000 students identified in the 1992-1993 National Postsecondary Student Aid Study (NPSAS) as having earned a bachelor's degree during the 1992-1993 academic year. These individuals were interviewed in 1994, 1997, and 2003 on a variety of topics, such as their education, job search activities, work experiences, further participation in degree and certificate programs, family formation, and other aspects of life after college. Initial B&B cohorts are a representative sample of graduating seniors in all majors. A third B&B cohort was recruited from 2008 graduates, and it has reached a third interview round, with full results forthcoming. Because it is larger than the previous two cohorts, this sample allows for more detail by industry sector.[18] These data can be useful in examining linkages between education and science, engineering, and health occupations, such as the educational background for these occupations and mobility across them for different academic areas.

A particular challenge is capturing the international mobility of doctorate holders. Although ample data are available on individuals that have earned a Ph.D. at a U.S. university and still reside in the United States, the same is not true for two other groups: individuals that have received a doctorate from a foreign university and U.S. doctorate holders that now reside abroad. Although there are no easy solutions to closing this gap, the panel strongly supports NCSES's efforts to identify these groups and include them in SESTAT survey populations, and notes that U.S. doctorate holders abroad have recently been included in the target population of the SDR.

The LEHD and B&B databases offer opportunities for tracking the mobility patterns within sectors and types of occupations and for specific kinds of education. Data on occupations and education can also be used to examine the demand for skills in key industries, an issue discussed in greater detail later in this chapter.

To carry out this activity, NCSES will have to navigate privacy issues as data from the SDR and the LEHD or B&B are linked. However, this activity offers high potential value, particularly for understanding the correspondence between supply and demand for skill sets in S&T sectors and worker mobility.

Social capital is also an important factor in the development and maintenance of human capital. For instance, similar levels of skills, knowledge, and networks developed through STEM education can produce different productivity outcomes in part as a result of differences in social capital. In cases in which skills and knowledge are fairly standardized, higher levels of social capital may lead to higher performance outcomes. Payne and colleagues (2011) synthesize the research on social capital published in the sociology and business management literatures. Although the literature contains a variety of definitions of social capital, Payne and colleagues note that "most scholars generally agree that social capital represents the resources an individual or a collective gains through a social structure or network of relationships. . . ." Saxenian (2002, p. 28) gives an illustra-

[16]See Abowd et al. (2004) for details on the characteristics and uses of LEHD data and Bjelland et al. (2011) for detail on the use of LEHD data for understanding labor mobility.

[17]For more information on the B&B study, see National Center for Health Statistics (2013).

[18]The second B&B cohort of about 10,000 students was drawn from the 2000 NPSAS and followed up in 2001. The third B&B cohort is the largest, with approximately 19,000 students sampled from the 2008 NPSAS and followed up in 2009 and 2012. See http://nces.ed.gov/surveys/b&b/about.asp [January 2014] for more detail on the types of data gathered by all three longitudinal surveys.

tion of how social capital can influence entrepreneurship and productivity in high-technology industries:

> First-generation immigrants, such as the Chinese and Indian engineers of SiliconValley, who have the language and cultural as well as the technical skills to function well in both the United States and foreign markets are distinctly positioned to play a central role in this environment. They are creating social structures that enable even the smallest producers to locate and maintain mutually beneficial collaborations across long distances and that facilitate access to Asian sources of capital, manufacturing capabilities, skills, and markets.

Payne and colleagues list several "key operational examples" that could inform indicators of social capital, a few of which may be relevant for the STEM workforce: an index measuring the extent of redundant connections for an individual or collective, an index measuring the shortest path between one node and other nodes in a network, affiliation of each actor with all other actors (e.g., membership in professional associations), access to information and resources (which may be correlated with the rank of the university attended by the individual), and embedded ties to external stakeholders. While these specific elements are not available in the LEHD or B&B, NCSES could develop proxies for these data from its SESTAT surveys. For example, the 2010 SDR and the 2010 NSCG include questions on professional associations: "During the past 12 months, did you attend any professional society or association meetings or professional conferences? Include regional, national, or international meetings." and "To how many regional, national, or international professional societies or associations do you currently belong?" NCSES could also consider obtaining information on social capital by bridging data on individuals who appear in both the LEHD and STAR METRICS[19] databases.

> **RECOMMENDATION 6-2: The National Center for Science and Engineering Statistics (NCSES) should draw on the Longitudinal Employer-Household Dynamics Program (occupations) and the Baccalaureate and Beyond Longitudinal Study (education levels) to create indicators of labor mobility. NCSES should focus in particular on industries that have been experiencing high growth and/or those in which the United States has a strong competitive advantage. Also relevant would be examining skill sets of firms with high growth.**

Supply of STEM Skills and Talent

As noted above, the current NCSES data and indicators provide extensive coverage of college graduates and doctorate holders in S&T. Much less information is available, however, on other important groups. These include individuals with master's degrees and those who hold degrees from community colleges or have attended community colleges and gone on to earn higher degrees. They also include recent doctorate recipients, a group for which more focus is needed on career choices in the first years after receiving the degree, when finding permanent employment may be difficult. Also needed is reliable, consistent information on STEM education and occupations by different demographic characteristics, such as gender, race/ethnicity, existence of disabilities, and age. Information on numbers and wages of postdoctorates with degrees from U.S. or foreign institutions, whether they work in the United States or abroad, is still incomplete. Needed as well is better information on foreign student stay rates, broken out by race/ethnicity and country of birth.

Although comprehensive statistics exist on college graduates in S&T, it would be helpful to be able to distinguish better among different levels of educational attainment. Master's degree holders currently are identified through the NSCG, and many of the statistics on college graduates published by NCSES distinguish between those with bachelor's and master's degrees. The panel views this distinction as important and worthy of expanding in NCSES's human capital indicators. To have a better understanding of the value of master's degrees, it is important to be able to track the career paths of those who receive them.

The same applies to community college degrees. Data are lacking on how many community college degrees are within STEM fields and to what extent their holders work in STEM occupations. Furthermore, because of the rising costs of college education, many students attend a community college for part of their bachelor's education. The question then arises of whether community college attendance affects the choice of field and the later choice of occupation. NCSES currently uses the National Survey of Recent College Graduates to identify community college degrees and attendance. As that survey is phased out, it will be important for NCSES to collect this information through other surveys, such as the NSCG and the ACS.

> **RECOMMENDATION 6-3: The National Center for Science and Engineering Statistics (NCSES) should enhance indicator coverage of individual science, technology, engineering, and mathematics groups such as early-career doctorate recipients, master's degree holders, and community college graduates. NCSES already distinguishes between bachelor's and master's degree holders in many of its statistics. Stay rates at different education levels by demographic characteristics such as gender, race/ethnicity, disability, and country of origin should be included.**

[19] STAR METRICS is an acronym that stands for Science and Technology for America's Reinvestment: Measuring the Effect of Research on Innovation, Competitiveness and Science. See https://www.starmetrics.nih.gov/ [June 2013].

RECOMMENDATION 6-4: The National Center for Science and Engineering Statistics should explore whether questions can be included in the National Survey of College Graduates and the American Community Survey that would allow the identification of community college graduates or of holders of higher university degrees that have attended a community college.

New Resources

The employment situation of early-career doctorate holders may be very different from that of more experienced researchers. Members of the former group must decide whether to pursue a research career in academia or employment with a private company. A number of interesting questions arise in this regard that cannot be examined with existing data, such as the main factors affecting the choice between academia and the private sector, types of occupations in the private sector and the Ph.D. skills most relevant for these occupations, and responses to potential difficulties in finding a permanent position at a university or public research institution.

Ph.D. dissertations themselves contain valuable information on the areas in which research is being conducted and thus also on the skills and research competencies being developed. Research topics do not always fit well within standard fields of study. They may transcend different fields or represent a new specialized subfield or a combination of topics normally not seen as closely related. To a large extent, a database and the necessary text mining tools with which to address these issues already exist, and with further adaptation and improvement could be used to develop a number of useful indicators on current doctoral work. It would clearly be advantageous if all Ph.D. dissertations were publically available for analysis. Text mining tools could then be developed to identify emerging research topics and to measure the closeness of key research areas. Privately managed services cover a large portion, but not all, of U.S. dissertations.[20]

RECOMMENDATION 6-5: The National Center for Science and Engineering Statistics should explore methods for exploiting the full-text resources of dissertation databases to create indicators on selected topics both within and across scientific fields and on the relatedness of different fields.

[20]The ProQuest Dissertations & Theses Database, for example, contains an extensive number of dissertations, and many, though not all, universities routinely submit accepted dissertations to this database.

Potential for Reducing the Number and Frequency of Surveys

Implementing the recommendations offered in this chapter, as well as others in this report, will require substantial resources. If their implementation is to be feasible, then savings will need to be achieved in NCSES's current activities. The panel identified in particular two areas for potential cost savings. The first concerns further exploitation of the ACS toward the production of S&T statistics. NCSES has adopted the ACS for drawing the sampling frame for the NSCG, an important element in this transition being the inclusion in the ACS of a question on field of study for bachelor's degree holders. Perhaps the ACS could be used not only to draw a sampling frame, but also as a data source to produce statistics on college graduates. A comparison of the NSCG and ACS questionnaires shows that they solicit fairly similar information on main employer and main job, albeit with some differences. For example, the NSCG contains additional information on college graduates that is not available in the ACS, including (1) the extent to which the principal job is related to the highest degree, (2) reasons for changing jobs, and (3) training and motivation for taking it. Hence, while there are differences in the coverage of relevant topics in the two surveys, there is also a significant degree of overlap. Use of the ACS as a source for statistics on college graduates could permit an increase in the frequency of indicators. Less immediate reliance on the NSCG could allow NCSES to reduce the frequency of data collection for that survey and potentially eliminate it altogether in the longer term. A reduction in frequency, for example, from every 2 to every 4 years would yield substantial cost savings that could be reallocated to many of the new indicators recommended in this report.

Second, as mentioned earlier, the panel also views the SED as a candidate for reduction in the frequency of data collection, for the following reasons. Primarily, an increased focus on early careers of doctorate holders would enable the production of statistics on the first job choices of recent graduates. Data on the first jobs of recent Ph.D.'s can be seen as a substitute for data from the SED on postgraduation plans. Data on actual outcomes (first jobs) may also be more useful than data on expectations of employment that can be gained from the SED. The sampling frame of the SDR is replenished by the population of newly earned Ph.D.'s from the SED. However, the SED is not needed to update the frame population of the SDR. Basic data on recent Ph.D.'s can be (and in effect already are[21]) supplied directly from the awarding universities. Finally, in terms of developing indicators on new Ph.D.'s, the panel believes alternative data sources, such as the ProQuest and WorldCat databases, can prove more useful for characterizing Ph.D.'s themselves. NCSES should conduct a benefit-cost analysis to determine the impact of the reduced frequency of data collection for the SED on

[21]Through the annual Graduate Student Survey.

NCSES's staff allocation and the cost of administering the survey. NCSES could analyze the requests of researchers to the NORC Enclave to determine how the SED has been used and the potential extent to consequences of changes in its frequency or coverage for NCSES's clients.

In sum, the panel believes NCSES would benefit from reducing the frequency of the NSCG and SED, instead producing statistics on college graduates based on the ACS. Regarding earned doctorates and the SED, data on the early careers of doctorate holders, particularly their first jobs, reduce the need for data on the plans of new doctorate recipients. In addition, data on the number of earned doctorates can be obtained using other data sources, freeing up resources needed for the development of the new indicators recommended in this report.

Demand for STEM Skills and Talent

Data are lacking on the demand for human capital. Which types of skills are, or will be, in greatest demand, and which are most important for innovation performance? In terms of S&T employment, the focus is typically on high-tech manufacturing, an industry that, while important, does not appear to be a large source of growth for U.S. jobs and the U.S. economy based on current statistics. A look at trade balances suggests an increasing reliance on imports within these industries. In contrast, strong growth is seen in exports of knowledge-intensive services. An examination of education and/or occupation profiles for these industries (or in general, industries with strong growth, particularly with respect to international competitiveness) might indicate what types of skills are in greatest demand and are generating value (and helping to generate new jobs). A number of other data sources can provide insights on the STEM skills in demand now and in the near future, including existing data sources such as BRDIS and nontraditional data sources such as help-wanted ads (see Chapter 4 and Chapter 7).

Wage levels provide an important indicator of the demand for occupations or skill sets. Wage levels may also influence individuals' choice of education. The panel thus views wage levels for science, engineering, and health occupations as useful indicators for gauging developments in both supply and demand for S&T-related skills.

> **RECOMMENDATION 6-6: The National Center for Science and Engineering Statistics should consider using American Community Survey data to produce indicators that can be used to track the salaries of science, technology, engineering, and mathematics occupations and/or college graduates receiving degrees in different fields and at different degree levels.**

BRDIS covers both R&D and innovation activities, including questions on the number of R&D personnel involved in R&D activities. R&D personnel include not only researchers but also other employees with a variety of other types of skills and educational backgrounds. This is particularly true when one considers businesses' innovation activities and not just their R&D. Little is known, however, about the skill sets needed by different types of businesses for their innovation activities. This is a central question entailed in examining whether the skill sets and competencies needed for the development of today's and tomorrow's innovations are being acquired.

Self-reporting on what industry wants has been shown to be an unreliable proxy for what industry actually does. To address this problem, NCSES could ask firms to identify their most recent innovation and for that innovation, the skill mix of the team responsible. Another way to obtain information on this issue would be to ask firms whether they have unfilled budgeted positions for people with specific skills. Lastly, firms could be asked what skills they would seek if they had to create a small team to develop an innovation they are currently considering. The main point is to ask a question that focuses on recent or current decisions on needed skill sets.

> **RECOMMENDATION 6-7: The National Center for Science and Engineering Statistics should consider adding questions to the Business Research and Development and Innovation Survey on the types of skill sets used by businesses to develop and implement innovations. The results would provide data on and indicators of innovative firms' demand for skills.**

The Growth of Online STEM Education

To inform STI indicators, NCSES could develop a series of case studies on massive open online courses (MOOCs), described in Box 6-3, to address the following questions: How will the financial structure of universities be affected by distance learning activities, particularly MOOCs? Will MOOCs allow for faster diffusion of innovation (e.g., trained workers in developing countries that can utilize new products and methods for local concerns)? Will this encourage more inter- or intracountry research collaboration? What types of courses are being offered online? And more important, what is the decay rate for people finishing the course, that is, the share of people who view the material online but do not finish the course, or registrations versus completions? Is there a higher decay rate for YouTube viewing than for sites where students actually sign up for a course, receive course credit, and may engage in on-site course-related activities?

NCSES could undertake three data-gathering efforts related to MOOCs. First, it would be helpful to understand the current inventory of these courses and track trends over time. A Google search shows that many courses and many free-standing lectures or lecture series are offered online.

BOX 6-3
Massive Open Online Courses: Steps in Progress

Massive open online courses (MOOCs) offered free over the Internet could revolutionize science, technology, engineering, and mathematics (STEM) education worldwide. They are descendants of the 1960s Sunrise Semester (United States) and Open University (United Kingdom) courses that offered distance learning to viewers on public television.

The Internet allows for worldwide exposure of academic material. The Khan Academy offers free short courses to students from K-12 to college without offline links, credit, or degrees. Academic organizations offer free lectures on the Internet, some on highly abstruse subjects. Many colleges and universities, some for-profit and some nonprofit, offer online courses for credit.

Only in the past 2-3 years, however, when leading universities around the world began offering their courses free over the Internet, did it strike many that MOOCs would likely alter the future of education. Some 160,000 people signed up for a Stanford University course in artificial intelligence. More than 90,000 people registered for the first MITx course, and some 100,000 registered for Harvard's first free courses a few months later. Most registrants drop out of free courses, but enough complete them to mark a massive expansion of the courses' reach.

The growth of MOOCs raises important questions about the best way to combine online and offline education. Online education readily fulfills its promise of great scale at low cost, but otherwise has not proven the panacea that many hoped it would be. Experimentation and research are needed to determine the appropriate mix of face-to-face offline education and MOOCs. Studies of STEM education find that lectures are less effective in getting students to understand science and math than working collaboratively and meeting with instructors and fellow students face to face. Much of the future of education may reside on the World Wide Web, but it appears that the greatest benefits will come from combining this form of learning with other, more traditional ways of approaching material.

Examples:

UDACITY: "Learn. Think. Do. Higher Education for Free"—formed by the creator of the Stanford artificial intelligence course. As of December 2012, 19 courses were being offered.

COURSERA: "Take the World's Best Courses, Online, For Free"—33 universities offering 210 courses, with student meet-ups in 1,020 cities.

edX: "The future of online education—for anyone, anywhere, anytime"—nine courses, with universities and courses increasing rapidly. The plan is to offer a sanctioned certificate at a "modest fee." Participating universities are Harvard; MIT; and the University of California, Berkeley. Students around the world who never dreamed of having the chance for an Ivy League-level education can take these courses. The portal for learning is edx.org.

SOURCES: Agarwal (2012); *The New York Times* (2012).

Most are free, without course credit being offered. For example, before edX at Harvard University[22] existed, the university offered extension course lectures free online; likewise, the Massachusetts Institute of Technology (MIT) had its open courseware available before its MOOCs initiative. To obtain an inventory of what exists and related trends, NCSES could conduct an online search similar to the Bureau of Labor Statistics' effort to construct price indices using web-based data. NCSES could then choose a few specific subjects in different STEM fields, say, differential equations in math, and count those courses on the Internet. NCSES then could obtain figures on downloads of courses that are for credit or not for credit. These numbers would be useful for understanding how pervasive MOOCs are and how important they will be in the future for generating credentialed STEM workers.

Another metric NCSES could develop is how frequently people access the online material. For example, MIT's differential equations course on YouTube had 490,783 views for the first lecture but 26,871 for the last lecture—a decay rate of 94.5 percent. A Khan Academy course registered 569,954 students at first, but 45,090 accessed the last lecture in the series—a decay rate of 92.0 percent. Lewin (2012) reports that the famous Stanford University course in artificial intelligence had 160,000 students from 190 countries enrolled at the beginning in 2011, but just 20,000 successfully completed the course. Stanford's Machine Learning course had 104,000 registered, and 13,000 completed the course; Introduction to Databases had 92,000 registered but was completed by just 7,000. NCSES could obtain these data from the universities or firms offering online courses. Data on decay rates would complement those on persistence rates for traditional college degrees. This metric would help

[22]See https://www.edx.org/ [December 2012].

university administrators understand the potential impact of distance learning courses on the financial structure of their institutions, or help policy makers understand the impact of innovation in the distribution of knowledge on knowledge diffusion and which countries are the likely benefactors of the distribution of knowledge in STEM fields.

NCSES could also link data from administrative records on people taking courses for credit online to LEHD earnings and employment data, NCSES's SDR data, other data on the STEM workforce, or even patent and publication data (see Chapter 5). For instance, it would be possible to develop a metric indicating the percentage of people who took MOOCs in engineering and changed their trajectory from or to a different field. One could also follow differences in earnings trajectories between traditional bricks-and-mortar and distance learning environments. Careful identification techniques would be necessary, but these data could yield informative analytical output on returns to different types of education investments and organizational structures. This is an example of a project that researchers could undertake using NCSES data along with data from other sources. A variety of users of NCSES's indicators would be interested in the findings from this research. Therefore, this would be an interesting topic for a box inserted item in NCSES's reports on indicators of human capital.

KEY OPPORTUNITIES

Because NCSES's indicators inform a rich set of topics on human capital, the panel chose to focus on a few of these in this chapter.

First, it is important to obtain better measures of rapid changes in the STI workforce, including job mobility. Worker mobility between jobs, occupations, and nations is important to measure because it indicates how knowledge flows and spillovers occur and how knowledge markets operate. For instance, it is important to know in what industries people with degrees in science—in both STEM fields and the social and behavioral sciences—work throughout their lifetimes. It is also of national interest to have information on the flows of undergraduate and graduate students, postdoctorates, and workers between states and countries.

Second, it is important to determine where datasets on STEM education and workforce statistics overlap among the federal statistical agencies. NCSES has long-term relationships with several statistical agencies that provide education data for S&E indicators. Some education indicators are collected through NCSES surveys, while others are reported by NCSES based on statistics generated by other agencies and organizations. In particular, NCSES and NCES both collect data on holders of higher education degrees. Recently, NCSES and NCES staff agreed to undertake a joint gap analysis to examine education data-gathering activities. The results of this analysis will enable both agencies, as well as others, to determine where there is useful overlap, where there are voids, and where efficiencies can be achieved by streamlining efforts. A series of workshops on potential linkages, interoperability, and rationalization of datasets on human capital will further improve efficiencies between the two agencies. NCSES will then be able to plan for a greater role than its current position of a data clearinghouse specifically on STI education and workforce data.

Third, NCSES has clear near-term opportunities to mine BRDIS for indicators related to the STEM labor force. The agency already has several activities under way or planned to update its portfolio of education and workforce statistics. For instance, NCSES is rethinking its collection of data on the S&E workforce as the NSCG transitions to a sampling frame built entirely from the ACS, implementing the International Survey of Doctorate Recipients and integrating it into the SDR, developing an early-career doctorate project, and considering earnings of STI workers as a potential new indicator. One possibility would be to develop an S&E wage index, which could facilitate international comparisons and become another explanatory variable for international flows of S&E students and workers. All of these efforts could yield new indicators that would address user needs.

SUMMARY

In this chapter, the panel has offered seven recommendations, which fall into four categories: (1) rationalization of existing human resource surveys, (2) measures of student and labor mobility that can be developed by using NCSES surveys alone or by linking NCSES data with data from other agencies, (3) measures of industry skill mix revealing demand and supply for STEM talent by sector, and (4) new datasets that can be developed without new surveys. The first priority should be efficiency principles for rationalization of SESTAT datasets along with the development of indicators from the existing, rich longitudinal database at NCSES. Over time, NCSES can collaborate with other agencies to deliver highly useful human capital indicators that link educational inputs to employment and wage outcomes.

7

A Paradigm Shift in Data Collection and Analysis

The National Center for Science and Engineering Statistics (NCSES) finds itself in the midst of a paradigm shift in the way data are gathered, manipulated, and disseminated. The agency's science, technology, and innovation (STI) indicators program must deal with several challenges:

- As discussed in previous chapters, traditional surveys face increasing expense, declining response rates (Williams, 2013), and lengthy time lags between when data are gathered and when derived indicators and other statistics can be published.
- Tools for data extraction, manipulation, and analysis are evolving rapidly.
- Repositories of STI measures that users demand are distributed among several statistical agencies and private repositories.
- Sources of knowledge generation and innovation are expanding beyond developed countries to emerging and developing economies.
- Users have rising expectations, and they are demanding more access to statistics that are closer to actual measures of what they want to know.

This chapter explores this changing landscape of data collection and analysis and its implications for NCSES's STI indicators program.

NEW METHODS, NEW DATA SOURCES

Standards and taxonomies for data collection and analysis are expected to change before the end of this decade. OECD's National Experts on Science and Technology Indicators are revising the *Frascati Manual* (OECD, 2002) and the *Oslo Manual* (OECD-Eurostat, 2005) on a rolling basis. The group plans to work on priority themes and to build a better bridge between the two manuals. The North American Industry Classification System (NAICS) codes and the Standard Occupational Codes may also undergo revision within the next decade.

NCSES and, indeed, other government statistical agencies confront a world of dizzying change in the way information technology is integrated into their data-gathering and data-management activities. The World Wide Web, in particular, has been transformational in enabling new forecasting and data collection methods that yield useful insights in almost real time. These tools provide information much more rapidly than is possible with traditional surveys, which entail up to multiple-year lags.

Other changes are occurring as well. In his November 2011 presentation at the annual meeting of the Consortium of Social Science Associations, Robert Groves (2011a) conveyed the status of U.S. surveys: "Threatened coverage of frames; falling participation rates; increasing reliance on nonresponse adjustments; and for surveys with high response rate targets, inflated costs." His proposed solution for agencies to address these issues is to develop an approach of a "blended data world by building on top of existing surveys."[1] Groves (2011b) envisions multimodal data acquisition and manipulation of data, including "Internet behaviors; administrative records; Internet self-reporting; telephone, face-to-face, paper surveys; real-time mode switch to fill in missing data; and real-time estimation."[2]

Some of these innovations are already being implemented at the Census Bureau. The agency's economic directorate has combined administrative data with survey data in inventive ways. It also handles multiple response modes—paper forms, Internet responses, and telephone interviews. To address the timeliness of economic indicators, it has devised workable decision rules for defining which estimates are preliminary and what information is required to revise them.

[1]For further comments on this point, see U.S. Census Bureau (2011a,b,c).

[2]See Chapter 4 for detail on how business practice data (which include administrative records and web-based data) can be used to obtain innovation indicators.

Perhaps the most challenging innovation in Groves' vision of the future of surveys is performing real-time estimation during data collection. Groves (2011b) envisions implementing real-time estimation routines—including imputation, nonresponse adjustment, and standard error estimation—after every 24 hours of data collection. Part of this progress would entail assessing whether the standard error increase due to imputation was acceptable or additional nonresponse follow-up was necessary. In this context, imputation can, in effect, be viewed as another mode of data collection. To make trade-off decisions about whether to terminate nonresponse efforts for a case using a particular mode, key statistics on the fully imputed estimates and measures of the imputation standard error and sampling standard error of the estimates would be actively tracked. Groves believes successfully implementing this real-time estimation and decision process at the Census Bureau would take at least 5 years.

In this vein, one issue that needs to be explored is the feasibility of blending the use of administrative records, scientometric tools, and survey techniques to improve the accuracy of data on STI human capital measures and other indicators that NCSES produces, such as research and development (R&D) input and performance measures. A multimodal approach would help in creating longitudinal series using existing and new information. In the near term, the topic could be explored through a workshop designed specifically to discuss the conceptual framework and feasibility of blending data acquisition techniques and using this mixed-methods approach to develop new indicators.[3] This approach could be useful for developing real-time maps of networked scholars while measuring return on investments from federal research funds as they are used and linking them to outputs (paper and patents). At the same time, this approach would include periodically assembling basic data on education, employment, work activities, and demographic characteristics.

Data from administrative records and web-based sources—termed "business practice data" (see Chapter 4)—have been used for many years at federal agencies with two purposes: to benchmark sample survey data and, along with sample survey data, to produce official statistics. Horrigan (2012, 2013) gives several examples of sources being used by the Bureau of Labor Statistics (BLS), including the Billion Prices Project data; retail scanner data; the J.D. Power and Associates used car frame; stock exchange bid and ask prices and trading volume data; universe data on hospitals from the American Hospital Association; diagnosis codes from the Agency for Healthcare Research and Quality, used to develop the producer price index; Energy Information Agency administrative data on crude petroleum for the International Price Program; Department of Transportation administrative data on baggage fees and the Sabre data, used to construct airline price indices; insurance claims data, particularly Medicare Part B reimbursements to doctors, used to construct health care indices; and many more sources of administrative records data from within the U.S. government, as well as web-based data. According to Horrigan (2013), in addition to the development of price indices, administrative records and web-scraping data are used to "improve the efficacy of estimates . . . the Current Employment Statistics (CES) Survey uses administrative data from the Quarterly Census of Employment and Wages (QCEW). . . ." BLS also is "using web-scraping techniques to collect input price information used to increase the sample of observations we use to populate some of our quality adjustment models" (Horrigan, 2013, p. 26). Horrigan cautions, however, that "the principle of constructing an inflation rate based on the rate of price increase for a known bundle of goods with statistically determined weights lies at the heart of what we do. While research may show the viability of using a web-scraped source of data for a particular item, it needs to be done within the framework of this methodology" (Horrigan, 2013, p. 27).

The statistical methodology related to sampling and weights must be developed if these multimodal techniques are to be fully relied upon to deliver bedrock STI indicators. The panel must stress, moreover, that business practice data must be regularly calibrated using sample survey data. Business practice data contain a wealth of detailed and rapidly changing information that is not practically acquired using surveys. However, businesses and government enterprises generally do not maintain the sort of consistency across organizations, categories, and time that would enable cross-sectional and longitudinal comparisons. In time, and with appropriate financial and human resources, NCSES and other statistical agencies should be able to publish indicators based on business practice data, but only if the raw data are adjusted using a well-designed program of sample surveys. Indeed, the challenge will be to design the most efficient combination—financially and statistically—of traditional sample surveys and administrative and web-based sources.

IMPLICATIONS FOR NCSES

NCSES needs to determine now how it will handle the above changes if they materialize and how the types and frequencies of various STI indicators will be affected. During the panel's July 2011 workshop, Alicia Robb of the Kauffman Foundation encouraged NCSES to explore the use of administrative records to produce STI indicators. She also cautioned, however, that ownership issues associated with the use of those data will have to be addressed before they can become a reliable complement to traditional survey data.

[3] *Statistical Neerlandica* has prepublication views of a series of articles on the use of administrative records for analytical purposes, including regression analysis; see http://onlinelibrary.wiley.com/journal/10.1111/(ISSN)1467-9574/earlyview [December 2011]. For theoretical foundations of combining information from multiple sources of data, see Molitor et al. (2009). Also see Eurostat (2003).

Also at the workshop, Stefano Bertuzzi of the National Institutes of Health (NIH) and the STAR METRICS Program at that time, in collaboration with Julia Lane from the National Science Foundation (NSF), presented techniques for using administrative records at universities to determine the impact of federal research funds on scientific outputs and the development of human capital in the physical and biological sciences. In follow-on discussions, George Chacko, who took the helm of the STAR METRICS Program at NIH in late 2011, described that $1.5 million per year activity. Level 1 data outputs (described by Bertuzzi) are in the data validation process. There are two potential sources of error (data entry and data transformation into readable files), but biases are not yet known. Chacko noted that further research is needed to determine the quality of the Level 1 data. He also described Level 2, which will allow the integration of research project data; that effort had not yet begun as of the writing of this report. Participants in Level 2 will include the U.S. Department of Agriculture, the Environmental Protection Agency, NIH, and NSF. Because each agency has different ways of reporting the same kinds of grants, one of the first tasks will be the creation of a data ontology and taxonomy before a database is developed. Sometime in the future, Chacko expects that STAR METRICS Level 3 will enable demographic identifiers for individuals, thereby allowing analysis of science, technology, engineering, and mathematics (STEM) outcomes by gender and race/ethnicity.

In May 2012, Christopher Pece gave a presentation at NSF on NCSES's Administrative Records Project (ARP), which is still in the feasibility testing stage. Pece cited the National Research Council (2010) report recommending that NCSES (Pece, 2012):

- develop R&D descriptors (tags) into administrative databases to better enable identification of R&D components of agency or program budgets;
- use administrative data to test new classification schemata by direct access to intramural spending information from agency databases; and
- develop several demonstration projects to test for the best method for moving to a system based at least partly on administrative records.

Accordingly, NCSES is working with a subset of agencies that have data reported in the Federal Funds Survey and Federal Support Survey to pilot methods for using administrative records to produce data comparable to the survey data. In addition to budgetary constraints and negotiation of interagency agreements, other challenges must be addressed, including the creation of data tags and R&D crosswalks between agency systems that use different data taxonomies, accounting practices, and information technology systems. The panel was impressed by NCSES's willingness to experiment with the use of administrative records to complement its survey-based datasets, but also recognized the need for NCSES to acquire increased resources—funding and staff—at least in the short term, with the potential ultimately for reduced survey costs, reduced data validation costs, and increased timeliness of data delivery.

During the July 2011 workshop, presentations by Erik Brynjolfsson of the Massachusetts Institute of Technology, Lee Giles of Pennsylvania State University, Carl Bergstrom of the University of Washington, and Richard Price of Academia.edu provided insights regarding tools that can be used to obtain up-to-date information on science and engineering networks and linkages between human capital investments and STI outcomes and impacts. These experts showed panel members how nowcasting, netometrics, CiteSeerX, Eigenfactor, and Academia.edu (similar to Lattes in Brazil) can be used to create scientometric[4] data for use in developing STI "talent" indicators. Such tools can be used, say, to track intangible assets and knowledge flows from online recruiting sites and social networks.

Many questions remain about the representativeness of data sources such as STAR METRICS Level 1 (which includes data from 80 universities) and the datasets drawn from web-based sources, primarily because they are nonrandom convenience samples. Recent work on medical care expenditures by Ana Aizcorbe at the Bureau of Economic Analysis (BEA)[5] and by Dunn and colleagues (2012) shows that insurance companies' patient claims data can be used to develop reliable price estimates, given the appropriate weighting strategy. Both projects use MarketScan data, which include sampling weights designed to provide population estimates from the MarketScan sample. This is a potentially cost-effective approach compared with the use of traditional household surveys (in this case, the Medical Expenditure Panel Survey [MEPS]). Clearly, the MarketScan data cannot address all of the questions that the MEPS was designed to address. However, Dunn and colleagues find that the MarketScan data "produce spending growth figures that are more aligned with other benchmark estimates of price and expenditure growth from national statistics" (Dunn et al., 2012, p. 26).

INDICATORS FROM FRONTIER TOOLS: EXAMPLE OF THE DATA SCIENCE DISCIPLINE

Consider the rise of data science, an emerging discipline that encompasses analysis, visualization, and management of large datasets. ("Large" in this context typically means many millions or billions of records.) The digitization of records, increasing numbers of sensors, and inexpensive storage have combined to produce enormous quantities of data in the sciences and business. Data scientists use specialized techniques to sift through these troves of information

[4]In practice, scientometrics often uses bibliometrics, a measurement of the impact of (scientific) publications and patents (see Chapter 5).

[5]See Aizcorbe et al. (2012) for more detail on the Health Care Satellite Accounts at BEA.

to discover new insights and create new value. Google's chief economist, Hal Varian, has characterized the statistical work of data scientists as "the sexy job in the next 10 years" (Lohr, 2009); *Forbes* magazine describes the data scientist's role as "the hot new gig in tech" (Lev-Ram, 2011); and *The Economist* (2011) says data science is "where the geeks go."

In line with perennial concerns about the supply of knowledge workers in the United States (Atkinson, 1990; Freeman and Aspray, 1999; Jackson, 2001; The Partnership for a New American Economy and The Partnership for New York City, 2012), data scientists are already projected to be in short supply in the near future. According to a 2011 McKinsey study (Manyika et al., 2011), "a shortage of people with the skills necessary to take advantage of the insights that large datasets generate is one of the most important constraints on an organization's ability to capture the potential from big data." Likewise, an EMC Corporation (2011) study foresees a "looming talent shortage." Access to talent is not an issue just for industry: 23 percent of respondents to a 2011 *Science* survey said their laboratories were lacking in data analysis skills.

Given that past projections of shortages of knowledge workers have proven controversial (Lowell and Salzman, 2007; Matloff, 2003; Stanford News Service, 1995; Weinstein, unpublished), it is worth examining the above claims more closely. Consider some of the questions a policy maker concerned about the future data scientist workforce might ask of NCSES:

- How many new data scientists are graduating each year?
  - How many in the United States?
  - How many in other parts of the world?
- Where were existing data scientists educated?
  - What schools?
  - What programs?
- Where are data scientists employed?
  - What fraction work in industry?
  - In government?
  - In academia?
- What range of salaries do data scientists command?
  - How much do salaries vary with degree level?
  - With sector?
- Is the United States producing enough or too many data scientists?

A funding agency director (such as the director of NSF) might want to know:

- Is NSF funding data science research?
  - Which directorates?
  - How much is NSF spending?
- What basic research does data science draw upon?

NCSES's existing STEM workforce surveys would be hard pressed to answer these questions. For one thing, "data science" is not in the taxonomy of fields used in the STEM degree surveys, so one cannot obtain data science degree counts directly from existing NCSES datasets. Similarly, the taxonomy of occupations used by the Current Population Survey/American Community Survey does not contain "data scientist," so NCSES datasets derived from these surveys will likewise miss this group. Fixed, slowly evolving taxonomies restrict the ability of existing surveys to provide insights about emerging fields.

One potential remedy might be a one-time custom survey. Given the cost of existing surveys, however, this would likely be a prohibitively expensive undertaking. A custom survey would entail the additional difficulty that there is no obvious, well-defined frame. An alternative might be for NCSES to update its taxonomy of fields for future surveys. This would be a slow process, however: turnaround time for the NCSES surveys is on the order of 2 years (National Science Foundation, 2012a), and additional time would be needed to formulate and validate a revised taxonomy. Even if taxonomic issues were resolved, the limitation would remain that NSF's datasets cover only the United States.

An Alternative Approach

Datasets exist that could shed light on questions about data science, but they are very different from those produced by NCSES. They are not among the datasets typically used to analyze the STEM workforce in part because, while they offer significant advantages over NCSES's data, they also come with significant challenges.

Consider, first, the task of counting data science doctoral degrees. Rather than using a survey to ask new doctorate recipients whether they did data science-related research, an expert could look at their dissertations to make this determination. The expert could then tally the number of new data science graduates. The expert could also identify the degree-granting institutions and programs from the dissertations. The idea is not merely theoretical: both ProQuest and WorldCat maintain large databases of dissertations (Online Computer Library Center, 2012; ProQuest, 2012). Although counting data scientists in this fashion is labor-intensive, it is potentially faster and less expensive than either conducting a custom survey or waiting for a taxonomy update in an existing survey. In addition, the approach has the benefit of providing global counts of data science degrees, not just U.S. counts.

For the task of learning about the number of data scientists in the current workforce, one could examine a database of resumés; count the number of people whose job titles or job descriptions include such phrases as "data scientist," "data mining," or "big data"; tally their educational backgrounds; and observe their sectors of employment. A large, global resumé database such as LinkedIn (http://www.linkedin.com;

BOX 7-1
Employment Shifts from LinkedIn Data

LinkedIn's data science team (The Noisy Channel, 2012) recently collaborated with the White House Council of Economic Advisors to identify the industries that grew and shrank the most during the 2008-2009 recession and the subsequent recovery. By following people who were site members in 2007 longitudinally through 2011, they were able to see the rapid growth in renewable energy and Internet companies, as well as sharp declines in newspapers, restaurants, and the retail sector. The cohort they followed numbered in the tens of millions, and LinkedIn contains detailed data on its members' educational backgrounds, so one can readily imagine conducting similar analyses restricted to workers with science, technology, engineering, and mathematics (STEM) degrees. Moreover, one of the study's authors says that in principle, LinkedIn could track such changes in real time.

SOURCES: Nicholson (2012); *The Economist* (2012).

see Box 7-1) or profiles of science professionals in science-related social networks such as ResearchGate (http://www.researchgate.net/), Academia.edu (http://academia.edu/), or BioMedExperts (http://www.biomedexperts.com/) could be used for this procedure. Again, the process of counting the resumés or profiles and classifying the associated educational backgrounds and employers would be labor-intensive, but it potentially could provide inexpensive and timely insights on the supply of data scientists in both the United States and international markets that would otherwise be unavailable or prohibitively expensive to generate.

To assess demand and salary levels for data scientists, one could turn to large databases of job listings such as Monster.com (http://www.monster.com/), Indeed.com (http://www.indeed.com/), or SimplyHired.com (http://www.simplyhired.com/). An expert could identify data science-related jobs and then tally offered salary levels as a function of the level of the job.

Mandel and Scherer (2012) recently used techniques of this sort to estimate the size and location of the "App Economy"—jobs related to smartphone and tablet applications and to Facebook plugins.[6] Because this is a very recently created category of jobs, existing labor market statistics could not provide useful information about these types of jobs. Mandel and Scherer turned to The Conference Board's Help Wanted OnLine data (The Conference Board, 2012), a collection of job listings from more than 16,000 Internet job boards. They counted job listings containing keywords associated with the App Economy (e.g., "iPhone," "iPad," "Android") and then used historical data on the ratio of job listings to jobs in the tech sector to estimate the total number of App Economy jobs. They were able to identify the geographic distribution of App Economy jobs from the job locations listed in the ads. One might apply a similar approach to data science jobs by looking for keywords related to this field (e.g., "data science," "data mining," "big data") and common data science tools and techniques (e.g., "map-reduce," "Hadoop," "Hive," "Pig").

Mandel (2012) and Mandel and Scherer (2012) analyzed online help-wanted ads to track the demand for "app-related" skills geographically, using key words such as "iOS" and "Android." This analysis made it possible to identify states with a higher-than-expected number of app-related want ads relative to their size (see Table 7-1). This procedure could be repeated for any innovation; one could identify clusters of green innovations, software innovations, or medical innovations in the same way, for example, at the state or metro level.

The data from help-wanted ads also can be combined with conventional survey data to provide a more complete and timely picture of innovation activities at the national or subnational level. Suppose, for example, one wanted to get a picture of R&D activities in biotechnology by state. The Business Research and Development and Innovation Survey (BRDIS) would provide some information, but many cells of the tables derived from its data would likely be suppressed to avoid disclosure. However, it would be possible to use biotechnology-related help-wanted ads to impute the missing state information without violating confidentiality. This analysis could even be taken down to the metro level, using help-wanted ads combined with state-level survey data to provide a benchmark. Using help-wanted ads to track the diffusion and use of innovation at both the national and subnational levels has several advantages: these ads are public and continuously refreshed; full databases of the ads already exist in digital form, available in near real time; the ads are semistructured—they are free text, but must include information about the skills and experience needed, using recognizable terms; and organizations such as The Conference Board already have procedures for automatically tagging them by occupation, industry, location, and so forth. As a result, the ads provide a continually changing real-time map of the skills employers need. In particular, as the content of the ads changes, one can see innovations diffusing through the economy.

Finally, to gauge existing support for data science research within NSF, an expert could read through existing NSF awards and identify those related to data science research. The expert could then identify the institutions and programs doing data science work, as well as tally the directorates supporting data science. To identify the basic research on which data science draws, the expert could compile a set of recent, important data science papers; follow their citations back, possibly multiple levels; and tally the journals and/or

[6]See also Box 4-8 in Chapter 4 on the use of this approach to develop innovation indicators.

TABLE 7-1 The App Leaders: States #1-15

State	App Intensity (U.S. average = 1)	App Economy Jobs (thousands)	App Economy Economic Impact (millions of dollars, annual rate)
1. Washington	4.47	49.8	$2,671
2. California	2.71	151.9	8,241
3. Massachusetts	1.71	21.4	1,143
4. Oregon	1.70	10.8	526
5. Georgia	1.56	24.0	1,062
6. New Jersey	1.29	19.5	1,087
7. New York	1.16	39.8	2,313
8. Virginia	1.04	15.0	788
9. Delaware	0.93	1.5	76
10. Colorado	0.90	8.1	429
11. Illinois	0.90	19.9	847
12. Connecticut	0.88	5.6	294
13. Minnesota	0.87	9.1	475
14. Utah	0.86	4.1	192
15. Maryland	0.84	8.4	436

NOTES: All figures estimated as of April 2012, includes spillover jobs. Data from The Conference Board, Bureau of Labor Statistics, and calculations from South Mountain Economics LLC.

SOURCE: Reprinted with permission from Mandel and Scherer (2012).

fields of research represented by frequently cited precursors to data science papers. Several large databases of papers and citations—such as Web of Science (Thomson Reuters, 2012) or Scopus (http://www.scopus.com/home.url)—could be used to facilitate the process of tracing papers back.

Advantages and Challenges

Using the datasets described above to learn about the state of data science offers several advantages over using surveys. First, the datasets have already been created for other purposes, so the incremental cost of using them to learn about data science is modest. Second, the databases are all updated continuously, so the lengthy delays associated with gathering survey data do not come into play. And third, because experts classify the data, there is no locked-in, limiting, pre-existing taxonomy that could lead to misclassification of data scientists (although this also presents its own issues).

Along with the benefits associated with these new datasets, however, come challenges:

- In many cases an expert is needed to assign data to categories because the datasets are *unstructured* (see the discussion of this issue in the next section). There will be uncertainty in this classification process, especially if multiple experts are used because they may not always agree.
- Classifying large collections of dissertations, resumés, awards, and papers by hand is labor-intensive for the expert—there is an issue of *scale*.
- Some of the datasets are commercial products, so one must pay for or negotiate *access* to the datasets.

In addition, some of the datasets are sensitive and require special handling.

- More generally, concerns have been raised about the inconsistency in the way R&D-related data are specified, making data sharing and linking by researchers difficult. A web-based infrastructure for data sharing and analysis, including clear data exchange standards, would be a useful first step in addressing this issue.[7]
- Finally, the question of *validation* arises. Many of the databases cited above are incomplete in that they do not include the entire population of interest. It is necessary to understand the portion that is missing in order to estimate or at least bound the possible bias in using the database of interest to characterize that population.[8] In addition to coverage and sampling biases, measurement, nonresponse, and interviewer biases should be examined to determine the validity of statistical indicators derived from such databases. Moreover, a process needs to be in place for measuring the reliability of the expert's classifications.
- As noted, most of the web-based datasets described here are neither representative samples nor complete censuses of the population of interest. That being the case, developing and implementing methods for

[7]See Haak et al. (2012, pp. 196-197) for a discussion of this problem and possible solutions.

[8]Sample surveys are used to draw inferences about well-defined populations. Survey methodologists have developed tools for measuring how valid and robust their inferences from surveys are. Conceptually, these methods can be applied to nonsurvey databases as well. See Groves et al. (2009), particularly Chapters 3 and 5.

using these datasets is largely tilling new ground for the staff of any statistical agency. Should NCSES decide to move in this direction, it will need to ensure that it has staff with the appropriate training and experience to develop and implement suitable analytic and data collection approaches in this new environment.

Considerable progress has been made in addressing all of these challenges, but many important open questions remain.

A NEW DIRECTION FOR NCSES

The general approach described above for learning quickly and inexpensively about an emerging field by repurposing existing datasets holds considerable promise for improving understanding of many aspects of innovation in science and engineering. At the same time, the approach entails methodological and logistical problems. The tasks of structuring unstructured data, dealing with challenges of scale, negotiating access to data and protecting sensitive data, and validating nontraditional data sources are common to many potentially useful but nontraditional datasets.

The panel proposes that NCSES explore and experiment with these new, nontraditional data sources. This section describes four core areas in which NCSES could contribute to improving the state of the art in this area, with the goal of improving outputs from its data development and indicators programs.

Identification of Data Sources

Plummeting prices for data storage, low-cost sensors, improvements in data collection mechanisms, and increases in Internet access have given rise to vast new collections of digital data (*The Economist*, 2010a), and the total amount of digital data is growing rapidly—a 10-fold expansion occurred between 2006 and 2011 (Gantz et al., 2008). A wide variety of datasets could be used to better understand STEM innovation—the ones mentioned above barely scratch the surface. Annex 7-1 at the end of this chapter lists some promising possibilities.

NCSES could help answer two key questions regarding these new data sources: What are the most promising new datasets for understanding the state of STEM? and What are effective ways to analyze these datasets? NCSES has historically used a combination of internally generated and third-party datasets in assembling science and engineering indicators and *InfoBriefs*. The agency could test the waters by adopting the goal of including in its publications and on its website analyses performed with nontraditional data in the areas of human resources, R&D, and innovation. Such analyses could be performed by external researchers funded by targeted awards.

RECOMMENDATION 7-1: The National Center for Science and Engineering Statistics (NCSES) should use research awards to support the development and experimental use of new sources of data to understand the broad spectrum of innovation activities and to develop new measures of science, technology, and innovation. NCSES should also support the development of new datasets to measure changing public perceptions of science, international trade in technological goods and services, new regions for entrepreneurial activity in science and technology, and precommercialized inventions.

Structuring of Unstructured Datasets

The data generated from NCSES's surveys are structured. Data are stored as tables of numbers, with each number having a well-defined meaning. As noted, many of the nontraditional datasets discussed above—perhaps the majority—are in unstructured form, such as free text. A traditional (but apocryphal) rule of thumb is that 80 percent of corporate data is unstructured (Grimes, 2011); a recent article in *The Economist* (2010b) estimates that 95 percent of new data is unstructured.

The databases of doctoral dissertations, resumés, job listings, and NSF grant proposals described above are vast and rich stores of information, but they are difficult to process by machine. The data science example given earlier assumes that a human expert is willing to spend weeks categorizing dissertations, job listings, and so forth. This role would likely be difficult to fill, as the work is tedious and repetitive. To tap the potential of unstructured datasets fully, new tools and techniques are needed.

The problem of extracting structured information from unstructured text is an active area of research, and several NSF directorates are funding work in this area (National Science Foundation, 2008). One approach is to use divide-and-conquer techniques: rather than having a single expert spend months on a repetitive task, one can use "crowdsourcing" (Wikipedia, 2012), a technique in which a large task is accomplished by being divided among a large collection of people. Services such as Amazon.com's Mechanical Turk (https://www.mturk.com/mturk/welcome) and CrowdFlower (http://crowdflower.com/) provide workers and infrastructure for distributing tasks to them.

Crowdsourcing can be used to extract information from unstructured data. For example, researchers have used the technique to identify people, organizations, and locations in tweets on Twitter (Finin et al., 2010) and to analyze and annotate charts (Willett et al., 2012). In the realm of STEM, crowdsourcing has been used to identify the correct authors of papers when there is ambiguity in names (e.g., identifying which of several John Smiths wrote a particular paper) (Brooks et al., 2011).

A second approach to structuring unstructured data is to use automated information extraction algorithms. Most of the tasks requiring an expert's input in the data science example involve extracting topics from documents ("Is this dissertation related to data science?") and extracting entities ("What university did the author of this dissertation attend?"). For other applications, it is also important to disambiguate entities ("Is the Susan Jones who wrote paper A the same as the Susan Jones who wrote paper B?") and to link entities from one dataset to another ("Is the Kim Lee who received patent A the same as the Kim Lee who wrote NSF award B?"). Automated algorithms exist that can perform all of these tasks with varying degrees of success, and research is ongoing on all of these problems.

Indeed, staff in NSF's Social, Behavioral, and Economics Directorate have started using some of these information extraction techniques in the construction of the directorate's portion of the STAR METRICS Program (National Science Foundation, 2012b). For example, latent Dirichlet allocation (Blei et al., 2003), a technique for automatically identifying topics in documents based on the frequency of keywords, was used to assign topics to NSF awards (Lane and Schwarz, 2012). Automated entity disambiguation techniques are being used to link NSF awards to subsequent patents (Lai et al., 2011). Improvements in text processing techniques and broader availability of tools for topic and entity extraction could open up rich new datasets that could shed light on STEM innovation. NCSES could catalyze progress by coordinating research and facilitating the dissemination of ideas.

NSF made use of text processing tools internally for STAR METRICS. These tools may have applicability beyond NSF, and NCSES should explore the possibility of making these tools more widely available. Although NCSES should not be in the business of supporting software products, it could explore the possibility of making parts of the text processing code available without support or available as an open-source project. More generally, NCSES could encourage the sharing of text processing tools produced by NSF-supported researchers. The Apache Software Foundation (http://www.apache.org/foundation/) could serve as a model and potential collaborator. A valuable role for NCSES would be to provide organizational and financial support for regularly developing open-source text processing projects. For example, NCSES could pay for a common repository for source code (through a provider such as GitHub), fund meetings among contributors to projects, and organize workshops to bring developers together with users. The broad value of these tools for NSF implies an opportunity for sharing across directorates the resources required to develop these activities.

Encouraging the sharing of extracted structured data would be valuable as well. For example, NSF has autogenerated topics for awards through text processing software developed for STAR METRICS and could start including these topics in its award database so that other researchers might benefit from them. Similarly, if one team of NSF-funded researchers were to link, say, papers in entomology journals to patents in pest control for one project, it might be useful for another team to use the same set of linkages. NCSES could provide a common repository for the sharing of extracted metadata about datasets and encourage NSF-funded researchers to contribute to it.

A fundamental question concerning the use of unstructured data for indicators requires more examination: What kind of statistical methodology should be applied to data derived from web scraping? There are other, related questions: What trade-offs are entailed in using web-based data instead of survey data? Is it possible to adjust web-based data accurately to represent a survey sample and to estimate sampling errors and statistical biases? Is applying modeling techniques to web-based data and traditional survey data a promising approach to achieving this end? How frequently must this be done? How frequently would NCSES want to publish new statistics? Would NCSES want to publish less reliable statistics if it meant publishing them more frequently at lower cost?

A company such as LinkedIn stores in its servers a social network representing all of its users and relationships among them, and techniques for accurately sampling this social network have been developed (see Maiya and Berger-Wolf, 2011; Mislove et al., 2007). To the panel's knowledge, however, researchers have not yet addressed how well this social network represents the larger population. For example, if one is interested in measuring how many chemical engineers are working in the United States, then some subset of these individuals is represented in LinkedIn's social network. Adjusting this subset accurately to represent the target population and estimating the error incurred in using this subset is a daunting challenge.[9] It is important to understand how the data collected from websites compare with traditional survey data, particularly because different websites have very different coverage. Facebook, for example, covers a large portion of the undergraduate population (at least for the next couple of years). However, sites such as Mendeley and Academia.edu clearly cover only a subset of the entire population of researchers.

It could prove useful to adopt a combination approach, in which web-based statistics would be calibrated periodically against a traditional survey. Of course, the survey would have to be administered less frequently than is currently the case, or there would be no cost or time savings. It could also be a useful experiment to run parallel efforts (collecting traditional indicators in addition to their possible replacements) for a period of time in order to examine the strengths and weaknesses of using certain nontraditional data sources for indicators. This period of time would also be important for

[9]LinkedIn and similar data could be quite useful for questions involving relative rather than absolute measures. For example, are there more chemical than electrical engineers? Do chemical engineers change jobs more frequently than other engineers? Where in the country are chemical engineers most highly concentrated?

assessing how well the newly constructed indicators identify trends and rates of change, particularly for policy purposes.

Because NCSES has reported that the response rates for some of its surveys are declining, questions arise about how well those data reflect the population sampled and how web-based data could be calibrated to those surveys. Calibrating to the Survey of Earned Doctorates (SED), which has a response rate above 90 percent, would be relatively straightforward, but only once and on questions asked by that survey. One solution to this dilemma would be for NCSES to devote resources to sampling for nonresponse follow-up,[10] that is, strive to achieve close to a 100 percent response rate from a small sample of nonrespondents to a standard survey, adjust the survey results for nonresponse, and use the adjusted survey data to calibrate information from web-based sources.[11] The calibration would be similar to what computer scientists and mathematicians do with compressed sensing of data on pixels and is a promising area of research.

Achieving a level of rigor comparable to that of a traditional survey with these methods may not be possible, and NCSES would need to consider its tolerance for publishing statistics that may not be as reliable as those it has previously published. The agency would need to balance reliability against timeliness: because little time is required for data collection with data mining techniques in comparison with traditional surveys, releasing statistics much more frequently would be possible.

In principle, nothing prevents statistics from being updated periodically or continuously. For example, the national unemployment rate, gross domestic product, and the consumer price index are updated periodically with no compromise to their importance. And the Billion Prices Project at the Massachusetts Institute of Technology uses an algorithm that collects prices daily from hundreds of online retailers worldwide, creating, among other things, a daily price index for the United States (see Massachusetts Institute of Technology, 2013).

RECOMMENDATION 7-2: The National Center for Science and Engineering Statistics (NCSES) should pursue the use of text processing for developing science, technology, and innovation indicators in the following ways:

- **explore synergies with National Science Foundation directorates that fund research on text processing; and**
- **enable NCSES staff to attend and develop workshops that bring together researchers working on text processing and on understanding the science, technology, engineering, and mathematics ecosystem.**

RECOMMENDATION 7-3: The National Center for Science and Engineering Statistics should use its grants program to encourage the sharing of new datasets and extracted metadata among researchers working on understanding innovation in science, technology, engineering, and mathematics.

Data Validation

Although the datasets discussed in the data science example offered earlier provide new windows into the state of the STEM workforce, the accuracy of statistics gleaned from some of these datasets is unknown. The ProQuest and WorldCat dissertation databases are both large, for example, but neither is complete. Do either or both contain biased subsets of new dissertations? If so, then can the biases be characterized in ways that can be understood and corrected systematically?

One way to better understand omissions in a dataset such as a dissertation database would be to compare it with a definitive source such as the SED (National Science Foundation, 2011a) or the Integrated Postsecondary Education Data System (IPEDS) (National Center for Education Statistics, 2012). If, say, the databases were less likely to contain dissertations from students at private institutions or from humanities students, then estimates based on those databases could be reweighted accordingly.

Assessing the accuracy of metrics based on other types of sources can be more difficult. For example, counts of Twitter mentions (tweets) have been proposed as an indicator of the impact of a paper (Priem and Costello, 2010), and journals such as *PLOS ONE* now report Twitter mentions for each article (*PLOS ONE*, 2012). How might one assess the validity of tweets as an indicator of impact?

NSF is supporting ongoing research in areas that could facilitate assessing nontraditional data sources. Techniques from sampling theory, approaches for handling missing data, and imputation algorithms could all prove useful. In addition, NCSES's own datasets could be used for calibrating new datasets.

RECOMMENDATION 7-4: The National Center for Science and Engineering Statistics should coordinate with directorates at the National Science Foundation in supporting exploratory research designed to validate new sources of data related to innovation in science, technology, engineering, and mathematics.

[10] This is one of several tools used by survey methodologists to address nonresponse in sample surveys. See Groves et al. (2009, Chapter 6) for a description of some of these methods.

[11] This approach would entail using the survey data as the dependent variable in a model that used information from the web-based data source to create the explanatory variables. That model could then be used to "nowcast" population values of interest directly from the web-based data.

Data Access

Many datasets that would be promising for better understanding STEM have restrictions on their usage. The ProQuest and WorldCat dissertation databases, Web of Science, Scopus, and The Conference Board's Help Wanted OnLine database are all commercial datasets for which the processes for obtaining access are well defined. Other datasets are more sensitive and may require carefully controlled access. For example, access to some types of census data entails stringent controls on how the data are handled. Likewise, some corporate datasets are zealously guarded by their owners and may be used only by employees.

NCSES has considerable experience with managing access to sensitive datasets—the Survey of Doctorate Recipients (SDR) and census data. The experience it has gained in the process may be useful in negotiating access to other sensitive datasets.

NCSES already has the infrastructure in place at NORC to house many of its survey data and allow licensed researchers access through remote dedicated computers.[12] In October 2012, data from the following surveys became available in the NORC Data Enclave: the SED, the National Survey of Recent College Graduates (NSRCG), the SDR, and the Scientists and Engineers Statistical Data System (SESTAT) integrated database (which includes the NSRCG, the SDR, and the National Survey of College Graduates [NSCG]). The panel heard from several people that NCSES sees the NORC Data Enclave as a way to build its community of licensed researchers while enabling its own staff to spend more time in helping researchers with the substance of the data rather than paperwork. Additionally, NCSES has worked with NORC to build an infrastructure that allows research teams to share their work in a secure environment, regardless of whether they are physically in one location.

There is strong interest in the dynamics of firm demographics, births, deaths, and employment contributions and the role of high-growth firms. The Census Bureau can develop these statistics by analyzing its business register. If these data were available to researchers—say, at the NORC Data Enclave—then a broad spectrum of evidence-based statistics and indicators could be made publicly available. One means by which such capability could be built is through NCSES's initiation of a research program. Such a program would energize the research community to use survey and other data as soon as the data arrived at the NORC Data Enclave. The program could also be designed to incentivize researchers to develop new, high-utility statistics that were based on linked data from several agencies and that related inputs to outputs, outcomes, and effects.

For datasets that cannot be used outside of a company, another approach NCSES could take would be to work with NSF directorates that sponsor industrial fellowships. For example, LinkedIn has an excellent Data Science team that could potentially provide mentorship for a graduate student or postdoctoral fellow. A program modeled after the NSF Division of Mathematical Sciences University-Industry Cooperative Research Programs in the Mathematical Sciences (National Science Foundation, 2004) could provide a way for researchers interested in the STEM labor market to collaborate with LinkedIn's Data Science team and explore LinkedIn's data while under close supervision.

RECOMMENDATION 7-5: The National Center for Science and Engineering Statistics should explore the use of university-industry exchanges as a mechanism for giving researchers access to promising datasets and industry teams access to new research techniques.

NEXT STEPS

The emerging field of data science is more than the motivating example for this chapter. The new approach to understanding STEM that the panel believes NCSES should explore is at its core a data science approach. Because the field of data science is new and the number of practitioners is relatively small, the panel proposes two concrete initiatives that would provide some opportunities for NCSES to gain experience with new data science tools and to collaborate with data scientists.

NSF has a long history of funding university-industry research collaborations. The model is typically that an industry partner with a problem to solve is paired with a university partner that has experience with techniques and tools relevant to the problem domain. A graduate student or postdoctoral fellow (or professor) splits his or her time between the university and the corporation and is mentored by people in both institutions. The student/postdoctoral fellow gains valuable real-world experience, the industry partner gains solutions to problems, and the university partner gains a better understanding of real problems and potentially relevant data. One example of this approach is the previously mentioned NSF Division of Mathematical Sciences University-Industry Cooperative Research Programs in the Mathematical Sciences (National Science Foundation, 2004).

NCSES could gain considerable experience in data science techniques by playing the role of the industry partner and collaborating with a university in this fashion. NCSES has a mandate to understand the state of STEM; access to interesting datasets; and a staff well versed in navigating federal research organizations, managing datasets, and conducting survey research. A collaboration with a university

[12] "The NORC Data Enclave exists to provide researchers with secure access to microdata and protect confidentiality, as well as index, curate and archive data. The NORC Data Enclave provides authorized researchers with remote access to microdata using the most secure methods to protect confidentiality." See the full description of the NORC Data Enclave in The University of Chicago (2013). BEA does not permit data migration to research data centers. Instead, it has a program whereby individuals can use the data in house under a special sworn employee arrangement.

laboratory focused on such matters as text processing, web mining, or Internet-oriented econometrics could yield valuable experience for both sides.

RECOMMENDATION 7-6: The National Center for Science and Engineering Statistics (NCSES) should collaborate with university researchers on the use of data science techniques to understand the science, technology, engineering, and mathematics ecosystem, using a mechanism similar to existing National Science Foundation university-industry partnerships. One or two graduate students or postdoctoral fellows could alternate between working at NCSES and at their home institution for up to 2 years, with the specific goal of contributing new findings to NCSES's data and indicators programs.

NCSES has considerable experience with managing structured data in the form of its survey products, but much less experience with unstructured data. Conveniently, NSF has a rich but relatively untapped dataset of unstructured data that could provide a wealth of new insights into STEM in the United States—its database of awards. This dataset is quite sensitive, but there is precedent for granting researchers access to the data: for the *Discovery in a Research Portfolio* report (National Science Foundation, 2010a), NSF's Science of Science and Innovation Policy Program recently granted 10 academic research laboratories access to a small set of recent award data. NCSES is well versed in managing researcher access to sensitive datasets and would doubtless be up to the task of ensuring responsible access to its award database.

The *Discovery in a Research Portfolio* report recommends that NSF make a number of improvements to its award data, several of which align well with the panel's recommendations. For example, the report recommends combining award data with internal and external data—a task that would benefit from automated techniques for extracting entities (people, laboratories, programs, institutions) from awards and linking them to related entities in other datasets. The report also recommends improving visualization techniques and understanding the interrelationships between people and topics—both of which would make promising research projects for a visiting graduate student or postdoctoral fellow.

Managing award data for research purposes would provide a useful test bed for several other recommendations offered in this chapter. For example:

- NCSES and NSF's Science of Science and Innovation Policy Program could formulate a set of key questions they believe NSF award data could help answer and then work with relevant directorates to fund this research.
- NCSES could work to share some of the tools used to add structure (in the form of automatically assigned topics) to awards. NCSES could also share the topics themselves, either as additions to the existing online awards database or as a separate metadata file.

RECOMMENDATION 7-7: The National Center for Science and Engineering Statistics (NCSES) should explore methods of mining the awards database at the National Science Foundation as one means of discovering leading pathways for transformational scientific discoveries. NCSES should engage researchers in this exploratory activity, using its grants program. NCSES should develop mechanisms for using the tools and metadata developed in the course of this activity for the development of leading indicators of budding science and engineering fields.

One way to develop these ideas further would be through a contest for research funding or prize competition. Several "open innovation" organizations, such as InnoCentive, the Innovation Exchange, and challenge.gov, facilitate this type of contest. Working with an outside entity to design and administer a contest would allow NCSES to focus on the problems it hoped to address rather than the implementation details of the contest. The National Research Council (2007) report *Innovation Inducement Prizes at the National Science Foundation* and NSF's Innovation Corps Program could also serve as useful models, although these resources are focused more specifically on technology commercialization.

If the contest were designed to address statistical questions related to the usefulness of web-based data sources, then it would be necessary to supply some sample data, and this might affect negotiations with companies. For example, LinkedIn might be willing to supply its data for NCSES to use but unwilling to allow use of the data in a public contest.

How can a federal statistical agency develop and rely on web-based and scientometric tools to produce gold-standard data for periodic publication? This is a basic question that needs to be considered in the current climate of rapidly changing technologies and increasing demands for data. There are numerous related questions, including: How can an agency overcome looming privacy and security issues? How many personnel internal to the agency will it take to develop and operate the tools to produce the indicators? These are good questions that will need to be fully addressed before NCSES or any other federal statistical agency implements the frontier methods described in this section.

One way to address these questions is by example. In 2011, NIH decided to sponsor a competition[13] to find improved methods for using the National Library of Medicine (NLM)

[13]The panel thanks Jerry Sheehan (National Institutes of Health) for providing information and materials on this competition (see http://showoffyourapps.challenge.gov/ [December 2011]).

to show knowledge flows from scientific exploration through to commercialized products. The agency also wanted to use the NLM resource for taxonomic development and for showing relationships among research activities. Knowledge spillovers and effects are difficult to measure. NIH determined that one way to mine millions of data entries would be to automate the process. Yet that was not the expertise of any specific department at NIH, and it was important to cast a broad net to obtain the best expertise for addressing the problem. The competition was announced on challenge.gov and was titled "The NLM Show Off Your Apps: Innovative Uses of NLM Information Challenge." The competition was open to individuals, teams of individuals, and organizations, and its purpose was to "develop innovative software applications to further NLM's mission of aiding the dissemination and exchange of scientific and other information pertinent to medicine and public health."[14] The competition ended August 31, 2011, and winners were announced on November 2.[15]

SUMMARY

In this chapter, the panel has presented seven recommendations. The focus is on exploratory activities that should enable NCSES to produce over time STI indicators that measure more accurately what users really want measured, and in a more timely fashion. Researcher engagement and incentives for exploratory activities are important aspects of these recommendations. While the recommendations in Chapters 4-6 take priority over the recommendations in this chapter, the panel views these exploratory efforts as important investments in the long-term viability of NCSES with respect to its ability to meet evolving user needs in changing technological and economic environments.

ANNEX 7-1: POTENTIAL DATA SOURCES TO EXPLORE

Measuring Research Impact

Considerable activity is focused on finding better measures than citations for the impact of papers (see Priem and Hemminger, 2010, for an overview). The approaches being used fall into three broad categories:

1. Impact measured through refinements of citation counts—The Eigenfactor algorithm (http://www.eigenfactor.org/) gauges impact by computing impact-weighted citation counts. Citations from high-impact papers count for more than citations from low-impact papers. The algorithm is related to Google's PageRank algorithm for determining the relevance of web pages.
2. Impact measured through aggregation of online indicators
   a. In addition to citations, Public Library of Science (PLOS) journals use article-level metrics to track usage statistics (article views and downloads), user feedback, and blog posts (Patterson, 2009).
   b. Total Impact is an application programming interface (API) that allows sites to display PLOS-style article-level metrics for arbitrary articles (http://impactstory.org/).
   c. Altmetric.com is a start-up that tracks mentions of scholarly articles in blogs, social media, newspapers, and magazines and provides scores for articles (http://altmetric.com/).
3. Impact gauged by expert raters—Faculty of 1000 is a subscription-based service that selects new publications deemed by expert reviewers to be important. Articles are assigned a numerical rating (http://f1000.com/).

Measuring Knowledge Diffusion

Diffusion Within the Research Literature

Citation flows from Thomson/Reuters have been analyzed to gauge the flow of knowledge within disciplines (see, e.g., Rosvall and Bergstrom, 2008). These flows can provide insight into the extent to which basic research in various fields propagates out into more applied disciplines.

Diffusion Within and Outside the Research Literature

The Kauffman-funded COMETS (Connecting Outcome Measures in Entrepreneurship Technology and Science) database (Ewing Marion Kauffman Foundation, 2012) and COMETSandSTARS database seek to shed light on the next stage of diffusion of ideas, from research to products (Zucker et al., 2011). The databases link awards from NSF/NIH, patents, papers from Web of Knowledge, doctoral dissertations, and more. The initial implementation of STAR METRICS at NSF involves similar types of linkages, initially linking research awards to patents and jobs (supported by awards), with ambitious future plans for tracking outputs, such as publications, citations, workforce outcomes, public health outcomes, and more (National Institutes of Health, 2012). The linking was accomplished using sophisticated text mining tools (Lai et al., 2011), in this case a variant of the Torvik-Smalheiser algorithm (Smalheiser and Torvik, 2009).

[14]See http://showoffyourapps.challenge.gov/ [December 2011].

[15]The U.S. Census Bureau ran a visualization competition in 2012 to develop a statistical model that could predict census mail return rates (see U.S. Census Bureau, 2012). Another example of a competition is the Netflix Prize, documented in the materials for the seminar of the Committee on Applied and Theoretical Statistics and Committee on National Statistics of the National Academy of Sciences titled "The Story of the Netflix Prize" (November 4, 2011) (see http://www.netflixprize.com/community/ and Gillick, 2012).

BOX 7-2
Tracking the Commercialization of Technologies Through Records of New Books

Michelle Alexopoulos and collaborators at the University of Toronto have been measuring the commercialization of technology using records of new books from the Library of Congress (Alexopoulos and Cohen, 2011). The idea is that publishers invest in new books on commercially promising innovations and stop investing when innovations are in decline. Hence, a count of new book titles on a particular technology provides an indicator of the success of that technology in the marketplace. Alexopoulos and Cohen trace the diffusion of such inventions as the Commodore 64 and Microsoft Windows Vista by searching for related keywords in new book titles.

One potential generalization of this work would be to attempt to trace the flow of ideas back to the research that preceded commercialization. This task would be considerably more difficult than tracking commercialization as it would be necessary to examine vastly more papers than books, but techniques such as automated topic extraction could make the task more feasible.

Diffusion Outside the Research Literature

Alexopoulos and Cohen (2011) have mined Machine Readable Cataloging (MARC) records (Library of Congress, 2012) from the Library of Congress to identify patterns of technical change in the economy (see Box 7-2). They argue that the book publication counts captured in these records correspond more closely than patent records to the commercialization of ideas. Other tools for mining data in books include

- Google's NGram viewer, a search engine for n-grams in Chinese, English, French, German, Hebrew, Russian, and Spanish books published between 1800 and 2008; and
- Culturomics for arXiv, a search engine for n-grams in scientific preprints published in ArXiv between 1992 and 2012.

Two of NSF's administrative datasets could potentially shed additional light on the translation of research work into commercial products:

- NSF proposals have a required section (National Science Foundation, 2011b) on the impact of past NSF-funded work.
- NSF-funded researchers submit Research Performance Progress Reports (National Science Foundation, 2012c) that document accomplishments, including inventions, patent applications, and other products.

The STEM Labor Market

Demand

Large job boards such as Monster.com or job board aggregators such as Indeed.com or SimplyHired.com could provide a rich source of information on demand for STEM professionals. The Conference Board's Help Wanted OnLine dataset includes jobs collected from 16,000 online job boards (The Conference Board, 2012).

One can collect data from a site such as Monster.com either by scraping information from the public website or by negotiating with the site for access to the data. Two reasons for preferring negotiation are legality and data structure. The legality of web scraping has been challenged several times in courts both in the United States and abroad,[16] and there appears to be no consensus on what is legal. However, all the cases to date that the panel found involved one for-profit company scraping data from another for-profit company's site for commercial use. For example, a travel company might use web scraping to collect information on ticket prices from an airline and then use those prices to facilitate customers' comparative shopping. During the course of this study, the panel found no example of a nonprofit or government organization or academic researcher being sued over web scraping.

Supply

Several new social networks for researchers could be used to learn more about career trajectories in the sciences, particularly nonacademic careers:

- ResearchGate—http://www.researchgate.net/
- Mendeley—http://www.mendeley.com/
- Academia.edu—http://academia.edu/

LinkedIn.com is a broader social network for professionals that had 175 million members as of June 2012.

Several initiatives may make new data on researchers available online:

[16]For example, Ryanair, a European airline, initiated a series of legal actions to prevent companies such as Billigfleuge and Ticket Point from scraping ticket price data from its website to allow for easier comparison shopping (see Ryanair, 2010). In a California 2000 case, *eBay v. Bidder's Edge*, eBay sued Bidder's Edge over price-scraping activities; see http://www.law.upenn.edu/fac/pwagner/law619/f2001/week11/bidders_edge.pdf [December 2011]. And in another California case, in 2009, *Facebook, Inc. v. Power Ventures, Inc.*, Facebook sued Power Ventures over scraping of personal user data from the Facebook site; see http://jolt.law.harvard.edu/digest/9th-circuit/facebook-inc-v-power-ventures-inc [December 2011].

- Vivo is a web platform for exposing semantic data on researchers and their work on the websites of research institutions. Vivo tools provide a way for institutions to create rich, structured datasets on their research activities.
- SciENCV is an NIH demonstration project for allowing researchers to create public research profiles. These profiles are designed to streamline the process of applying for NIH and other grants, but will also generate useful structured datasets on researchers.
- Brazil's Lattes Platform is a database of all Brazilian researchers and their work. It extends the ideas in Vivo and SciENCV, and participation is mandatory.
- The ORCID project seeks to provide researchers with unique identifiers that will be used as author identifiers for publications, awards, and so on. The goal is to facilitate linking of datasets involving individual researchers. ORCID will serve as a registry rather than a data provider, but the use of these identifiers can help structure existing unstructured datasets. (Some researchers [Smalheiser and Torvik, 2009] have expressed skepticism about the utility of such identifiers, however.)
- The U.S. Department of Labor issues quarterly foreign labor certification data for H-1B visa holders (U.S. Department of Labor, 2012a). The dataset contains job titles and employers for new H-1B holders, and degree level can be inferred for some broad categories of jobs (e.g., "postdoctoral scholar"). The data are imperfect in that not all Ph.D.'s are on H-1B visas, there will be some overlap between SED respondents and those receiving H1-B visas, and job title is an imperfect predictor of degree status, but one may be able to see useful year-to-year changes in numbers of foreign postdoctoral fellows.

Finally, there are several databases of dissertations:

- ProQuest,
- WorldCat, and
- OpenThesis.

8

Informing the Strategic Planning Process

Collectively, the recommendations on science, technology, and innovation (STI) indicators offered in Chapters 2-7 constitute a program of work for the National Center for Science and Engineering Statistics (NCSES) that accords with NCSES's obligations as a statistical agency and the requirements of the America Creating Opportunities to Meaningfully Promote Excellence in Technology, Education, and Science (America COMPETES) Reauthorization Act of 2010 (see Chapter 1). The panel understands that resource constraints will not permit all of its recommendations to be pursued to the same extent or on the same time table, so that it will be necessary for NCSES to develop a strategic plan (as recommended in Recommendation 3-1 in Chapter 3), followed by an implementation plan.

This chapter presents five recommendations intended to inform NCSES's strategic planning process for the future of its program of STI indicators. The first strategic recommendation is to accord priority to data quality, broadly defined. Then come recommendations for four strategic processes or pathways for development: data linkage and sharing, methodological research, user access to NCSES data and collaborative research on new and revised STI indicators, and establishment of a chief analyst position at NCSES. Although data quality is the principal priority for a statistical agency, the setting of relative priorities among the remaining four strategic recommendations must be done by NCSES. Once strategic priorities have been developed, NCSES can use the recommendations from Chapters 2-7 to lay out the requirements and priorities for the implementation plan that should accompany the strategic plan.

This chapter begins by noting the many strengths of the current program of STI indicators at NCSES. It then outlines pathways for moving the program forward. The final section of the chapter maps the recommendations in Chapters 2-7 to the five strategic recommendations to form a program of work that will enable NCSES to develop policy-relevant STI indicators to better meet user needs.

NCSES'S ACCOMPLISHMENTS

NCSES has displayed several areas of strength in its data and statistical products, its interactions with domestic and foreign statistical agencies, and its outreach to the research community. The panel offers the following specific observations about NCSES's accomplishments:

- NCSES provides comprehensive information to the National Science Board, whose members find this service highly useful. A broad range of users (academic and government) also rely on STI indicators produced by NCSES that these users believe they cannot obtain reliably from other sources.
- The panel identified no little-known, proven STI indicators and methodologies used by other countries that could easily and inexpensively be adopted by NCSES. The panel identified new and revised indicators to recommend, but essential indicators are covered in the National Science Board's *Science and Engineering Indicators* biennial volume prepared by NCSES and other NCSES publications.
- NCSES is recognized worldwide by statistical agencies and organizations in other countries as a leader in the development and curation of human resources statistics. This is a well-deserved reputation. NCSES might consider publishing periodic highlights (similar to highlights published by the research arm of the National Science Foundation [NSF]), with specific references to data series and statistics that have been shown to be pivotal in policy decisions. These highlights might include lists of peer-reviewed published articles that use NCSES's microdatasets or published statistics.
- NCSES has undertaken numerous productive collaborative efforts with other federal statistical agencies, including the use of data on global multinational research and development (R&D) activities from

the U.S. Bureau of Economic Analysis (BEA) and a wide range of statistics from the National Center for Education Statistics.

- The recent development of the Business Research and Development and Innovation Survey (BRDIS) instrument has produced needed statistics on national and international R&D expenditures and performance. BRDIS is a first step toward obtaining data on the diffuse nature of innovation and R&D in the United States and around the world.
- NCSES took the initiative to rationalize its human resources surveys by eliminating the National Survey of Recent College Graduates (NSRCG), relying instead on a sampling frame from the American Community Survey to incorporate information on recent science and engineering graduates into the National Survey of College Graduates (NSCG). This was a necessary and well-designed cost reduction strategy.[1]
- The U.S. Census Bureau established a microdata Survey Sponsor Data Center at NCSES's location, giving NCSES staff ready access to microdata collected for NCSES by the Census Bureau, including from BRDIS and the NSCG. This should reduce the coordination costs of accessing the data and, once the process for accessing the data is fully operational, enable NCSES staff to accomplish more timely analysis and reporting of analytical statistics and indicators. NCSES staff have also worked to facilitate greater access by outside researchers to its surveys of science and engineering personnel through a secure data enclave operated by NORC at the University of Chicago.
- NCSES staff have been leaders in various international forums, especially OECD, where NCSES officials have consistently provided leadership to the Committee on Scientific and Technological Policy's expert statistical working party, the National Experts on Science and Technology Indicators (NESTI). In this role, NCSES staff have led work to improve the international comparability of a variety of STI indicators, including patent-based indicators, the careers of doctorate holders, statistics on Chinese STI performance, and cognitive testing of surveys on R&D and innovation.

PROCESSES FOR CHANGE

NCSES products, especially the biennial *Science and Engineering Indicators* volume for the National Science Board and the agency's *InfoBriefs*, are highly regarded and widely used. However, they reflect a fraction of the information present in NCSES's microdata holdings, let alone what would be available if these microdata were linked to the holdings of other federal agencies to expand their analytical potential. NCSES thus has an opportunity to produce much more output without having to invest in major new data collection. Taking advantage of this opportunity would enable NCSES to increase its contribution to the public good and provide users of its data, expert or novice, with deeper insights into the workings of the science and engineering enterprise in the United States.

For these new products to be produced, decisions will have to be made and priorities set, as accomplishing this will likely require a somewhat different skill mix from that currently present in NCSES, and there are other resource implications as well. Given that NCSES is limited in the number of staff it can employ, the production of new products will require a clear vision and a commitment to resource reallocation over a significant period of time. The necessary decisions cannot be made in isolation from other decisions that are part of managing the agency and the expectations of the users of its data. These decisions must be part of a strategic plan that provides a vision for the organization and lays out how the goals that support that vision are to be achieved.

This panel cannot provide a strategic plan for NCSES. Formulating such a plan requires priority setting, resource reallocation, and monitoring processes that only NSF can undertake using information on likely resource levels and other factors to which NSF alone is privy. However, the panel can, and does, make recommendations (presented later in this chapter) that can inform the strategic planning process. As outlined earlier, these recommendations deal with data quality (Recommendation 8-1), which is multidimensional but fundamental for a statistical agency; data linkage and sharing with other agencies, both inside and outside of government (Recommendation 8-2); the need for a program of methodological research (Recommendation 8-3); the building of a community of practice engaged in the use of data managed by NCSES, both its own and from other sources, which can support not only methodological but also substantive research on new and revised STI indicators (Recommendation 8-4); and the establishment of the position of a chief analyst who would interface with the users of NCSES's products to monitor changes in demand and would make recommendations on methods and data sources for the STI indicators program (Recommendation 8-5).

It should be made clear that the panel's recommendations in this chapter are hardly new in the sense that most of these areas have been touched on by previous National Research Council (NRC) reports. NCSES is encouraged to review the NRC (2005) report *Measuring Research and Development Expenditures in the U.S. Economy*. That report, although focused on R&D expenditures, deals with many of the issues that are addressed in this chapter and quotes earlier reports examining similar issues.

What makes the recommendations in this chapter even more important now is that the global position of the United States has changed, particularly with the emergence of

[1]The last year of data collection using the NSRCG was 2010.

competencies in R&D and innovation and of human capital skills in science and technology within emerging economies around the world. As the United States competes in a global economy, its policy makers, business managers, educators, and residents need to understand what is happening to the science and engineering enterprise so they have the informed capacity to advise, consent, and act to improve it. That understanding can come, in part, from more and better use of NCSES data and statistics, as well as other metrics produced by federal agencies and other organizations.

Data Quality

As NCSES is a statistical agency, it is bound by data quality considerations, and that is where the recommendations in this chapter begin. The above-referenced NRC report notes four components of data quality—accuracy, relevance, timeliness, and accessibility (National Research Council, 2005, p. 11). That report focuses on accuracy; this panel regards all four elements as important and interrelated.[2] Accuracy is essential for all indicators, but relevance can be lost if the data release is not timely. Accessibility matters as it allows researchers from other institutions to work with the data, leveraging NCSES's limited analytical capacity, but also imposing an obligation for NCSES to protect the confidentiality of the data and the need to train external users in how to access and manipulate the data.

A special emphasis on timeliness is warranted because users whose views were solicited for this study were emphatic that certain types of indicators lose relevance if they are not made available within a short time after the date of observation—providing indicators that are 2-3 years out of date is not helpful. For NCSES, greater timeliness could be achieved in several ways, including (1) the release of preliminary data and statistics, which is common for leading economic indicators from other statistical agencies, although users will need to be cognizant of the greater uncertainty of the estimates; (2) nowcasting and other techniques discussed in Chapter 7 of this report, although their use will require changes in skill sets or contractual relationships at NCSES; and (3) allocating resources to more timely release of final products by reengineering every component of the process from survey development and sourcing of raw data through publication processes and media for distribution. NCSES's strategic planning process for its STI indicators program should specifically address methods to be used to respond to users' needs for timeliness.

The panel is aware that NCSES surveys are contracted to other organizations, but the production of indicators to meet NCSES's data quality requirements, including timeliness, should be a contractual obligation. Data quality indicators are necessary to enable the identification of quality deficiencies and to develop methods for data improvement. As earlier NRC reports provide specific advice about what kinds of quality indicators NCSES should monitor and disseminate for its data collection programs, including unit nonresponse, item nonresponse, and population coverage, the following recommendation does not go into further detail.[3]

RECOMMENDATION 8-1: Given the fundamental importance of data quality and reporting of quality measures, the National Center for Science and Engineering Statistics should review its data quality framework, establish a set of indicators for all of its surveys, and publish the results at regular intervals and at least annually.

Data Linkages and Sharing

Once data are seen to conform to quality assurance indicators, analysis of the data can follow, leading to new, relevant, and timely products addressing issues of importance to the policy community and to other users of the information. Datasets need not come only from NCSES; other agencies produce data that fall within the charge given to NCSES by the America COMPETES Act, and there are data from nongovernmental sources as well.

The usefulness of a single dataset can be increased if it is linked to other datasets of comparable quality, whether within NCSES or in collaboration with other agencies. Collaboration with other agencies on data linkages activities, which in some cases is already taking place, would lead to working-level knowledge of how other agencies manage data. In due course, the ongoing exchange of information would enable NCSES to assume the clearinghouse function required under the America COMPETES Act, which mandates that NCSES "shall serve as a central Federal clearinghouse for the collection, interpretation, analysis, and dissemination of objective data on science, engineering, technology, and research and development."

Better integration of data sources is needed to develop more robust STI indicators. At the panel's July 2011 workshop, John Haltiwanger of the University of Maryland described how infrastructure datasets could be fully integrated to track the career histories of innovators and entrepreneurs and the relationships between start-up, young, high-growth firms and large, mature firms (see Chapter 4 for more detail). These infrastructure datasets could be fully integrated with all existing Census Bureau business surveys and other data. For example, one could integrate economic censuses and annual surveys to measure productivity, business practices, and R&D, linked to patent, citation, and

[2]See also National Research Council (2013b).

[3]One example of a useful tool for communicating various elements of the quality of data and statistics to users is the U.S. Census Bureau's website for the American Community Survey (U.S. Census Bureau, 2013). The panel recognizes that web-based links like this are rare, and that NCSES and other statistical agencies typically give this type of information in technical notes that accompany data releases.

other information about innovators from the U.S. Patent and Trademark Office.

Any new STI indicators that are developed will need to be integrated into the existing infrastructure (if not at the person/business level, then at some level of disaggregation). Data sharing and synchronization would permit even richer integration of Bureau of Labor Statistics (BLS) and BEA firm data. At the panel's July 2011 workshop, Matthieu Delescluse of the European Commission remarked that the European Union (EU) is commissioning the linking of patent data with company databases to support the development of new indicators. For example, this type of linking will make it possible to track the relationship between small and medium firms and the number of patents over time. The EU is also using data from Community Innovation Survey Business Registers for member countries to determine the international sourcing of R&D. This statistic could also be developed in the United States through the linking of Census Bureau and BEA data. Employment dynamics, including worker mobility trends in science and engineering occupations, could be developed by linking Census Bureau, BLS, and BEA data. Existing research data centers or data enclaves could facilitate platforms for data integration, potentially making the data comparable with those of other nations that have similar data administration policies while protecting the confidentiality of the information.

RECOMMENDATION 8-2: The National Center for Science and Engineering Statistics should work with other federal agencies bilaterally and in interagency statistical committees to share data, link databases where feasible, and produce products that would not be possible if the agencies worked independently. The use of data from outside the federal system, where appropriate, should be part of this process.

Methodological Research

The production of datasets that conform to quality standards and meet user needs will almost certainly necessitate research on methodological questions within NCSES and in the wider community to provide optimum solutions. A critical way in which to identify important methodological issues is to work with the data, which typically leads to the discovery of errors, inconsistencies, and gaps. This is why data analysis by staff responsible for the data matters—it is a quality issue.

Not all of the methodological issues raised by data analysis—whether by staff or outside researchers—can be solved by staff or by methodologists in other agencies. NCSES has an opportunity, through its own and other NSF granting programs, to support research on the methodology of data production, including survey methods, data linkage, estimation, and quality control. NCSES could also issue contracts to address specific requirements, but the advantage of an NSF methodological research program that met the immediate needs of NCSES is that it would, over the years, result in a community of methodologists familiar with the agency's work and serve as an invaluable resource for staffing and peer review of new initiatives.

RECOMMENDATION 8-3: The National Center for Science and Engineering Statistics should use its existing Grants and Fellowships Program and related programs at the National Science Foundation to support methodological research that addresses the agency's specific needs.

Building of a Community of Practice in the Use of NCSES and Other Data

If NCSES is to advance its analytical capacity beyond the interpretation of tables and management of its biennial indicators report, then it will have to make greater use of its own data. In addition, these data can and should be used by researchers outside of NCSES to produce more analytical material and to build a community of practice that is knowledgeable about NCSES datasets. Such a community of practice could expand the range of outputs based on these datasets, including new and revised STI indicators; contribute to methodological improvements; and enhance the evident public good of NCSES's surveys. Including other federal agencies in such collaboration would contribute to the knowledge and relationships NCSES needs to move toward assuming the clearinghouse function mandated by the America COMPETES Act.

Researchers would have to be trained in the use of NCSES's datasets and their work monitored to ensure that there would be no breach of confidentiality or privacy. Again, this would require resource allocation and a commitment to working with outside communities. The panel stresses the importance of greater use of the data, or greater accessibility, to use the data quality term. A good start has been made in that direction for the science and engineering personnel surveys.

RECOMMENDATION 8-4: The National Center for Science and Engineering Statistics should make its data holdings available to external researchers to facilitate their research while protecting confidentiality and privacy.

Development of a Chief Analyst Position

The four recommendations offered thus far in this chapter relate to data quality, broadly defined, as a bedrock of NCSES's STI indicators program, data linkages and sharing to permit the development of new and improved indicators, methodological research on indicators and the underlying data, and the building of a community of practice to leverage the work of NCSES staff. The need to manage these activities suggests the need for a separate unit within NCSES led by a senior staff member who would be responsible for these activities, including building the internal team to carry them out and managing the necessary links with other organizations. This unit would monitor new indicators emerging across the globe and would have the capacity to absorb the knowledge needed to bring these new indicators into NCSES and present them to users for their evaluation. A close relationship and regular communication between members of this analytical unit and the survey units in NCSES would be necessary.

The new unit should include a position of chief analyst, analogous to the current position of chief statistician but with a different portfolio. The chief analyst should not manage the unit but instead provide substantive leadership and communication across units within NCSES and with other agencies and organizations to develop NCSES's STI indicators program in ways that are most useful for the policy and research communities.

Taking these steps would facilitate NCSES's ability to more fully embody Practice 10 in *Principles and Practices for a Federal Statistical Agency, Fifth Edition*. Practice 10 calls for a federal statistical agency to have an active research program, including not only methodological research, but also research "on the substantive issues for which the agency's data are compiled, . . . [for] the identification and creation of new statistical measures, . . . [and] to understand how the agency's information is used, in order to make the data more relevant to policy concerns, and more useful for policy research and decision making" (National Research Council, 2013b, p. 22).

RECOMMENDATION 8-5: The National Center for Science and Engineering Statistics (NCSES) should establish a unit to manage data quality assurance, cooperation with other institutions, and analysis for its science, technology, and innovation indicators and related programs. NCSES should develop a new position of chief analyst within this unit whose role would be to (1) interface periodically with users of NCSES data and statistics, including indicators, so the agency can remain up to date on changing demand for its products; (2) provide NCSES staff with periodic updates on areas of change that are likely to have an impact on the agency's statistical operations; and (3) assess the utility of new types of datasets and tools that NCSES could use either in house or by contractual arrangement to develop and improve indicators.

A PROGRAM OF WORK

Once the strategic planning proposed in Recommendation 3-1 has been undertaken and informed by the five recommendations offered in this chapter, those five recommendations can be linked to the recommendations in the previous chapters to constitute a program of work for NCSES for its STI indicators. Implementing the panel's recommendations would support the development of the capacity within NCSES to meet the requirements of the America COMPETES Act, especially SEC. 505 (b) (2) and (3) and (c). Table 8-1 shows which recommendations in Chapters 2-7 fall under the recommendations in this chapter; the full text of the recommendations in Chapters 2-7 is provided in the boxes at the end of this chapter.

Considerations for Prioritization

The panel believes that all pathways outlined in this chapter are important for NCSES to include in its strategic plan for its STI indicators program, although the agency will undoubtedly need to prioritize them with respect to the pace and specifics of implementation. As part of implementing the strategic plan, NCSES will also need to assign priorities for the detailed program of work along each pathway. Prioritization requires gauging which policy issues are likely to be important in the near, medium, and longer terms; assessing which policy issues might be informed by specific STI indicators; estimating the benefit versus the cost of developing indicators in house or obtaining them through contractual relationships or from scholarly research; and focusing on those indicators that would cost-effectively shed light on important policy questions. The key to this prioritization process is the recognition that not all indicators need to be sourced through traditional means such as surveys.

While, as noted earlier, the panel is not in a position to set priorities for NCSES's STI indicators program, Table 8-1 reveals that some recommendations are pertinent for progress along more than one strategic pathway, which suggests a relative importance. Likewise, some of the recommendations are contingent on the implementation of others, again suggesting a relative importance. Linkages between the pathways and the programmatic recommendations are summarized below.

Pathways and Programmatic Recommendations

As is appropriate for a statistical agency, all the panel's recommendations relate to data quality, **Recommendation 8-1**. At the same time, the recommendations in Chapters 2-7 entail actions by NCSES along four strategic pathways, corresponding to **Recommendations 8-2 through 8-5**, respectively.

Data Sharing and Linkage (Recommendation 8-2)

First is the development of new policy-relevant, internationally comparable indicators that are based on existing NCSES survey data and on data collections at other statistical

TABLE 8-1 Linking the Recommendations in Chapters 2-7 to the Strategic Pathways in Chapter 8 to Form a Strategic Program of Work

Chapter 8	Chapter 2	Chapter 3	Chapter 4	Chapter 5	Chapter 6	Chapter 7
Recommendation 8-2—Data Linking and Sharing	Rec. 2-2	Rec. 3-1	Recs. 4-2 4-3, 4-4	Recs. 5-1, 5-2, 5-3, 5-4	Recs. 6-2, 6-4, 6-5, 6-6	Rec. 7-5
Recommendation 8-3—Methodological Research	Rec. 2-1	Rec. 3-1	Recs. 4-1, 4-2, 4-4, 4-5	Recs. 5-1, 5-2, 5-4	Recs. 6-1, 6-2, 6-3, 6-4, 6-5, 6-6, 6-7	Recs. 7-1, 7-2, 7-4, 7-6, 7-7
Recommendation 8-4—Use of NCSES and Other Data	Rec. 2-1	Rec. 3-1	Rec. 4-2		Rec. 6-1	Recs. 7-3, 7-5
Recommendation 8-5—Chief Analyst		Rec. 3-1	Rec. 4-1		Recs. 6-2, 6-3	Recs. 7-2, 7-7

NOTE: All recommendations link to Recommendation 8-1 on data quality indicators. NCSES = National Center for Science and Engineering Statistics.

agencies, both inside and outside the government, using data sharing and linkage techniques as appropriate.

- A critical aspect of this effort is the development of integrating processes to leverage synergies within NCSES and collaborations outside NCSES at the same time that new data extraction and management methods for generating statistics are explored. A prerequisite for linking data from different sources is the development of a consistent taxonomy of science and engineering fields and occupations (**Recommendation 2-2**).
- NCSES has many data series that have not been analyzed but have great potential to help answer questions posed by users. **Recommendation 3-1** stresses the importance of developing new measures from existing data. This is very apparent for the innovation indicators.
- BRDIS data already provide the input that could allow for analysis of comparative statistics on innovation by the same cutoffs for firm size used by the OECD countries (**Recommendation 4-2**); for cross-tabs on R&D and innovation spending in the United States and abroad, by firm characteristics (**Recommendation 4-2**); and for analysis of yield measures of activities by high-growth firms and of firm dynamics in terms of births and deaths of businesses linked to innovation outputs (**Recommendation 4-4**). Although some of the results require linkage with data from other agencies, these types of outputs would be welcomed by the user community. Better access to BRDIS data by NCSES staff is imperative for the timely distribution of such statistics to the users of STI indicators (**Recommendation 4-3**).

Methodological Research (Recommendation 8-3)

Second is the need for NCSES to build a community of practice around existing and emerging methodological issues if it is to update its data acquisition and analysis techniques.

- Given resource constraints, and to avail itself of methodological best practices, NCSES needs to use existing grants programs and input from researchers and practitioners (**Recommendations 2-1 and 3-1**, respectively) to learn about the usefulness of its own data and new techniques for developing more useful indicators.
- New measures are needed, including specific cutoffs for innovation statistics; measures of organizational and market innovations, as well as innovations in training and design; measures for understanding hindrances in the innovation process; and the use of business practice data (e.g., administrative records, web-based data) to produce new, timely indicators of innovation (**Recommendations 4-1, 4-2, 4-4, and 4-5**). Users of indicators also want to see improved measures of knowledge networks (**Recommendation 5-1**) and payments and receipts for R&D services (**Recommendation 5-2**), as well as indicators that track the development and diffusion of general-purpose technologies (**Recommendation 5-4**).
- New and revised human capital indicators for labor mobility, career paths, stay rates for students at various levels of education, wages and salaries by skill set, and demand and supply of skill sets in various industries (**Recommendations 6-1 through 6-7**) also will depend on the cultivation of existing datasets and the development of new techniques for using business practice data (**Recommendations 7-1, 7-2, 7-4, 7-6, and 7-7**).

Wider Use of Data (Recommendation 8-4)

The development of new measures and methods for STI indicators will require research and implementation strategies that should be developed collaboratively among NCSES staff and the research and practitioner communities. NCSES will need to expand its mechanisms for providing researchers access to data so it can leverage its own limited staff resources. NCSES will also need to address directly the issue of timeliness, given that in many cases, the utility of indicators and the underlying databases to user communities is inversely related to the lag between when the data were sourced and when the indicators are publicly released. Building another community of practice engaged in the use of data managed by NCSES, both its own and from other sources, is important to address this timeliness issue, as well as to contribute to the methodological research described under Recommendation 8-3 above.

- The timeliness with which NCSES delivers indicators to user communities depends on its own access to data resources, primarily from surveys, but increasingly from other sources as well (**Recommendations 2-1, 3-1, 4-2, 6-2, 7-3, and 7-5**).
- Nontraditional methods hold promise for delivering information to users more quickly, but the panel recognizes that many of these methods are still in the exploratory stage. NCSES could play an important role in supporting research to advance the utility of these methods (see in particular **Recommendations 7-3 and 7-5**).

Chief Analyst (Recommendation 8-5)

Fourth and last, establishment of a chief analyst position would improve NCSES's interface with the users of indicators, allowing the agency to monitor changes in demand. The chief analyst would also engage with users to observe and make recommendations on methods and data sources of particular relevance for the STI policy community (**Recommendations 3-1, 4-1, 6-2, 6-3, 7-2, and 7-7**). This role implies the forging of a balanced approach to governance of NCSES activities that span data collection and analytical processes. The goal is a process of feedback and improvement for NCSES's STI indicators program—and the data collections on which it draws—to assist policy makers and other users in understanding the evolving U.S. science and engineering enterprise.

PANEL RECOMMENDATIONS: CHAPTERS 2-7

Chapter 2: Concepts and Uses of Indicators—Outreach Recommendations

RECOMMENDATION 2-1: The National Center for Science and Engineering Statistics should continue its Grants and Fellowships Program for using its datasets, maintaining the high National Science Foundation standards for peer-reviewed award decisions.

RECOMMENDATION 2-2: The National Center for Science and Engineering Statistics should engage with other statistical agencies, including but not limited to the Bureau of Labor Statistics, the U.S. Census Bureau, the National Center for Education Statistics, and the National Institutes of Health, to develop a consistent taxonomy of science and engineering fields and occupations (including the health and social sciences). There should also be an established process for performing updates of this taxonomy as needed.

Chapter 3: Data Resources for Indicators—Prioritization Recommendation

RECOMMENDATION 3-1: In the near term, the National Center for Science and Engineering Statistics (NCSES) should work to produce new and revised science, technology, and innovation indicators in a few key areas, using existing data from the Business Research and Development and Innovation Survey and the Scientists and Engineers Statistical Data System or from productive collaborations with other statistical agencies in the United States and abroad. Over time, NCSES should build capacity in house and through its Grants and Fellowships Program to develop measures that are high priority for users but that require deeper knowledge to obtain statistically valid data or to use frontier methods appropriately. NCSES should also develop a strategic plan for adding new indicators or case studies because doing so may require curtailing the frequency of some of its current measures.

Chapter 4: Measuring Innovation—Recommendations

RECOMMENDATION 4-1: The National Center for Science and Engineering Statistics should develop additional indicators for measuring innovation outcomes that would complement existing data on patents, inputs to innovation

activities, and broader measures of economic performance.

RECOMMENDATION 4-2: The National Center for Science and Engineering Statistics should build on its Business Research and Development and Innovation Survey (BRDIS) to improve its suite of innovation indicators in the following ways:

- **tabulate the results from BRDIS using the same cutoffs for firm size (as well as comparable industry sectors) that are used by OECD countries in order to facilitate international comparisons;**
- **fund research exploring precisely what companies mean when they report an innovation or report no innovation on BRDIS—such research would help inform current policy debates;**
- **broaden the innovations tracked by BRDIS to encompass organizational and marketing innovations, as well as new data algorithms;**
- **consider adding a section to BRDIS on unmarketed innovations, giving respondents the opportunity to cite the main reason these innovations have not yet been marketed or implemented;**
- **as funds permit, extend BRDIS to gather information on innovation-related expenditures in such areas as training and design; and**
- **publish more results from BRDIS that link innovation to business characteristics, including the amount of research and development spending by U.S.-based companies outside of the United States. Production and distribution of such cross-tabulations should be timely, and they should address contemporary policy questions.**

RECOMMENDATION 4-3: The Survey Sponsor Data Center at the National Science Foundation should house the Business Research and Development and Innovation Survey data, improving access to the data for National Center for Science and Engineering Statistics staff who develop the research and development statistics.

RECOMMENDATION 4-4: The National Center for Science and Engineering Statistics (NCSES) should begin a project to match its Business Research and Development and Innovation Survey data to data from ongoing surveys at the U.S. Census Bureau and the Bureau of Labor Statistics. It should use the resulting data linkages to develop measures of activities by high-growth firms, births and deaths of businesses linked to innovation outputs, and other indicators of firm dynamics, all of which should be tabulated by geographic and industry sector and by business size and business age to facilitate comparative analyses. NCSES should conduct a sensitivity analysis to fine-tune meaningful age categories for high-growth firms.

RECOMMENDATION 4-5: The National Center for Science and Engineering Statistics should make greater use of business practice data to track research and development spending and innovation-related jobs at a more detailed geographic and occupational level than is possible with government survey data.

Chapter 5: Measuring the Three K's: Knowledge Generation, Knowledge Networks, and Knowledge Flows—Recommendations

RECOMMENDATION 5-1: The National Center for Science and Engineering Statistics should expand its current set of bibliometric indicators to develop additional measures of knowledge flows and networking patterns. Data on both coauthorship and citations should be exploited to a greater extent than is currently the case.

RECOMMENDATION 5-2: The National Center for Science and Engineering Statistics (NCSES) should make greater use of data from its Business Research and Development and Innovation Survey to provide indicators of payments and receipts for research and development services purchased from and sold to other countries. For this purpose, NCSES should continue collaboration with the U.S. Bureau of Economic Analysis on the linked dataset.

RECOMMENDATION 5-3: The National Center for Science and Engineering Statistics (NCSES) should continue to report statistics on knowledge-based capital and intangible assets obtained from other agencies as part of its data repository function. In addition, NCSES should seek to use data from the Business Research and Development and Innovation Survey on research and development and potentially also on innovation-related expenditures as valuable inputs to ongoing work in this area.

RECOMMENDATION 5-4: The National Center for Science and Engineering Statistics (NCSES) should develop a suite of indicators that can be used to track the development and diffusion of general-purpose technologies, including information and communication technologies, biotechnology, nanotechnology, and green technologies. NCSES should attempt to make greater use of data from the Business Research and Development and Innovation Survey for this purpose while also exploring the use of other sources, such as patent and bibliometric data.

Chapter 6: Measuring Human Capital—Recommendations

RECOMMENDATION 6-1: The National Center for Science and Engineering Statistics (NCSES) should do more to exploit existing longitudinal data. Specifically, NCSES should exploit the longitudinal panel structure of the Survey of Doctorate Recipients (SDR) in the following ways:

- **create indicators of researcher mobility over time, by constructing longitudinal weights for the SDR that take account of changes in the sample and target population over time—these weights should be constructed both for subsequent survey cycles and for existing data;**
- **create a dynamic database for researcher use in which data from the SDR over time would be linked at the level of the individual; and**
- **enhance coverage of recent doctorate recipients to better track their initial employment and career path in the first years after they receive their Ph.D, which could potentially be accomplished by including an additional module in the SDR or by exploiting that survey's longitudinal capacities or both.**

RECOMMENDATION 6-2: The National Center for Science and Engineering Statistics (NCSES) should draw on the Longitudinal Employer-Household Dynamics Program (occupations) and the Baccalaureate and Beyond Longitudinal Study (education levels) to create indicators of labor mobility. NCSES should focus in particular on industries that have been experiencing high growth and/or those in which the United States has a strong competitive advantage. Also relevant would be examining skill sets of firms with high growth.

RECOMMENDATION 6-3: The National Center for Science and Engineering Statistics (NCSES) should provide indicators for individual science, technology, engineering, and mathematics groups such as early-career doctorate recipients, master's degree holders, and community college graduates. NCSES already distinguishes between bachelor's and master's degree holders in many of its statistics. Stay rates at different education levels by demographic characteristics such as gender, race/ethnicity, disability, and country of origin should be included.

RECOMMENDATION 6-4: The National Center for Science and Engineering Statistics should explore whether questions can be included in the National Survey of College Graduates and the American Community Survey that would allow the identification of community college graduates or of holders of higher university degrees who have attended a community college.

RECOMMENDATION 6-5: The National Center for Science and Engineering Statistics should explore methods for exploiting the full-text resources of dissertation databases to create indicators on selected topics both within and across scientific fields and on the relatedness of different fields.

RECOMMENDATION 6-6: The National Center for Science and Engineering Statistics should consider using American Community Survey data to produce indicators that can be used to track the salaries of science, technology, engineering, and mathematics occupations and/or college graduates receiving degrees in different fields and at different degree levels.

RECOMMENDATION 6-7: The National Center for Science and Engineering Statistics should consider adding questions to the Business Research and Development and Innovation Survey on the types of skill sets used by businesses to develop and implement innovations. The results would provide data on and indicators of innovative firms' demand for skills.

Chapter 7: A Paradigm Shift in Data Collection and Analysis—Recommendations

RECOMMENDATION 7-1: The National Center for Science and Engineering Statistics (NCSES) should use research awards to support

the development and experimental use of new sources of data to understand the broad spectrum of innovation activities and to develop new measures of science, technology, and innovation. NCSES should also support the development of new datasets to measure changing public perceptions of science, international trade in technological goods and services, new regions for entrepreneurial activity in science and technology, and precommercialized inventions.

RECOMMENDATION 7-2: The National Center for Science and Engineering Statistics (NCSES) should pursue the use of text processing for developing science, technology, and innovation indicators in the following ways:

- **explore synergies with National Science Foundation directorates that fund research on text processing; and**
- **enable NCSES staff to attend and develop workshops that bring together researchers working on text processing and on understanding the science, technology, engineering, and mathematics ecosystem.**

RECOMMENDATION 7-3: The National Center for Science and Engineering Statistics should use its grants program to encourage the sharing of new datasets and extracted metadata among researchers working on understanding innovation in science, technology, engineering, and mathematics.

RECOMMENDATION 7-4: The National Center for Science and Engineering Statistics should coordinate with directorates at the National Science Foundation in supporting exploratory research designed to validate new sources of data related to innovation in science, technology, engineering, and mathematics.

RECOMMENDATION 7-5: The National Center for Science and Engineering Statistics should explore the use of university-industry exchanges as a mechanism for giving researchers access to promising datasets and industry teams access to new research techniques.

RECOMMENDATION 7-6: The National Center for Science and Engineering Statistics (NCSES) should collaborate with university researchers on the use of data science techniques to understand the science, technology, engineering, and mathematics ecosystem, using a mechanism similar to existing National Science Foundation university-industry partnerships. One or two graduate students or postdoctoral fellows could alternate between working at NCSES and at their home institution for up to 2 years, with the specific goal of contributing new findings to NCSES's data and indicators programs.

RECOMMENDATION 7-7: The National Center for Science and Engineering Statistics (NCSES) should explore methods of mining the awards database at the National Science Foundation as one means of discovering leading pathways for transformational scientific discoveries. NCSES should engage researchers in this exploratory activity, using its grants program. NCSES should develop mechanisms for using the tools and metadata developed in the course of this activity for the development of leading indicators of budding science and engineering fields.

References

Abowd, J.M., J. Haltiwanger, and J. Lane. (2004). Integrated longitudinal employer-employee data for the United States. *The American Economic Review* 94(2):224-229.

Abramovitz, M. (1956). Resource and output trends in the United States since 1870. *The American Economic Review* 46:5-23.

Acemoglu, D., P. Aghion, L. Bursztyn, and D. Hemous. (2012). The environment and directed technical change. *American Economic Association* 102(1):131-166.

Adams, J., and Z. Griliches. (1996). Measuring science: An exploration. *Proceedings of the National Academy of Sciences* 93(12):12644-12670.

Agarwal, A. (2012). *Reinventing Education*. Presented at the Government-University-Industry Research Roundtable meeting, October 10, National Academy of Sciences, Washington, DC.

Aizcorbe, A.M., C.E. Moylan, and C.A. Robbins. (2009). Toward better measurement of innovation and intangibles. *Survey of Current Business* 89(1):10-23. Available: http://www.bea.gov/scb/pdf/2009/01%20January/0109_innovation.pdf [September 2012].

Aizcorbe, A.M., E. Liebman, D. Cutler, and A. Rosen. (2012). Household consumption expenditures for medical care: An alternate presentation. *Survey of Current Business* 92(6):35-48. Available: http://www.bea.gov/scb/pdf/2012/06%20June/0612_healthcare.pdf [September 2012].

Alexopoulos, M., and J. Cohen. (2011). Volumes of evidence: Examining technical change in the last century through a new lens. *Canadian Journal of Economics/Revue Canadienne D'Économique* 44(2):413-450.

Anderson, G. (2011). *Federal Technology Transfer Data 1987-2009*. Available: http://www.nist.gov/tpo/publications/upload/Tech_Trans_Data_Source_2011_10_4.pdf [January 2013].

Arundel, A., K.R. O'Brien, and A. Torugsa. (2010). *Errors in the Interpretation of "Innovation": Preliminary Results from a 2007 Innovation Survey in Australia*. Presented at working group meeting on statistics on science, technology, and innovation, November 22-23, Luxemburg. Available: http://eprints.utas.edu.au/10813/ [July 2013].

Atkinson, R.C. (1990). Supply and demand for scientists and engineers: A national crisis in the making. *Science* 248(4954):425-432. Available: http://www.sciencemag.org/content/248/4954/425.abstract [December 2011].

Audretsch, D.B., E.E. Lehmann, and S. Warning. (2005). University spillovers and new firm location. *Research Policy* 34(7):1113-1122. Available: http://www.sciencedirect.com/science/article/pii/S0048733305001071 [December 2012].

Auriol, L. (2010). *Careers of Doctorate Holders: Employment and Mobility Patterns*. Paris, France: OECD. Available: http://dx.doi.org/10.1787/5kmh8phxvvf5-en [January 2013].

Baldwin, J.R., W. Gu, A. Lafrance, and R. Macdonald. (2008). *Investment in Intangible Assets in Canada: R&D, Innovation, Brand, and Mining, Oil, and Gas Exploration Expenditures*. The Canadian Productivity Review Series, Catalogue No.15-206-X—No. 026. Ottawa, Ontario, Canada: Statistics Canada Economic Analysis Division.

Barnes, P. (2010). *Investments in Intangible Assets and Australia's Productivity Growth: Sectoral Estimates*. Productivity Commission staff Working Paper. Available: http://www.pc.gov.au/__data/assets/pdf_file/0007/99907/intangible-investment-sectoral-estimates.pdf [November 2012].

Barnes, P., and A. McClure. (2009). *Investments in Intangible Assets and Australia's Productivity Growth*. Productivity Commission staff Working Paper. Available: http://www.pc.gov.au/__data/assets/pdf_file/0006/86496/intangible-investment.pdf [November 2012].

Barro, R., and X. Sala-I-Martin. (1997). Technological diffusion, convergence, and growth. *Journal of Economic Growth* 2(1):1-27.

Becker, R. (2010). International branch campuses: New trends and directions. *International Higher Education* 58(Winter):2-3.

Belleflamme, P., P. Picard, and J.-F. Thisse. (2000). An economic theory of regional clusters. *Journal of Urban Economics* 48(1):158-184.

Bernanke, B. (2011). *Promoting Research and Development: The Government's Role*. Presented at the Conference on New Building Blocks for Jobs and Economic Growth, May 16, Washington, DC.

Bjelland, M., B. Fallick, J. Haltiwanger, and E. McEntarfer. (2011). Employer-to-employer flows in the United States: Estimates using linked employer-employee data. *Journal of Business and Economic Statistics* 29(4):493-505.

Blei, D.M., A.Y. Ng, and M.I. Jordan. (2003). Latent Dirichlet allocation. *Journal of Machine Learning Research* 3:993-1022. Available: http://www.cs.princeton.edu/~blei/papers/BleiNgJordan2003.pdf [January 2012].

Borjas, G.J. (2005). The labor-market impact of high-skill immigration. *American Economic Review* 95(2):56-60.

Bornmann, L., and W. Marx. (2013). How good is research really? *EMBO Reports* 14(3):226-230.

Borousch, M. (2010, October). *NSF Releases New Statistics on Business Innovation. InfoBrief* NSF 11-300. Arlington, VA: National Science Foundation. Available: http://www.nsf.gov/statistics/infbrief/nsf11300/ [December 2012].

Boyd, D., and K. Crawford. (2011). *Six Provocations for Big Data*. A decade in Internet time: Symposium on the dynamics of the Internet and society. Available: http://ssrn.com/abstract=1926431 [December 2011].

Braunerhjelm, P., and M. Henrekson. (2013). Entrepreneurship, institutions, and economic dynamism: Lessons from a comparison of the United States and Sweden. *Industrial and Corporate Change* 22(1):107-130.

Bresnahan, T.F., and M. Trajtenberg. (1995). General purpose technologies "engines of growth." *Journal of Econometrics, Annals of Econometrics* 65(1):83-108.

Brooks, T.C., S. Carli, S., Dallmeier-Tiessen, S., Mele, S., and H. Weiler. (2011). Authormagic in INSPIRE in author disambiguation in scholarly communication. *Proceedings of the ACM Web Science* 11:1-2. Koblenz, Germany. Available: http://www.websci11.org/fileadmin/websci/Posters/158_paper.pdf [December 2011].

Burrelli, J. (2010, July). *Foreign Science and Engineering Students in the United States. InfoBrief* NSF 10-324. Arlington, VA: National Science Foundation. Available: http://www.nsf.gov/statistics/infbrief/nsf10324/nsf10324.pdf [December 2011].

Card, D. (2009). How Immigration Affects U.S. Cities. In *Making Cities Work: Prospects and Policies for Urban America*, R.P. Inman, ed., pp. 158-200. Princeton, NJ: Princeton University Press.

Carson, C.S., B.T. Grimm, and C.E. Moylan. (1994). A satellite account for research and development. *Survey of Current Business* 74(11):37-71.

Chesbrough, H., W. Vanhaverbeke, and J. West, eds. (2008). *Open Innovation: Researching a New Paradigm*. United Kingdom: Oxford University Press.

Clark, G.L., M.S. Gertler, and M.P. Feldman, eds. (2003). *The Oxford Handbook of Economic Geography*. United Kingdom: Oxford University Press. Available: http://EconPapers.repec.org/RePEc:oxp:obooks:9780199250837 [December 2012].

Coates, J.F. (1985). Foresight in federal government policy making. *Futures Research Quarterly* 1(2):29-53.

The Conference Board. (2012). *Global Business Cycle Indicators*. Available: http://www.conference-board.org/data/bcicountry.cfm?cid=1 [December 2012].

Cooke, P., and O. Memedovic. (2003). *Strategies for Regional Innovation Systems: Learning Transfer and Applications*. Policy papers. Vienna: United Nations Industrial Development Organization.

Cornell University, INSEAD, and World Intellectual Property Organization. (2013). *The Global Innovation Index 2013: The Local Dynamics of Innovation*, S. Dutta and B. Lanvin eds. Geneva, Switzerland: World Intellectual Property Organization.

Corrado, C.A., C. Hulten, and D. Sichel. (2005). Measuring capital and technology: An expanded framework. In *Measuring Capital in the New Economy*, C. Corrado, J. Haltiwanger, and D. Sichel, eds., pp. 11-46. Chicago, IL: University of Chicago Press.

Corrado, C.A., C. Hulten, and D. Sichel. (2006). *Intangible Capital and Economic Growth*. Finance and Economics Discussion Series No. 24. Washington, DC: Division of Research and Statistics and Monetary Affairs, Federal Reserve Board.

Corrado, C.A., J. Haskel, C. Jona-Lasinio, and M. Iommi. (2012). *Intangible Capital and Growth in Advanced Economies: Measurement Methods and Comparative Results*. IZA Discussion Paper No. 6733. Available: http://ssrn.com/abstract=2114913 [September 2012].

Crépon, B., E. Duguet, and J. Mairesse. (1998). Research and development, innovation, and productivity: An econometric analysis at the firm level. *Economics of Innovation and New Technology* 7(2):115-158.

Dahlman, C.J. (2012). *The Changing Geography of Innovation: The Rise of the BICs—Challenges and Opportunities*. Policy brief, No.6. Available: http://www.growthdialogue.org/sites/default/files/publication/documents/GD_Policy_Brief_No._6_5-23-12.pdf [July 2013].

David, P.A. (1990). The dynamo and the computer: An historical perspective on the modern productivity paradox. *The American Economic Review* 80(2):355-361.

David, P. A. (2010). Introductory comments on innovation incentives, institutions, and economic growth. In *The Rate and Direction Inventive Activity Revisited*, J. Lerner and S. Stern, eds. University of Chicago Press (forthcoming).

David, P.A., D. Foray, and B. Hall. (2011). *Measuring Smart Specialisation: The Concept and the Need for Indicators*. Available: http://cemi.epfl.ch/files/content/sites/cemi/files/users/178044/public/Measuring%20smart%20specialisation.doc [April 2014].

Davis, T., and D.M. Hart. (2010). International cooperation to manage high-skill migration: The case of India-U.S. relations. *Review of Policy Research* 27(4):509-526.

Defoort, C. (2008). Long-term trends in international migration: An analysis of the six main receiving countries. *Population-E* 63(2):285-318.

Delbecque, V., and A. Bounfour. (2011). *Intangible Investment: Contribution to Growth and Innovation Policy Issues*. The European chair on intellectual capital management Working Paper Series, No. 2011-1A. Available: http://survey.peterboswell.net/media/504/delbeque_bonfour_2010.pdf [November 2012].

Department of Business Innovation and Skills. (2011). *Innovation and Research Strategy for Growth*. Available: http://www.bis.gov.uk/assets/biscore/innovation/docs/e/11-1386-economics-innovation-and-research-strategy-for-growth.pdf [May 2012].

Deutsche Bank Research. (2011). *International Division of Labour in R&D: Research Follows Production*. Available: http://www.dbresearch.de/PROD/DBR_INTERNET_EN-PROD/PROD0000000000269346.pdf [December 2011].

Dunn, A., E. Liebman, and A. Shapiro. (2012). *Developing a Framework for Decomposing Medical-Care Expenditure Growth: Exploring Issues of Representativeness*. Bureau of Economic Analysis Working Paper No. 0086. Available: http://www.bea.gov/papers/pdf/OverTimeRepresent.pdf [December 2012].

The Economist. (2010a). *Data, Data Everywhere*. Available: http://www.economist.com/node/15557443 [December 2011].

The Economist. (2010b). *All Too Much. Monstrous Amounts of Data*. Available: http://www.economist.com/node/15557421 [December 2012].

The Economist. (2011). *Where the Geeks Go*. Available: http://www.economist.com/blogs/babbage/2011/09/big-data [December 2011].

The Economist. (2012). *A Pixelated Portrait of Labor. LinkedIn Offers a New Way to Look at Employment*. Available: http://www.economist.com/node/21549948 [December 2012].

Edelman, B., M. Ostrovsky, and M. Schwarz. (2007). Internet advertising and the generalized second-prize auction: Selling billions of dollars worth of keywords. *American Economic Review* 97(1):242-259.

Edquist, C. (2005). Systems of Innovation: Perspectives and Challenges. In *The Oxford Handbook of Innovation*, J. Fagerberg, D.C. Mowery, and R.R. Nelson, eds., pp. 181-208. Oxford: Oxford University Press.

EMC Corporation. (2011). *New Global Study: Only One-Third of Companies Making Effective Use of Data*. Available: http://www.emc.com/about/news/press/2011/20111205-02.htm?pid=home-datascientist pr-120511 [December 2011].

European Commission. (2009). *System of National Accounts 2008*. United Nations, New York. Available: http://unstats.un.org/unsd/national account/docs/SNA2008.pdf [March 2012].

European Commission. (2010). *Community Innovation Statistics*. Available: http://epp.eurostat.ec.europa.eu/portal/page/portal/microdata/cis [December 2012].

European Union. (2013). *Innovation Union Scoreboard, 2013*. European Commission. Available: http://ec.europa.eu/enterprise/policies/innovation/files/ius-2013_en.pdf [July 2013].

Eurostat. (2003). *Quality Assessment of Administrative Data for Statistical Purposes*. Presented at working group "Assessment of Quality in Statistics," sixth meeting, Luxembourg, October 2-3.

Eurostat. (2010). *Sixth Community Innovation Survey: More than Half of EU27 Enterprises are Innovative*. One in nine innovative enterprises cooperate with European partners. News release 166/2010, November 10. Available: http://epp.eurostat.ec.europa.eu/cache/ITY_PUBLIC/9-10112010-AP/EN/9-10112010-AP-EN.PDF [May 2012].

Eurostat. (2012). *The Community Innovation Survey 2012 (CIS 2012): The Harmonized Survey Questionnai*re. Luxemburg: Eurostat.

Eurostat. (2013). *Eurostat 2010 Statistics*. Available: http://epp.eurostat.ec.europa.eu/portal/page/portal/statistics/themes [December 2011].

Ewing Marion Kauffman Foundation. (2012). *Connecting Outcomes Measures in Entrepreneurship, Technology, and Science (COMETS)*. Available: http://www.kauffman.org/comets/ [December 2012].

Falck, O., and S. Heblich. (2008). Modern location factors in dynamic regions. *European Planning Studies* 16(10):1385-1403.

Federal Statistics. (2007). *List of Statistical Agencies*. Available: http://www.fedstats.gov/agencies/agencies.html [January 2013].

Feldman, M. (2013a). *Patent Dataverse Network*. Available: http://maryannfeldman.web.unc.edu/data-sources/rd-and-innovation/patent-dataverse-network/ [January 2013].

Feldman, M. (2013b). *STAR METRICS*. Available: http://maryannfeldman.web.unc.edu/data-sources/business-creation-development/star-metrics/ [December 2012].

Feser, E.J., and E.M. Bergman. (2000). National industry cluster templates: A framework for applied regional cluster analysis. *Regional Studies* 34(1):1-19. Available: http://www.tandfonline.com/doi/abs/10.1080/00343400050005844#preview [January 2013].

Finin, T., W. Murnane, A. Karandikar, N. Keller, J. Martineau, and M. Dredze. (2010). *Annotating Named Entities in Twitter Data with Crowdsourcing*. Proceedings of the NAACLHLT 2010 Workshop on Creating Speech and Language Data with Amazon's Mechanical Turk, CSLDAMT, pp. 80-88, Stroudsburg, PA. Association for Computational Linguistics. Available: http://www.cs.jhu.edu/~mdredze/publications/amt_ner.pdf [December 2011].

Franzoni, C., G. Scellato, and P. Stephan. (2012). Foreign-born scientists: Mobility patterns for 16 countries. *Nature Biotechnology* 30(12):1250-1253.

Fraumeni, B., and S. Okubo. (2005). R&D in the National Income and Product Accounts: A first look at its effect on GDP. In *Measuring Capital in the New Economy*, C. Corrado, J. Haltiwanger, and D. Sichel, eds., pp. 275-316. Chicago, IL: The University of Chicago Press.

Freeman, C. (1995). The "National System of Innovation" in historical perspective. *Cambridge Journal of Economics* 19(1):5-24.

Freeman, P., and W. Aspray. (1999). *The Supply of Information Technology Workers in the United States*. Washington, DC: Computing Research Association. Available: http://archive.cra.org/reports/wits/it_worker_shortage_book.pdf [December, 2011].

Fukao, K., S. Hamagata, T. Miyagawa, and K. Tonogi. (2007). *Intangible Investment in Japan: Measurement and Contribution to Economic Growth*. Discussion Paper No.07-E-034. Tokyo, Japan: Research Institute of Economy, Trade, and Industry.

Fukao, K., T. Miyagawa, K. Mukai, Y. Shinoda, and K. Tonogi, K. (2009). Intangible investment in Japan: Measurement and contribution to economic growth. *Review of Income and Wealth* 55(3):717-736.

Fukao, K., S. Hisa, and T. Miyagawa. (2012). *The Measurement of Intangible Investment by Industry and Its Role in Productivity Improvements in Japan*. Paper presented at the Workshop on Intangible Investment, Innovation, and Productivity at the National Institute of Science and Technology Policy (NISTEP), Tokyo, Japan.

Furman, J.L., M.E. Porter, and S. Stern. (2002). The determinants of national innovative capacity. *Research Policy* 31(6):899-933.

Gantz, J.F., C. Chute, A. Manfrediz, S. Minton, D. Reinsel, W. Schlichting, and A. Toncheva. (2008). *The Diverse and Exploding Digital Universe. An Updated Forecast of Worldwide Information Growth Through 2011*. IDC White Paper. Available: http://www.emc.com/collateral/analyst-reports/diverse-exploding-digital-universe.pdf [December 2011].

Gault, F. (2010). *Innovation Strategies for a Global Economy: Development, Implementation, Measurement, and Management*. Northampton, MA: Edward Elgar.

Gertler, M.S., P. Oinas, M. Storper, and P. Scranton. (1995). Discussion of regional advantage: Culture and competition in Silicon Valley and Route 128, by AnnaLee Saxenian. *Economic Geography* 71(2):199-207.

Gil, V., and J. Haskel. (2008). *Industry-level Expenditure on Intangible Assets in the UK*. Working Paper. Available: http://www.ralfmartin.me.uk/pub/CoInvest/CoinvestGilHaspaper/Intang_Report_for_DTI_Nov08-1.pdf [July 2011].

Gillick, D. (2012). *Predicting Movie Ratings: The Math That Won the Netflix Prize*. Available: http://www.science20.com/random_walk/predicting_movie_ratings_math_won_netflix_prize [January 2012].

Godin, B. (2008). *The Making of Statistical Standards: The OECD and the Frascati Manual, 1962-2002*. Working Paper No. 39. Available: http://www.csiic.ca/PDF/Godin_39.pdf [January 2013].

The Government of Abu Dhabi. (2008). *The Abu Dhabi Economic Vision 2030*. Available: http://www.tdic.ae/en/media/get/20110814_economic-vision-2030-executive-summary-mandate2propertypdf.pdf [January 2013].

Griliches, Z. (1958). Research cost and social returns: Hybrid corn and related innovations. *Journal of Political Economy* 66(5):419-431.

Griliches, Z. (1974). Errors in variables and other unobservables. *Econometrica* 42(6):971-998.

Griliches, Z. (1986). Economic data issues. In *Handbook of Econometrics*, Z. Griliches and M. Intriligator, eds., pp. 1465-1514. Amsterdam and New York: Elsevier.

Griliches, Z. (1998). Issues in assessing the contribution of research and development to productivity growth. In *R&D and Productivity: The Econometric Evidence*, pp. 17-45. Chicago, IL: University of Chicago Press.

Grimes, S. (2011). *Unstructured Data and the 80 Percent Rule*. Available: http://clarabridge.com/default.aspx?tabid=137&ModuleID=635&ArticleID=551 [December 2012].

Groves, R.M. (2011a). *A Possible Data Future for the Observational Social Sciences*. Paper presented at the COSSA 30th anniversary meeting, Washington, DC. Available: http://www.cossa.org/annualmtg/2011/Robert-Groves.pdf [December 2011].

Groves, R.M. (2011b). Data futures for the observational social sciences. *COSSA Update*, 30(20B), 13-14. Available: http://www.cossa.org/volume30/Update_Volume_30_Issue_20-B.pdf [December 2011].

Groves, R.M., F.J. Fowler, Jr., M.P. Couper, J.M. Lepkowski, E. Singer, and R. Tourangeau. (2009). *Survey Methodology* (Second Edition). Hoboken, NJ: John Wiley & Sons. Available: http://www.wiley.com/WileyCDA/WileyTitle/productCd-0470465468.html [December 2012].

Haak, L.L., D. Baker, D.K. Ginther, G.J. Gordon, M.A. Probus, N. Kannankutty, and B.A. Weinberg. (2012). Standards and infrastructure for innovation data exchange. *Science* 338(6104):196-197. Available: http://www.sciencemag.org/content/338/6104/196.summary [January 2013].

Hall, B.H. (2011). *Innovation and Productivity*. NBER Working Paper No.17178. Cambridge, MA: National Bureau of Economic Research.

Hall, B.H., and A.B. Jaffe. (2012). *Measuring Science, Technology, and Innovation: A Review*. Report commissioned for the Panel on Developing Science, Technology, and Innovation Indicators for the Future, Committee on National Statistics, Division of Behavioral and Social Sciences and Education, Washington, DC.

Hall, B.H., J. Mairesse, and P. Mohnen. (2010). Measuring the returns to R&D. In *Handbook of the Economics of Innovation*, B.H. Hall and N. Rosenberg, eds., pp. 1032-1080. Amsterdam and New York: Elsevier.

Hill, C.T. (2007). The post-scientific society. *Issues in Science and Technology* Fall:78-88. Available: http://www.issues.org/24.1/c_hill.html [September 2011].

Hollanders, H. (2013). Measuring regional innovation: A European perspective. In *Global Innovation Index 2013: The Local Dynamics of Innovation*, S. Dutta and B. Lanvin, eds., pp. 79-86. Geneva, Switzerland: World Intellectual Property Organization.

Horrigan, M.W. (2012). *Big Data and Official Statistics*. Presented at National Association for Business Economics Annual Meeting, New York.

Horrigan, M.W. (2013). *Big Data: A Perspective from the BLS*. Amstat News. Available: http://magazine.amstat.org/blog/2013/01/01/sci-policy-jan2013/ [January 2013].

Huang, N., and E. Diewert. (2011). Estimation of R&D depreciation rates: A suggested methodology and preliminary application. *Canadian Journal of Economics/Revue Canadienne D'Économique* 44(2):387-412.

Hulten, C.R., and J.X. Hao. (2012). *The Role of Intangible Capital in the Transformation and Growth of the Chinese Economy*. Working Paper No.w18405, Cambridge, MA: National Bureau of Economic Research.

Hunt, J., and M. Gauthier-Loiselle. (2010). How much does immigration boost innovation? *American Economic Journal: Macroeconomics* 2(2):31-56.

Hunter, P. (2013). Brain drain, brain gain, or brain sharing? New studies of the migration routes of scientists show that international mobility benefits all parties including countries that are net exporters of researchers. *EMBO Reports* 14(4). Available: http://embor.embopress.org/content/14/4/315 [April 2014].

International Monetary Fund. (2011). *Sixth Edition of the IMF's Balance of Payments and International Investment Position Manual*. Available: http://www.imf.org/external/pubs/ft/bop/2007/bopman6.htm [December 2011].

Jackson, S.A. (2001). *The Quiet Crisis: Falling Short in Producing American Scientific and Technical Talent*. BEST, San Diego, CA. Available: http://www.rpi.edu/homepage/quietcrisis/Quiet_Crisis.pdf [December 2011].

Jalava, J., P. Aulin-Ahmavaara, and A. Alanen. (2007). *Intangible Capital in the Finnish Business Sector, 1975-2005*. Discussion Paper No.1103. Helsinki, Finland: The Research Institute of the Finnish Economy.

Jankowski, J. (2012). *Business Use of Intellectual Property Protection Documented in NSF Survey. InfoBrief* NSF 12-307. Arlington, VA: National Science Foundation. Available: http://www.nsf.gov/statistics/infbrief/nsf12307/nsf12307.pdf [March 2012].

Jaumotte, F., and N. Pain. (2005). *From Ideas to Development: The Determinants of R&D and Patenting*. Economics Department Working Paper No. 457. Paris, France: OECD Publishing. Available: http://search.oecd.org/officialdocuments/displaydocumentpdf/?doclanguage=en&cote=ECO/WKP(2005)44 [March 2012].

Jenkins, T., B. Dahlby, A. Gupta, M. Leroux, D. Naylor, and N. Robinson. (2011). *Innovation Canada: A Call to Action*. Review of Federal Support to Research and Development—Expert Panel Report. Available: http://rd-review.ca/eic/site/033.nsf/eng/h_00287.html [January 2013].

Jorgenson, D.W., and F. Gollop. (1992). Productivity growth in U.S. agriculture: A postwar perspective. *American Journal of Agricultural Economics* 74(3):745-750.

Jorgenson, D.W., and Z. Griliches. (1967). The explanation of productivity change. *Review of Economic Studies* 34(3):249-283.

Jovanovic, B., and P.L. Rousseau. (2005). General purpose technologies. In *Handbook of Economic Growth*, P. Aghion and S. Durlauf, eds., pp. 1181-1224. Amsterdam and New York: Elsevier.

Kerr, W.R. (2008). Ethnic scientific communities and international technology diffusion. *The Review of Economics and Statistics* 90(3):518-537.

Khan, M., and H. Dernis. (2006). *Global Overview of Innovative Activities from the Patent Indicators Perspective*. STI Working Paper No. 2006/3. Available: http://www.oecd-ilibrary.org/docserver/download/5l9pscs8bmzp.pdf?expires=1360364065&id=id&accname=guest&checksum=672CE85DB36737CC5198BE447F1C3A93 [March 2012].

Lai, R., A. D'Mour, A., Yu, Y. Sun, V. Torvik, and L. Fleming. (2011). Disambiguation and Co-authorship Networks of the U.S. Patent Inventor Database. Available: http://scienceofsciencepolicy.net/system/files/usptodisamb8_18_2011.pdf [November 2011].

Lane, J., and L. Schwarz. (2012). *Creating New Administrative Data to Describe the Scientific Workforce: The STAR METRICS Program*. IZA Discussion Paper No. I6600. Bonn, Germany: Institute for the Study of Labor. Available: http://ftp.iza.org/dp6600.pdf [November 2012].

Lev, B. (2001). *Intangibles: Management, Measurement, and Reporting*. Washington, DC: The Brookings Institution Press.

Lev-Ram, M. (2011). *Data Scientist: The New Hot Gig in Tech*. CNN Money. Available: http://tech.fortune.cnn.com/2011/09/06/data-scientist-the-hot-new-gig-in-tech/ [December 2011].

Lewin, T. (2012). Introduction for masses knocks down campus walls. *The New York Times*, March, 4. Available: http://www.nytimes.com/2012/03/05/education/moocs-large-courses-open-to-all-topple-campus-walls.html?pagewanted=all&_r=0 [November 2012].

Library of Congress. (2012). *MARC Standards*. Available: http://www.loc.gov/marc [January 2012].

Link, A.N., and J.T. Scott (2003). U.S. science parks: The diffusion of an innovation and its effects on the academic missions of universities. *International Journal of Industrial Organization* 21(9):1323-1356.

Lipsey, R.G., K.I. Carlaw, and C.T. Bekar. (2005). *Economic Transformations: General Purpose Technologies and Long-Term Economic Growth*. United Kingdom: Oxford University Press. Available: http://ukcatalogue.oup.com/product/9780199290895.do [March 2012].

Litan, R.E. (2010). *In Defense of Much, But Not All, Financial Innovation*. Washington, DC: The Brookings Institution. Available: http://www.brookings.edu/research/papers/2010/02/17-financial-innovation-litan [January 2013].

Lohr, S. (2009). For today's graduate, just one word: Statistics. *The New York Times*. Available: http://www.nytimes.com/2009/08/06/technology/06stats.html?_r=1& [December 2012].

Lonmo, C. (2005). *Innovation in Selected Professional, Scientific, and Technical Services: Results from the Survey of Innovation 2003*. Ottawa, Ontario: Statistics Canada. Available: http://www.statcan.gc.ca/pub/88f0006x/88f0006x2005013-eng.pdf [January 2013].

Lowell, B.L., and H. Salzman. (2007). *Into the Eye of the Storm: Assessing the Evidence on Science and Engineering Education, Quality, and Workforce Demand*. Washington, DC: The Urban Institute. Available: http://www.urban.org/UploadedPDF/411562_Salzman_Science.pdf [December 2011].

Lundvall, B.-Å. (1988). Innovation as an interactive process: From user-producer interaction to the national innovation systems. In *Technology and Economic Theory*, G. Dosi, C. Freeman, R.R. Nelson, G. Silverberg, and L. Soete, eds., pp. 349-370. London, Pinter Publishers.

Lundvall, B.-Å. ed. (1992). *National System of Innovation. Towards a Theory of Innovation and Interactive Learning*. London: Pinter Publishers.

Lundvall, B.-Å. (2007). National innovation systems: Analytical concept and development tool. *Industry and Innovation* 14(1):95-119.

MADAR Research and Development. (2006). *United Arab Emirates Knowledge Economy 2006*. Available: http://www.madarresearch.com/default.aspx [January 2013].

Mairesse, J., and P. Mohnen. (2010). *Using Innovations Surveys for Econometric Analysis*. NBER Working Paper No. 15857. Cambridge, MA: National Bureau of Economic Research.

Maiya, A.S., and T.Y. Berger-Wolf. (2011). *Benefits of Bias: Towards Better Characterization of Network Sampling*. Available: http://arun.maiya.net/papers/maiya_etal-netbias.pdf [December 2011].

Mandel, M. (2012). *Beyond Goods and Services: The (Unmeasured) Rise of the Data-Driven Economy*. Washington, DC: Progressive Policy Institute. Available: http://www.progressivepolicy.org/2012/10/beyond-goods-and-services-the-unmeasured-rise-of-the-data-driven-economy/ [January 2013].

Mandel, M., and J. Scherer (2012). *The Geography of the App Economy*. Available: http://southmountaineconomics.files.wordpress.com/2012/11/the_geography_of_the_app_economy-f.pdf [November 2012].

Manyika, J., M. Chui, B. Brown, J. Bughin, R. Dobbs, C. Roxburgh, and A.H. Byers. (2011). *Big Data: The Next Frontier for Innovation, Competition, and Productivity*. Available: http://www.mckinsey.com/insights/mgi/research/technology_and_innovation/big_data_the_next_frontier_for_innovation [December 2011].

Marrano, M.G., J. Haskel, and G. Wallis. (2009). What happened to the knowledge economy? ICT, intangible investment, and Britain's productivity record revisited. *Review of Income and Wealth* 55(3):686-716.

Martin, B.R., and R. Johnston. (1999). Technology foresight for wiring up the national innovation system: Experiences in Britain, Australia, and New Zealand. *Technological Forecasting and Social Change* 60(1):37-54.

Marx, M., and L. Fleming. (2012). Non-compete agreements: Barriers to entry . . . and exit? *Innovation* 12(1):39-64.

Massachusetts Institute of Technology. (2013). *The Billion Prices Project*. Available: http://bpp.mit.edu/ [January 2012].

Matloff, N. (2003). On the need for reform of the H1-B non-immigrant work visa in computer-related occupations. *University of Michigan Journal of Law Reform* 36(4):1-99. Available: http://heather.cs.ucdavis.edu/Mich.pdf [December 2011].

Minh-Tam, L., J. Sweeney, B.M. Russett, and S.W. Zucker. (2012). *Structural Inference in Political Datasets*. Presented at the 2012 IEEE International Conference, Arlington ,Virginia.

Mislove, A., M. Marcon, K.P. Gummadi, P. Druschel, and B. Bhattacharjee. (2007). *Measurement and Analysis of Online Social Networks*. Available: http://conferences.sigcomm.org/imc/2007/papers/imc170.pdf [December 2012].

Miyagawa, T., and S. Hisa. (2012). *Estimates of Intangible Investment by Industry and Productivity Growth in Japan*. Paper presented at the 32nd General Conference of the International Association for Research in Income and Wealth, Boston, MA.

Molitor, N-T., N. Best, C. Jackson, and S. Richardson. (2009). Using Bayesian graphical models to model biases in observational studies and to combine multiple sources of data: Applications to low birth weight and water disinfection by-products. *Journal of the Royal Statistical Society: Series A* 72(3):615-637.

Mooney, G.M., and D.J. Foley. (2011). *Community Colleges: Playing an Important Role in the Education of Science, Engineering, and Health Graduates. InfoBrief* NSF 11-317. Arlington, VA: National Science Foundation. Available: http://www.nsf.gov/statistics/infbrief/nsf11317/nsf11317.pdf [December 2012].

Moris, F. (2009). R&D exports and imports: New data and methodological issues. In *International Flows of Invisibles: Trade in Services and Intangibles in the Era of Globalization*, M. Reinsdorf and M. Slaughter, eds., pp. 175-202. Chicago, IL: University of Chicago Press.

Moris, F., and N. Kannankutty, N. (2010). *New Employment Statistics from the 2008 Business R&D and Innovation Survey. InfoBrief* NSF 10-326. Arlington, VA: National Science Foundation. Available: http://www.nsf.gov/statistics/infbrief/nsf10326/ [December 2011].

Mowery, D.C. (1992). The U.S. national innovation system: Origins and prospects for change. *Research Policy* 21(2):125-144.

Mowery, D.C. (2010). Handbook of the economics of innovation. *Handbook of the Economics of Innovation* 2:1219-1256.

Muller, E., and A. Zenker. (2001). Business services as actors of knowledge transformation: The role of KIBS in regional and national innovation systems. *Research Policy* 30(9):1501-1516.

Myers, R.E., III, and J.T. Fouts. (2004). A cluster analysis of high school science classroom environments and attitude toward science. *Journal of Research in Science Training* 29(9):929-937.

Nakamura, L. (1999). *Intangibles: What Put the 'New' in the New Economy?* Available: http://www.phil.frb.org/research-and-data/publications/business-review/1999/july-august/brja99ln.pdf [November 2012].

NASSCOM, and Booz & Company. (2010*). Global ER&D: Accelerating Innovation with Indian Engineering*. Available: http://www.booz.com/media/uploads/NASSCOM_Booz_ESR_Report_2010.pdf [December 2011].

National Center for Education Statistics. (2012). *Integrated Postsecondary Education Data System*. Available: http://nces.ed.gov/ipeds/ [January 2013].

National Center for Health Statistics. (2013). *Baccalaureate and Beyond Longitudinal Study (B&B)*. Available: http://nces.ed.gov/surveys/b&b/ [January 2013].

National Center for Science and Engineering Statistics. (2013a). *Industrial Research and Development Information System*. Available: http://www.nsf.gov/statistics/iris/start.cfm [November 2012].

National Center for Science and Engineering Statistics. (2013b). *NSF Survey of Earned Doctorates: Tabulation Engine*. Available: https://ncses.norc.org/NSFTabEngine/#WELCOME [November 2012].

National Institutes of Health. (2012). *STAR METRICS*. Available: https://www.starmetrics.nih.gov/ [November 2012].

National Research Council. (2005). *Measuring Research and Development Expenditures in the U.S. Economy.* Panel on Research and Development Statistics at the National Science Foundation. Committee on National Statistics, Division of Behavioral and Social Sciences and Education. L.D. Brown, T.J. Plewes, and M.A. Gerstein, eds. Washington, DC: The National Academies Press.

National Research Council. (2007). *Innovation Inducement Prizes at the National Science Foundation*. Committee on the Design of an NSF Innovation Prize. Board of Science, Technology, and Economic Policy, and Policy and Global Affairs. Washington, DC: The National Academies Press.

National Research Council. (2008). *Using the American Community Survey for the National Science Foundation's Science and Engineering Statistics Programs*. Panel to Assess the Benefits of the American Community Survey for the National Science Foundation's Engineering Workforce Statistics Programs. Committee on National Statistics, Division of Behavioral and Social Sciences and Education. Washington, DC: The National Academies Press.

National Research Council. (2010). *Data on Federal Research and Development Investments: A Pathway to Modernization*. Panel on Modernizing the Infrastructure of the National Science Foundation Federal Funds Survey. Committee on National Statistics. Division of Behavioral and Social Sciences and Education. Washington, DC: The National Academies Press.

National Research Council. (2012). *Communicating Science and Engineering Data in the Information Age*. Panel on Communicating National Science Foundation Science and Engineering Information to Data Users. Committee on National Statistics and Computer Science and Telecommunications Board, Division of Behavioral and Social Sciences and Education. Washington, DC: The National Academies Press.

National Research Council. (2013a). *Nonresponse in Social Science Surveys: A Research Agenda*. Panel on a Research Agenda for the Future of Social Science Data Collection, Committee on National Statistics. Division of Behavioral and Social Sciences and Education. R. Tourangeau, and T.J. Plewes, eds. Washington, DC: The National Academies Press.

National Research Council. (2013b). *Principles and Practices for a Federal Statistical Agency: Fifth Edition*. Committee on National Statistics, Division of Behavioral and Social Sciences and Education. Washington, DC: The National Academies Press.

National Science Board. (2012a). *Science and Engineering Indicators 2012*. NSB 12/01. Arlington, VA: National Science Foundation. Available: http://www.nsf.gov/statistics/seind12/ [December 2012].

National Science Board. (2012b). *Science and Engineering Indicators 2012 Digest*. Arlington, VA: National Science Foundation. Available: http://www.nsf.gov/statistics/digest12/intro.cfm [December 2012].

National Science Board. (2012c*). Research and Development, Innovation, and the Science and Engineering Workforce*. Arlington, VA: National Science Foundation. Available: http://nsf.gov/nsb/publications/2012/nsb1203.pdf [December 2012].

National Science Foundation. (2004). *University-Industry Cooperative Research Programs in the Mathematical Sciences (UICRP)*. Available: http://www.nsf.gov/pubs/2005/nsf05504/nsf05504.txt [March 2012].

National Science Foundation. (2008). *Award Database*. Available: http://www.nsf.gov/awardsearch/piSearch.do?SearchType=piSearch&page=1&QueryText=%22text+mining%22+or+%22text+processing%22+or+%22entity+extraction%22+or+%22topic+extraction%22&PIFirstName=&PILastName=&PIInstitution=&PIState=&PIZip=&PICountry=&Restriction=2&Search=Search#results [December 2012].

National Science Foundation. (2010a). *Discovery in a Research Portfolio: Tools for Structuring, Analyzing, Visualizing, and Interacting with Proposal and Award Portfolios*. Available: http://www.nsf.gov/sbe/advcomm11_2010/CISE_SBE_ACResearchPortfolioAnalysis_FinalReport.pdf [November 2012].

National Science Foundation. (2010b). Research and development: National trends and international linkages. In *Science and Engineering Indicators: 2010*, Ch. 4. Arlington, VA: National Science Foundation. Available: http://www.nsf.gov/statistics/seind10/c4/c4s5.htm [June 2011].

National Science Foundation. (2011a). *Survey of Earned Doctorates*. Available: http://www.nsf.gov/statistics/srvydoctorates/ [November 2012].

National Science Foundation. (2011b). *Chapter II—Proposal Preparation Instructions*. Available: http://www.nsf.gov/pubs/policydocs/pappguide/nsf11001/gpg_2.jsp#IIC2diii [December 2011].

National Science Foundation. (2012a). *Fiscal Year 2013 Budget Request to Congress*. Available: http://www.nsf.gov/about/budget/fy2013/pdf/10-SBE_fy2013.pdf [October 2012].

National Science Foundation. (2012b). Appendix: Methodology and statistics. In *Science and Engineering Indicators 2012*. Arlington, VA: National Science Foundation. Available: http://www.nsf.gov/statistics/seind12/appmeth/appmeths1.htm [January 2013].

National Science Foundation. (2012c). Appendix tables. In *Science and Engineering Indicators 2012*. Arlington, VA: National Science Foundation. Available: http://www.nsf.gov/statistics/seind12/appendix.htm [January 2013].

National Science Foundation. (2012d). *Survey of Doctorate Recipients*. Available: http://www.nsf.gov/statistics/srvydoctoratework/ [December 2012].

National Science Foundation. (2012e). *WebCASPAR. Integrated Science and Engineering Resources Data System*. Available: https://webcaspar.nsf.gov/ [December 2012].

National Science Foundation. (2013a). *Business R&D and Innovation Survey*. Available: http://www.nsf.gov/statistics/srvyindustry/about/brdis/ [December 2012].

National Science Foundation. (2013b). *International Investment and R&D Data Link*. Available: http://www.nsf.gov/statistics/rdlink/ [December 2012].

National Science Foundation. (2013c). *National Patterns of R&D Resources: 2010-11 Data Update*. Available: http://www.nsf.gov/statistics/nsf13318/pdf/nsf13318.pdf [July 2013].

National Science Foundation. (2013d). *Scientists and Engineers Statistical Data System*. Available: http://www.nsf.gov/statistics/sestat/ [November 2012].

Nelson, R.R. (1993). *National Innovation Systems: A Comparative Analysis*. University of Illinois at Urbana-Champaign's Academy for Entrepreneurial Leadership Historical Research Reference in Entrepreneurship. Available: http://ssrn.com/abstract=1496195 [December 2012].

Newby, P.K., and K.L. Tucker. (2008). Empirically derived eating patterns using factor or cluster analysis: A review. *Nutrition Reviews* 62(5):177-203.

The New York Times. (2012). *The Big Three, at a Glance*. Available: http://www.nytimes.com/2012/11/04/education/edlife/the-big-three-mooc-providers.html?_r=0 [November 2012].

Nicholson, S. (2012). *LinkedIn Industry Trends: Winners and Losers during the Great Recession*. Available: http://blog.linkedin.com/2012/03/08/economic-report/ [December 2012].

Nishimura, M., D. Maurer, and X. Gao. (2009). Exploring children's face-space: Multidimensional scaling analysis of the mental representation of facial identity. *Journal of Experimental Child Psychology* 103(3):366-376.

The Noisy Channel. (2012). *Data Science at LinkedIn: My Team*. Available: http://thenoisychannel.com/2012/05/17/data-science-at-linkedin-my-team/ [December 2012].

OECD. (1992). *Oslo Manual: Guidelines for Collecting and Interpreting Innovation Data, First Edition*. Paris, France: OECD Publishing.

OECD. (1995). *Canberra Manual: The Measurement of Human Resources Devoted to S&T*. Paris, France: OECD Publishing. Available: http://www.tubitak.gov.tr/tubitak_content_files/BTYPD/kilavuzlar/Canberra.pdf [December 2012].

OECD. (2002). *Frascati Manual 2002: Proposed Standard Practice for Surveys on Research and Experimental Development*. Paris, France: OECD Publishing. Available: http://www.oecd.org/document/6/0,374,en_2649_34451_33828550_1_1_1_1,00.html [December 2012].

OECD. (2006). *KISA: Innovation and Knowledge-Intensive Service Activities*. Paris, France: OECD Publishing.

OECD. (2007). *Working Party of National Experts on Science and Technology Indicators*. Paris, France: OECD Publishing. Available: http://www.oecd.org/science/innovationinsciencetechnologyandindustry/38235147.pdf [January 2013].

OECD. (2008). *Compendium of Patent Statistics 2008*. Available: http://www.oecd.org/science/innovationinsciencetechnologyandindustry/37569377.pdf [March 2012].

OECD. (2009). *Innovation in Firms: A Microeconomic Perspective*. Paris, France: OECD Publishing. Available: http://www.keepeek.com/oecd/media/science-and-technology/innovation-in-firms_9789264056213-en [January 2013].

OECD. (2010). *Measuring Innovation: A New Perspective*. Paris, France: OECD Publishing. Available: http://www.oecd.org/document/22/0,3746,en_41462537_41454856_44979734_1_1_1_1,00.html [December 2011].

OECD. (2011). *OECD Science, Technology, and Industry Scoreboard 2011: Innovation and Growth in Knowledge Economies*. Paris, France: OECD Publishing. Available: http://www.oecd.org/document/10/0,3746,en_2649_33703_39493962_1_1_1_1,00.html#toc [December 2012].

OECD. (2012a). *New Sources of Growth Knowledge-Based Capital Driving Investment and Productivity in the 21st Century*. Interim project findings. Paris, France: OECD Publishing. Available: http://www.oecd.org/science/innovationinsciencetechnologyandindustry/50498841.pdf [November 2012].

OECD. (2012b). *Main Science and Technology Indicators (MSTI)*. Paris, France: OECD Publishing. Available: http://www.oecd.org/science/innovationinsciencetechnologyandindustry/mainscienceandtechnologyindicatorsmsti20122edition.htm [January 2013].

OECD. (2012c). *Knowledge Networks and Markets in the Life Sciences*. Paris, France: OECD Publishing.

OECD. (2013a). *Knowledge Flows in the Innovation System: Selected Indicators and Key Trends*. IPR strategies and policies in the era of knowledge networks and markets (KNM), OECD, Directorate for Science, Technology and Industry (OECD DSTI).

OECD. (2013b). *Metadata Application*. Available: http://webnet.oecd.org/rd_gbaord_metadata/Default.aspx [December 2012].

OECD-Eurostat. (1997). *Oslo Manual: Guidelines for Collecting and Interpreting Innovation Data, Second Edition*. Paris, France: OECD Publishing.

OECD-Eurostat. (2005). *Oslo Manual: Guidelines for Collecting and Interpreting Innovation Data, Third Edition*. Paris, France: OECD Publishing.

OECD-Eurostat. (2008). *Manual on Business Demography Statistics*. Paris, France: OECD Publishing.

Offit, P.A. (2005). Why are pharmaceutical companies gradually abandoning vaccines? *Health Affairs* 24(3):622-630. Available: http://content.healthaffairs.org/content/24/3/622.full.html [May 2013].

Okubo, S. (2012). *Payments and Receipts for R&D Services: Outsourcing US R&D*. Paper presented to the Panel on Developing Science, Technology, and Innovation Indicators for the Future, Washington, DC.

Online Computer Library Center. (2012). *OCLC WorldCat and Dissertations and Theses (WorldCatDissertations)*. Available: http://www.oclc.org/support/documentation/firstsearch/databases/dbdetails/details/WorldCatdissertations.htm [December 2012].

The Partnership for a New American Economy and The Partnership for New York City. (2012). *Not Coming to America: Why the U.S. Is Falling Behind in the Global Race for Talent*. Available: http://www.renewoureconomy.org/sites/all/themes/pnae/not-coming-to-america.pdf [December 2011].

Patterson. (2009). *Article-level Metrics at PLoS—Addition of Usage Data*. Available: http://blogs.plos.org/plos/2009/09/article-level-metrics-at-plos-addition-of-usage-data/ [November 2012].

Payne, G.T., C.B. Moore, S.E. Griffis, and C.W. Autry. (2011). Multilevel challenges and opportunities in social capital research. *Journal of Management* 37(2):491-520.

Pece, C. (2012). *NCSES Administrative Records Project (ARP)*. Presented to the Advisory Committee for Social, Behavioral, and Economic Sciences, National Science Foundation, Arlington, VA.

Petersen, D.R., and N. Ahmad. (2007). *High-Growth Enterprises and Gazelles—Preliminary and Summary Sensitivity Analysis*. Paris, France: OECD Publishing. Available: http://www.oecd.org/dataoecd/47/4/39639605.pdf [December 2011].

Phelps, C., R. Heidl, and A. Wadhwa. (2012). Knowledge, networks, and knowledge networks: A review and research agenda. *Journal of Management* 38(4):1115-1166.

Pianta, M. (2005). Innovation and employment. In *The Oxford Handbook of Innovation*, J. Fagerberg, D. Mowery, and R. Nelson, eds., pp. 568-598. New York: Oxford University Press.

PLOS ONE. (2012). *Article-Level Metrics Information*. Available: http://www.plosone.org/static/almInfo.action [November 2012].

Popp, D. (2010). *Innovation and Climate Policy*. NBER Working Paper Series. Working Paper No. 15673. Available: http://www.nber.org/papers/w15673 [January 2013].

Porter, M. (1990). *The Competitive Advantage of Nations*. New York: The Free Press, Macmillan, Inc.

Priem, J., and K.L. Costello. (2010). *How and Why Scholars Cite on Twitter*. Available: https://docs.google.com/present/view?id=0ASyDkfrsAcUjZGRmZzc4N2NfMjIwZ2N6NXRrYzg [March 2012].

Priem, J., and B.M. Hemminger. (2010). Scientometrics 2.0: Toward new metrics of scholarly impact on the social web. *First Monday* 15(7-5). Available: http://firstmonday.org/htbin/cgiwrap/bin/ojs/index.php/fm/article/viewArticle/2874/2570 [December 2011].

ProQuest. (2012). *ProQuest Dissertations and Theses Database*. Available: http://www.proquest.com/en-US/catalogs/databases/detail/pqdt.shtml [December 2012].

Regets, M. (2010). *Employment in Science and Engineering Occupations Reached 5.8 Million in 2008. InfoBrief* NSF 10-315. Arlington, VA: National Science Foundation. Available: http://www.nsf.gov/statistics/infbrief/nsf10315/nsf10315.pdf [January 2013].

Rogoff, K. (2011). *Processed Food Very Tasty for Producers*. Available: http://www.thenational.ae/thenationalconversation/industry---insights/economics/processed---food---very---tasty---for---producers [May 2013].

Rogoff, K. (2012). Is modern capitalism sustainable? *International Economy* 26(1):60. Available: http://relooney.fatcow.com/0_New_12809.pdf [December 2012].

Romer, P.M. (1990). Endogenous technological change. *Journal of Political Economy* 98(5):S71-S102.

Rosenberg, N. (1982). *Inside the Black Box: Technology and Economics*. Cambridge: Cambridge University Press.

Rosvall, M., and C. Bergstrom. (2008). Maps of random walks on complex networks reveal community structure. *Proceedings of the National Academy of Sciences* 105(4):1118-1123. Available: http://octavia.zoology.washington.edu/publications/RosvallAndBergstrom08.pdf [March 2012].

The Royal Society. (2011). *Knowledge, Networks and Nations: Global Scientific Collaboration in the 21st Century.* United Kingdom. Available: http://royalsociety.org/policy/projects/knowledge-networks-nations/report/ [June 2013].

Ruegg, R., and I. Feller. (2003). *A Toolkit for Evaluating Public R&D Investment: Models, Methods, and Findings from ATP's First Decade*. Gaithersburg, MD: U.S. Department of Commerce, Technology Administration, National Institute of Standards and Technology.

Ryanair. (2010). *Ryanair Wins Screenscraper High Court Action*. Available: http://www.ryanair.com/en/news/ryanair-wins-screenscraper-high-court-action [December 2011].

Saka, A., M. Igami, and T. Kuwahara. (2010). *Science Map 2010: Study on Hot Research Areas (2003-2008) by Bibliometric Method*. Japan: National Institute of Science and Technology Policy (NISTEP), Ministry of Education, Culture, Sports, Science, and Technology (MEXT).

Sampat, B.N. (2011). The impact of publicly funded biomedical and health research: A review. In National Research Council, *Measuring the Impacts of Federal Investments in Research: A Workshop Summary* (Appendix D). Washington, DC: The National Academies Press.

Sampat, B.N., and R. Lichtenberg. (2011). What are the respective roles of the public and private sectors in pharmaceutical innovation? *Health Affairs* 30(2):332-339.

Samuelson, P. (1962). The gains from international trade once again. *The Economic Journal* 72(288):820-829.

Saxenian, A. (1996). *Regional Advantage: Culture and Competition in Silicon Valley and Route 128*. Cambridge, MA: Harvard University Press.

Saxenian, A. (2002). Silicon Valley's new immigrant high-growth entrepreneurs. *Economic Development Quarterly* 16(1):20-31.

Schumpeter, J.A. (1934). *Theory of Economic Development. 1st ed. 1911*. Cambridge, MA: Harvard University Press.

Shachar, A. (2006). The race for talent: Highly skilled migrants and competitive immigration regimes. *New York University Law Review* 81(1):148-206.

Shanks, S., and S. Zheng. (2006). *Econometric Modelling of R&D and Australia's Productivity*. Available: http://www.pc.gov.au/__data/assets/pdf_file/0004/37183/economicmodelling.pdf [December 2012].

Smalheiser, N.R., and V.I. Torvik. (2009). *Author Name Disambiguation*. Available: http://arrowsmith.psych.uic.edu/arrowsmith_uic/tutorial/ARIST_preprint.pdf [March 2012].

Soete, L. (2012). *Maastricht Reflections on Innovation*. Working Paper No. 2012-001. Maastricht, Netherlands: UNU-MERIT.

Soloveichik, R.H. (2010a). Artistic originals. *American Economic Review* 100(2):110-114.

Soloveichik, R.H. (2010b). *Miscellaneous Artwork as a Capital Asset*. Washington, DC: Bureau of Economic Analysis.

Soloveichik, R.H. (2010c). *Music as a Capital Asset*. Washington, DC: Bureau of Economic Analysis.

Soloveichik, R.H. (2010d). *Theatrical Movies as Capital Assets*. Washington, DC: Bureau of Economic Analysis.

Soloveichik, R.H. (2011a). *Research Spotlight: Artistic Originals as Capital Assets*. Washington, DC: Bureau of Economic Analysis. Available: http://www.bea.gov/scb/pdf/2011/06%20June/0611_artistic.pdf [April 2012].

Soloveichik, R.H. (2011b). *Television Programs as Capital Assets*. Washington, DC: Bureau of Economic Analysis.

Solow, R.M. (1956). A contribution to the theory of economic growth. *The Quarterly Journal of Economics* 70(1):65-94.

Solow, R.M. (1957). Technical change and the aggregate production function. *Review of Economics and Statistics* 39(3):312-320.

Solow, R.M. (1994). Perspectives on growth theory. *The Journal of Economic Perspectives* 8(1):45-54.

Stanford News Service. (1995, June 5). *Doctorate Surplus in Science, Engineering Is Ongoing*. News Release. Available: http://www.stanford.edu/dept/news/pr/95/950605Arc5192.html [December 11, 2011].

Stephan, P.E. and S.G. Levin. (2001). Exceptional contributions to U.S. science by the foreign-born and foreign-educated. *Population Research and Policy Review* 20(1-2):59-79.

Stone, A., S. Rose, B. Lal, and S. Shipp. (2008). *Measuring Innovation and Intangibles: A Business Perspective*. Washington, DC: Science and Technology Policy Institute, Institute for Defense Analyses.

Tassey, G. (2011). *Beyond the Business Cycle: The Need for a Technology-Based Growth Strategy*. Gaithersburg, MD: Economic Analysis Office, National Institute of Standards and Technology.

Tett, G. (2009). *Fool's Gold: The Inside Story of J.P. Morgan and How Wall St. Greed Corrupted Its Bold Dream and Created a Financial Catastrophe*. New York: Free Press.

Thomson Reuters. (2012). *Web of Science*. Available: http://thomsonreuters.com/products_services/science/science_products/a-z/web_of_science/ [December 2012].

Tödtling, F., and M. Trippl. (2005). One size fits all? Towards a differentiated regional innovation policy approach. *Research Policy* 34(8):1203-1219.

Travis, I. (2013). *Association for Information Science and Technology*. Bulletin, editor's desktop. April/May. Available: http://www.asis.org/Bulletin/Apr-13/AprMay13_Editor.html [July 2013].

United Nations Educational, Scientific and Cultural Organization. (1978). *Recommendation Concerning the International Standardization of Statistics on Science and Technology*. Paris, France: United Nations Educational, Scientific and Cultural Organization. Available: http://www.uis.unesco.org/Libary/Documents/STSManual78_en.pdf [January 2013].

United Nations Educational, Scientific and Cultural Organization. (1984). *UNESCO Manual for Statistics on Scientific and Technological Activities*. Paris, France: United Nations Educational, Scientific and Cultural Organization. Available: http://www.uis.unesco.org/Library/Documents/STS Manual84_en.pdf [December 2012].

U.S. Bureau of Economic Analysis. (2010). *2010 Survey of Current Business. Research and Development Satellite Account Update: Estimates for 1959-2007*. Available: http://www.bea.gov/scb/pdf/2010/12%20 December/1210_r-d_text.pdf [March 2012].

U.S. Census Bureau. (2011a). *The Future of Producing Social and Economic Statistical Information, Part I*. Available: http://blogs.census.gov/directorsblog/2011/09/the-future-of-producing-social-and-economic-statistical-information-part-i.html [December 2011].

U.S. Census Bureau. (2011b). *The Future of Producing Social and Economic Statistical Information, Part II*. Available: http://directorsblog.blogs.census.gov/2011/09/20/the-future-of-producing-social-and-economic-statistical-information-part-ii/ [December 2011].

U.S. Census Bureau. (2011c). *The Future of Producing Social and Economic Statistical Information, Part III*. Available: http://directorsblog.blogs.census.gov/2011/10/18/the-future-of-producing-social-and-economic-statistical-information-part-iii/ [December 2011].

U.S. Census Bureau. (2012). *U.S. Census Return Rate Challenge (Visualization Competition)*. Available: https://www.kaggle.com/c/us-census-challenge [December 2012].

U.S. Census Bureau. (2013). *American Community Survey*. Available: http://www.census.gov/acs/www/methodology/sample_size_and_data_quality/ [December 2012].

U.S. Department of Commerce. (2011). *2010 Business R&D and Innovation Survey*. Available: http://www.nsf.gov/statistics/srvyindustry/about/brdis/surveys/srvybrdis_2010.pdf [December 2011].

U.S. Department of Commerce. (2012). *2011 Business R&D and Innovation Survey*. Available: http://www.nsf.gov/statistics/srvyindustry/about/brdis/surveys/srvybrdis_2011.pdf [December 2012].

U.S. Department of Homeland Security. (2011a). *H-1B Specialty Occupations, DOD Cooperative Research and Development Project Workers, and Fashion Models*. Available: http://www.uscis.gov/portal/site/uscis/menuitem.eb1d4c2a3e5b9ac89243c6a7543f6d1a/?vgnextoid=73566811264a3210VgnVCM100000b92ca60aRCRD&vgnextchannel=73566811264a3210VgnVCM100000b92ca60aRCRD [January 2012].

U.S. Department of Homeland Security. (2011b). *Appendix Table 3-19. Occupations of New H1B Visa Recipients: FY 2009*. Available: http://www.nsf.gov/statistics/seind12/append/c3/at03-19.pdf [January 2013].

U.S. Department of Homeland Security. (2012). *Characteristics of H-1B Specialty Occupation Workers, Fiscal Year 2011*. Annual report to Congress, October 1, 2010-September 30, 2011. Available: http://www.uscis.gov/USCIS/Resources/Reports%20and%20Studies/H-1B/h1b-fy-11-characteristics.pdf [July 2013].

U.S. Department of Labor. (2012a). *OFLC Case Disclosure Data*. Available: http://www.foreignlaborcert.doleta.gov/quarterlydata.cfm [November 2012].

U.S. Department of Labor. (2012b). *OFLC Annual Report, October, 1 2010-September 30, 2011*. Available: http://www.foreignlaborcert.doleta.gov/pdf/2011AR_FINAL.pdf [July 2013].

U.S. House of Representatives. (2010). *America COMPETES Reauthorization Act of 2010*. Available: http://www.gpo.gov/fdsys/pkg/BILLS-111hr5116enr/html/BILLS-111hr5116enr.htm [January 2013].

The University of Chicago. (2013). *Population Research Center*. Available: http://popcenter.uchicago.edu/data/ [December 2012].

Updegrove, A. (2012). *A Trip through the Looking Glass: Valuing Standards (and SEPs)*. Presented at OECD's Workshop on A Policy Framework for Knowledge-Based Capital, Washington, DC. Available: http://www.oecd.org/sti/industryandglobalisation/Updegrove.pdf [December 2012].

Van Reenen, J. (1997). Employment and technical innovation: Evidence from U.K. manufacturing firms. *Journal of Labor Economics* 15(2):255-284.

van Rooijen-Horsten, M., D. Van den Bergen, M. de Haan, A. Klinkers, and M. Tanriseven. (2008). *Intangible Capital in the Netherlands: Measurement and Contribution to Economic Growth*. Discussion Paper No. 08016. The Hague: Statistics Netherlands.

Wadhwa, V. (2009). *Foreign-Born Entrepreneurs: An Underestimated American Resource*. Available: http://www.kauffman.org/uploaded Files/WadhwaTBook09.pdf [July 2013].

Wehrmeyer, W., A. Clayton, and K. Lum. (2002). Foresighting for development. *Journal of Corporate Environmental Strategy and Practice* 37(2):24-36.

Weinstein, E. (unpublished). *How and Why Government, Universities, and Industry Create Domestic Labor Shortages of Scientists and High-Tech Workers*. Available: http://users.nber.org/~peat/PapersFolder/Papers/SG/NSF.html [December 2011].

West, J., and C. Bergstrom. (2010). How to improve the use of metrics. *Nature* 465:870-872.

Wikipedia. (2012). *Crowdsourcing*. Available: http://en.wikipedia.org/wiki/Crowdsourcing [December 2012].

Wildavsky, B. (2010). *The Great Brain Race: How Global Universities Are Reshaping the World*. Princeton, NJ: Princeton University Press.

Wilkinson, L., and M. Friendly. (2009). The history of the cluster heat map. *American Statistician* 63(2):179-184.

Willett, W., J. Heer, and A. Agrawala. (2012). Strategies for crowdsourcing social data analysis. In *ACM Human Factors in Computing Systems*, pp. 227-236. Berkeley, CA: University of California. Available: http://bid.berkeley.edu/files/papers/CrowdAnalytics-CHI2012(Preprint).pdf [December 2012].

Williams, D. (2013). Explaining rising nonresponse rates in cross-sectional surveys. *The ANNALS of the American Academy of Political and Social Science* 645(1):36-59.

The World Bank. (1997). *Expanding the Measure of Wealth: Indicators of Environmentally Sustainable Development*. Environmentally Sustainable Development Studies and Monographs Series No. 17. Washington, DC: The World Bank.

Young, F.W. (2013). *Multidimensional Scaling*. Available: http://forrest.psych.unc.edu/teaching/p208a/mds/mds.html [January 2013].

Young, F.W., and R.M. Hamer. (1987). *Multidimensional Scaling: History, Theory, and Applications*. Hillsdale, NJ: Lawrence Erlbaum Associates.

Zucker, L., and M.R. Darby. (2011). Legacy and new databases for linking innovation to impact. In *The Science of Science Policy: A Handbook*, K.H. Fealing, J. Lane, J.H. Marburger, III, and S. Shipp, eds. Stanford, CA: Stanford University Press.

Zucker, L.G., M.R. Darby, and J. Fong. (2011). *Communitywide Database Designs for Tracking Innovation Impact: COMETS, STARS, and Nanobank*. NBER Working Paper No. 17404. Available: http://www.nber.org/papers/w17404 [December 2011].

Appendix A

Biographical Sketches of Panel Members and Staff

Robert E. Litan (*Cochair*) is director of research at Bloomberg Government in Washington, DC. He assumed this position after serving for 9 years as vice president for research and policy at the Kauffman Foundation, where he oversaw the foundation's program for data collection and research on entrepreneurship and economic growth. For more than two decades, he was a senior fellow in economic studies at the Brookings Institution, where he served from 1996 to 2003 as vice president and as director of the Economic Studies Program. Dr. Litan has published widely in academic journals, as well as in magazines and newspapers. Since the onset of the nation's recent financial crisis, he has published a number of essays on financial reforms for the Brookings website. Previously, he served in several capacities in the federal government, including associate director of the Office of Management and Budget, deputy assistant attorney general in the Antitrust Division of the Department of Justice, and staff economist at the Council of Economic Advisers. He received a B.S. in economics (summa cum laude) from the Wharton School of Finance at the University of Pennsylvania, a J.D. from Yale Law School, and an M.Phil. and a Ph.D. in economics from Yale University.

Andrew W. Wyckoff (*Cochair*) is director of the OECD Directorate for Science, Technology, and Industry. Previously, he held other positions in OECD, including head of the Information, Computer and Communications Policy Division, which supports the organization's work on information as well as consumer policy issues, and head of the Economic Analysis and Statistics Division, which has responsibility for developing OECD's methodology and data on science, technology, and innovation activity. His experience prior to OECD includes serving as program manager of the Information, Telecommunications, and Commerce Program of the Office of Technology Assessment of the U.S. Congress and as an economist at the National Science Foundation. He holds an undergraduate degree from the University of Vermont and a degree in public policy from the Kennedy School of Government at Harvard University.

Carter Bloch is research director at the Danish Centre for Studies in Research and Research Policy, Department of Political Science and Government, at Aarhus University in Denmark. His research focuses on impacts of funding on research careers and performance; innovation measurement; knowledge spillovers; and the relationships among research and development, innovation, and economic performance. He has led a number of international and national projects concerning research evaluation, the development of innovation indicators, the measurement of innovation in public sector organizations, and microbased data analysis. He holds an undergraduate degree in economics from the University of Wisconsin–Madison and a Ph.D. in economics from Aarhus University in Denmark.

Nicholas R. Chrisman has been discipline head of geospatial sciences at RMIT University, Melbourne, Australia, since January 2013. For the prior 8 years, he was professor of geomatics sciences at the Université Laval, Canada, his principal assignment being scientific director of the GEOIDE Network. His research has concentrated on time in geographic information systems (GIS), data quality testing, and the social and institutional aspects of GIS. From 1987 to 2004, he was professor of geography at the University of Washington. From 1982 to 1987, he was assistant professor in the Department of Landscape Architecture at the University of Wisconsin–Madison. From 1972 to 1982, he was a programmer at the Harvard Lab for Computer Graphics. Dr. Chrisman participated in the design of prototype GIS software. He holds a Ph.D. from the University of Bristol (United Kingdom) based on research on error and statistics for categorical maps. For 30 years, his writing has focused on connecting the technical details of GIS to larger issues of philosophy and culture.

Carl J. Dahlman is Henry R. Luce professor of international relations and information technology at Edmund A. Walsh School of Foreign Service at Georgetown University. Previously, he worked at The World Bank, where he served as senior adviser to The World Bank Institute and managed an initiative providing training on the strategic use of knowledge for economic and social development to business leaders and policy makers in developing countries. He also has conducted extensive analytical work in major developing countries, including Argentina, Brazil, Chile, China, India, Korea, Malaysia, Mexico, Pakistan, the Philippines, Russia, Thailand, Turkey, and Vietnam, and he is currently working on studies on China, Finland, Japan, Korea, and Mexico. Dr. Dahlman holds a B.A. in international relations from Princeton University and a Ph.D. in economics from Yale University.

Geoff M. Davis is an analyst in the Quantitative Insights Group at Google. Previously, he held positions in the Mathematics Department at Dartmouth College; in the Electrical and Computer Engineering Department at Rice University; with the Signal Processing Group at Microsoft Research; as a developer at a start-up company; and at Sigma Xi, the Scientific Research Society. He also was a Wertheim fellow in the Labor and Worklife Program at Harvard Law School. Dr. Davis's mathematical research centers on representations of information, with a particular focus on wavelets and related transforms. He is a recipient of the Leon K. Kirchmayer Prize Paper Award of the Institute of Electrical and Electronics Engineers. He has had a long-standing interest in science education and policy issues and is a past member of the Science and Engineering Workforce Project of the National Bureau of Economic Research. He holds a Ph.D. in applied mathematics from the Courant Institute at New York University.

Katharine G. Frase is vice president, industries research, at IBM. She is responsible for working across IBM Research on behalf of IBM clients to create transformational industry-focused solutions, including the application of "Watson" technologies to business applications and the realization of Smarter Planet solutions. Prior to assuming this role, she was vice president, technical and business strategy, for the IBM Software Group, where she was responsible for technical strategy, business strategy, business development, standards, competitive analysis, and the application of advanced technologies across IBM's software business. Her past roles in IBM include corporate assignments in technology assessment and strategy, as well as roles in IBM Microelectronics; management of process development; design/modeling methodology; and the production of chip carriers, assemblies, and test. She is a member of the National Academy of Engineering and the IBM Academy of Technology and sits on numerous external committees and boards. Dr. Frase received an A.B. in chemistry from Bryn Mawr College and a Ph.D. in materials science and engineering from the University of Pennsylvania.

Barbara M. Fraumeni is chair of the Public Policy and Management Masters Program and the Ph.D. in Public Policy Program and professor of public policy at the Muskie School of Public Service, University of Southern Maine. She is also a special-term professor at the China Center for Human Capital and Labor Market Research of the Central University for Finance and Economics in Beijing and a research associate of the National Bureau of Economic Research. She previously served as chief economist of the Bureau of Economic Analysis at the U.S. Department of Commerce and was a research fellow of the Program on Technology and Economic Policy at the John F. Kennedy School of Government at Harvard University. Her research interests include measurement issues and national income accounting; human and nonhuman capital, productivity, and economic growth; market and nonmarket accounts; investment in education and research and development; and measurement of highway capital stock and the real output of government by function. She holds a B.A. from Wellesley College and a Ph.D. in economics from Boston College.

Richard B. Freeman holds the Herbert Ascherman chair in economics at Harvard University and is currently serving as faculty director of the Labor and Worklife Program at Harvard Law School. He also directs the National Bureau of Economic Research/Sloan Science Engineering Workforce Projects and is senior research fellow in labor markets at the London School of Economics' Centre for Economic Performance. He is a fellow of the American Academy of Arts and Sciences, and he is currently serving as a member of the Initiative for Science and Technology of the American Association for the Advancement of Science. Dr. Freeman is a recipient of the Mincer Lifetime Achievement Prize of the Society of Labor Economics and of the IZA Prize in Labor Economics from the Institute for the Study of Labor. He holds a B.A. from Dartmouth College and a Ph.D. from Harvard University.

Fredrick D. Gault is a professorial fellow with the United Nations University-Maastricht Economic and Social Research and Training Centre on Innovation and Technology (UNU-MERIT). He is also a professor extraordinaire at the Tshwane University of Technology in South Africa and a member of the university's Institute for Economic Research on Innovation. He worked with OECD as a member of the management team coordinating the OECD Innovation Strategy. Previously, he held a visiting fellowship at the Canadian International Development Research Centre, spent some years at Statistics Canada, and was a senior lecturer in theoretical physics at the University of Durham in the United Kingdom. At Statistics Canada, Dr. Gault directed the division responsible for the development of statistics on all aspects of research, development, invention, innovation, and the diffusion of technologies, as well as on related human resources. He was chair of the OECD Working Party of National Experts on Science and Technology Indicators and of the Working Party on Indicators for the Information Society. He is a fellow of the Institute of Physics and a member of the British Computer Society. He holds a Ph.D. in theoretical physics and a B.Sc. (economics) from the University of London.

David Goldston is director of government affairs at the National Resources Defense Council. He was a visiting lecturer at the Harvard University Center for the Environment and in the Science, Technology and Environment Program at Princeton University's Woodrow Wilson School of Public and International Affairs. Previously, he was chief of staff of the U.S. House of Representatives' Committee on Science, which has jurisdiction over much of the federal research and development budget, and legislative director for Representative Sherwood Boehlert of New York. He wrote the monthly column "Party of One" on Congress and science policy for the journal *Nature*. He graduated from Cornell University and completed the course work for a Ph.D. in U.S. history at the University of Pennsylvania.

Kaye Husbands Fealing *(Study Director)* is a member of the staff of the Committee on National Statistics. During the study, she was on leave from the Hubert H. Humphrey School of Public Affairs of the University of Minnesota, where she is a professor in the Center for Science, Technology and Public Policy. Previously, she was William Brough professor of economics at Williams College. At the National Science Foundation, she initiated and developed the agency's Science of Science and Innovation Policy Program, cochaired the Science of Science Policy Interagency Task Group, and served as an economics program director. Her research has included a study of the impact of the North American Free Trade Agreement on the Mexican and Canadian automotive industries and on strategic alliances between aircraft contractors and their subcontractors. She holds a B.A. in mathematics and economics from the University of Pennsylvania and a Ph.D. in economics from Harvard University.

Michael Mandel is chief economic strategist at the nonpartisan Progressive Policy Institute in Washington and a senior fellow at the Mack Center for Technological Innovation at the Wharton School, University of Pennsylvania. His main areas of study include the economic impact of the data-driven economy, the impact of regulation on innovation, and measurement issues connected with globalization and innovation. His current research focuses on new methodologies for tracking job creation in innovative industries and on whether new regulatory institutions can improve economic performance. He formerly served as chief economist at *Businessweek*, where he directed the magazine's coverage of the domestic and global economies, and he has received multiple awards for his articles on economic growth and innovation. He holds a Ph.D. in economics from Harvard University.

John E. Rolph is professor of statistics (emeritus) at the Marshall School of Business, University of Southern California, where he also holds appointments in the mathematics department and the law school. Previously, he was a statistician at the RAND Corporation and served as head of the statistical research and consulting group. His major areas of research include statistics and public policy and empirical Bayes estimation. He is an elected member of the International Statistical Institute, a fellow of the American Statistical Association, a fellow of the Institute of Mathematical Statistics, and a lifetime national associate of the National Academies. He is a past editor of *CHANCE* magazine and has served in many other editorial capacities. He holds an A.B. and a Ph.D. in statistics from the University of California, Berkeley.

Leland Wilkinson is vice president of data visualization at Skytree Inc. and adjunct professor of computer science at the University of Illinois at Chicago. Previously, he was executive vice president of SYSTAT Software Inc., a company he founded, and adjunct professor of statistics at Northwestern University. Dr. Wilkinson is a fellow of the American Statistical Association, an elected member of the International Statistical Institute, and a fellow of the American Association for the Advancement of Science, and served as vice-chair of the board of the National Institute of Statistical Sciences. He was also a member of the Committee on Applied and Theoretical Statistics of the National Academy of Sciences. His innovation indicators include books, journal articles, the original SYSTAT statistics package, and patents in visualization and distributed computing. He holds an A.B. from Harvard University, an S.T.B. from Harvard Divinity School, and a Ph.D. from Yale University.

Appendix B

Users of Science, Technology, and Innovation (STI) Data and Indicators and Their Questions and Requests for STI Indicators

USERS*

- Ana Aizcorbe (Virginia Tech, formerly at the Bureau of Economic Analysis)
- Shinichi Akaike (Hitotsubashi University)
- Jeff Alexander (SRI International)
- Michelle Alexopoulos (University of Toronto)
- Howard Alper (Canada's Science Technology and Innovation Council)
- Rob Atkinson (Information Technology and Innovation Foundation)
- B.K. Atrostic, Cheryl Grim, Richard Hough, Dave Kinyon, Erika McEntarfer, and Mary Potter (U.S. Census Bureau)
- Asha Balikrishnan (Institute for Defense Analysis–Science, Technology Policy Institute)
- Carl Bergstrom (University of Washington)
- Stefano Bertuzzi, George Checko, and Jerry Sheehan (National Institutes of Health)
- Maria Borga (Bureau of Economic Analysis)
- Ray Bowen, Kelvin Froegemeier, Jose-Marie Griffiths, Arthur Reilly, and Arnold Stancell (National Science Board)
- Erik Brynjolfsson (Massachusetts Institute of Technology)
- Susan Butts (National Academy of Sciences, Government-University-Industry Research Roundtable)
- Jayanta Chatterjee (Indian Institute of Technology)
- Aaron Chatterji (Duke University, formerly of the Council of Economic Advisers)
- Cynthia Clark (National Agricultural Statistics Service)
- Patrick Clemins (American Association for the Advancement of Science)
- Alessandra Colecchia, Gili Greenberg, and Fernando Galindo-Rueda (OECD)
- Carol Corrado (The Conference Board)
- Gustavo Crespi (Inter-American Development Bank)
- Matthieu Delescluse (European Commission)
- Mark Doms (U.S. Department of Commerce)
- Nicholas Donofrio (IBM)
- Maryann Feldman (University of North Carolina)
- Changlin Gao (Chinese Academy of Science and Technology)
- Matthew Gerdin (U.S. Department of State)
- Lee Giles (Penn State University)
- Donna Ginther (University of Kansas)
- Martin Grueber (Battelle Foundation)
- Bronwyn Hall (University of California, Berkeley)
- John Haltiwanger (University of Maryland)
- Amber Hartman Scholz (President's Council of Advisors on Science and Technology)
- Jonathan Haskel (Imperial College Business School)
- Christopher Hill (George Mason University)
- Hugo Hollanders (United Nations University-Maastricht Economic and Social Research Institute on Innovation and Technology)
- Matthew Hourihan (American Association for the Advancement of Science)
- Charles Hulten (University of Maryland, College Park)
- Adam Jaffe (Brandeis University)
- Tom Kalil and Kei Koizumi (U.S. Office of Science and Technology Policy)
- Bhavya Lal (Institute for Defense Analysis–Science, Technology Policy Institute)
- Julia Lane (American Institutes for Research)
- Brian MacAulay (National Endowment for Science, Technology and the Arts, United Kingdom)

*All listed affiliations are as of February 2014.

- David McGranahan and Tim Wojan (U.S. Department of Agriculture)
- Daniel McGrath, Jessica Shedd, Matthew Soldner, and Tom Weko (National Center for Education Statistics)
- Christine Matthews (Congressional Research Service)
- Philippe Mawoko (The New Partnership for Africa's Development)
- OECD-National Experts on Science and Technology Indicators (NESTI) workshop participants (30 member states, regional representatives, OECD staff)
- Richard Price (Academia.edu)
- Andrew Reamer (George Washington University)
- Alicia Robb (Kauffman Foundation)
- Carol Robbins (Bureau of Economic Analysis)
- Laurie Salmon, Jim Spletzer, and David Talan (Bureau of Labor Statistics)
- Robert Samors and David Winwood (Association of Public and Land-grant Universities)
- Walter Schaffer (National Institutes of Health)
- Jerry Sheehan (National Institutes of Health)
- Stephanie Shipp (Institute for Defense Analysis-Science, Technology Policy Institute)
- Dahlia Sokolov (U.S. House of Representatives)
- Gregoy Tassey (National Institute of Standards and Technology)
- Katherine Wallman, Rochelle Martinez, and colleagues (Office of Management and Budget)

KEY ISSUES AND QUESTIONS FOR STI INDICATORS

- **Growth, competitiveness, and jobs:** What is the contribution of science, technology, and innovation (STI) activity to productivity, employment, and growth? What is the relative importance of technological innovation and nontechnological innovation for economic growth? What are the current advances and vulnerabilities in the global STI system? Is the United States falling behind with respect to innovation, and what are the effects on socioeconomic outcomes? Where is leadership in science and technology (S&T) trending?
- **STI activities:** What are the drivers of innovation? How important are the following for advancing innovation: small businesses, large businesses, strategic alliances, technology transfer between universities and firms, academic researchers, government laboratories and procurement activities, and nonprofit organizations? How influential is research and development (R&D) for innovation and growth (by sector)? What is the role of intangibles in affecting productivity? How do government investments in S&T contribute to innovation? What is the return on investment in basic research, and what good does it do for society? What would constitute a "balance" between the biological and physical sciences? On what basis could that be determined? Does biological science depend on physical science for advancement? What are the emerging innovative sectors, and what is unique about them? What is the international connectivity of science, technology, engineering, and mathematics (STEM) activities?
- **STI talent:** What is the status of STEM education around the world? How much knowledge capital does the United States have? How many people, possessing what kinds of skills, are needed to achieve a robust STI system? What additional sources of "talent" can best be tapped—especially among immigrants, women, and minorities? What are the career paths of foreign-born STEM-educated or foreign-trained individuals? What fraction of STEM-degree holders have STEM jobs? What is the return on investment for individuals who obtain STEM education or training? How many science and engineering (S&E) doctorate holders took nontraditional pathways into the STEM workforce? Did this vary by race/ethnicity, gender, or the existence of a disability? How important are community colleges in developing human resources for STEM talent? What is the trend in online learning and massive open online courses (MOOCs) in the United States and abroad? Is the United States falling behind in STEM workers? What fields other than STEM are important for advances in STI? What are the fields that contribute expertise to advances in clean energy?
- **Private investment, government investment and procurement:** What impact does federal research spending have on innovation and economic health, and over what time frame? How large should the federal research budget be? How should policy makers decide where to put additional research dollars or reallocate existing funding streams—information and communication technology (ICT), biotechnology, physical science, nanotechnology, environmental technology, social science, etc.? Does government investment crowd out or energize private investment in STI activities? What is the role of entrepreneurship in driving innovation?
- **Institutions, networks, and regulations:** What impacts are federal research programs (including federally funded research and development centers) having on entrepreneurial activities in S&E sectors? Where are the key gaps in the transfer of scientific and technological knowledge that undercut the performance of the STI system? Where is the supposed "valley of death" in innovation? In which industries is the valley of death most prevalent? What part of the process is underfunded for specific sectors? What is the nature and impact of intellectual property

protection with respect to scientific and innovation outputs? How do incentives for innovation activities work (national and state levels)?

- **Global STI activities and outcomes:** What can the United States learn from other countries, and what are other countries learning from the United States? In what technological areas are other countries accelerating? What impact does the international flow of STI have on U.S. economic performance? What is the relative cost of innovation inputs in the United States versus other countries? Where are multinational corporations sourcing R&D? What are the institutional differences that affect innovation activities among nations, and how are they changing?
- **Subnational STI activities and outcomes:** How does innovation activity in a given firm in a given place contribute to that firm's productivity, employment, and growth, and perhaps also to these characteristics in the surrounding area? How are those innovation supply chains working within a state? Are firms outsourcing new knowledge principally from customers or from universities?
- **Systemic changes on the horizon:** How is the global STI ecosystem changing or evolving? What sectors, regions, and people will rise in prominence in the near future? How will demographic shifts affect the STEM workforce, nationally and internationally? Will they alter the locus of the most highly productive regions? Will global financial crises slow innovation activities or merely change the locus of activities? When will emerging economies be integrated into the global ecosystem of innovation, and what impact will that have on the system? What changes are expected in the following sectors: clean energy, agriculture, biotechnology, nanotechnology, information technology, cyber technology, weapons, "big data," and others? How are public views of S&T changing over time? What is the culture of innovation (e.g., entrepreneurship, willingness to take risks) in U.S. regions and around the world? What are the general perceptions about science and the public value of science in the general population (United States and abroad)?

KEY INDICATORS SUGGESTED BY MAJOR USERS OF STI INDICATORS

Activities

R&D

- National R&D expenditures
  - Federal and state funds for basic research
  - Public-sector R&D (focus on advanced manufacturing, green technologies, energy-related R&D, nanotechnology, agriculture, weapons)
  - Public R&D spending as a share of gross domestic product (GDP)
  - Business R&D spending
  - Business R&D as a share of GDP
  - Industry support for R&D in universities
  - Social science R&D
- National R&D performance (by type of industry and source of funds)
- Trends in size of grants to universities
- Number of R&D centers in the United States and other countries

Innovation

- Direct measures of innovation (Community Innovation Survey-like data)
  - Propensity-to-innovate ratings
  - Subject matter experts (SMEs) innovating in house as a share of SMEs
  - Firms (<5, 5+, 10+, 20+ employees) introducing new or significantly improved products or processes as a share of all firms
  - Firms (<5, 5+, 10+, 20+ employees) introducing new or significantly improved goods or services as a share of all firms
  - Firms (<5, 5+, 10+, 20+ employees) introducing marketing or organizational innovations as a share of all firms
- Number and types of new products per year, by region (Thomasnet.com)
- Drug and other approvals per year, by region
- Sale of new-to-market and new-to-firm innovations as a share of turnover
- Non-R&D expenditures on innovation activities; non-R&D innovation spending as a share of turnover
- Inclusive innovation for development (case studies)
- Capital expenditures related to the introduction of new processes
- Marketing expenditures related to new products
- Expenditures on design and technical specifications
- Expenditures on service-sector innovation; investment in new ICT hardware and software
- Innovation inhibitors (case studies)

Market Capital Investments

- Venture capital investments in S&T (early-stage, expansion, and replacement); venture capital in S&T as a share of GDP
- Number of initial public offerings in S&T
- Number of S&T spinoffs

- Expenditures in later phases of development/testing that are not included in R&D

Outputs and Outcomes

Commercial Outputs and Outcomes

- Performance of "gazelles," small firms and small business units within large firms
- High-growth enterprises as a share of all enterprises
- Medium- and high-tech manufacturing exports as a share of total product exports
- Exports of knowledge-intensive services as a share of total service exports
- Value added in manufacturing
- Value added in technical services
- Trade flows of S&T products and services
- ICT output and sales (intermediate and final)
- Other intermediate inputs
- Technology balance of trade (especially intellectual property)
- Contracts to S&T firms
- Advanced manufacturing outputs (information technology-based processes)
- Market diffusion activities
- Emerging industries (based on universities, government laboratories, firms, value chains, key occupations, and individuals)
- Business practice data (e.g., help-wanted ads, "how to" books)

Knowledge Outputs

- U.S. receipts and royalty payments from foreign affiliates
- U.S. patent applications and grants by country, technology
- U.S. trademark applications and grants by country, technology
- Patent citations
- License and patent revenues from abroad as a share of GDP
- Triadic Patent Families by country
- Percentage of patent applications per billion GDP
- Percentage of patent applications related to societal challenges (e.g., climate change mitigation, health) per billion GDP
- Intangible assets
- Average length of a firm's products' life cycles or how often the firm usually introduces innovations
- Births and deaths of businesses linked to innovation outputs; firm dynamics by geography, industry, business size, and business age
- Knowledge depreciation
- Knowledge stocks and flows in specific sectors, including nanotechnology; information technology; biotechnology and agriculture research (local foods, organic foods, biofuels, environment, nutrition, health); oil and gas production; clean/green energy; space applications; weapons; health care technologies; educational technologies (MOOCs); mining

STEM Education

- Expenditures, direct and indirect costs, investments, revenues, financing on STEM education
- Percentage of faculty in nonteaching and nonresearch roles at universities
- Enrollment data by STEM at various levels (e.g., associate's, bachelor's, master's, doctoral degrees) and for various types of institutions
- New degrees (e.g., associate's, bachelor's, master's, doctoral); new doctoral graduates per 1,000 population aged 25-34
- Stock of degrees (e.g., associate's, bachelor's, master's, doctoral)
- Share of population aged 30-34 having completed tertiary education
- Share of youth aged 20-24 having attained at least upper-secondary-level education
- Persistence and dropout rates in education by geographic and demographic distinctions
- Number of high school students pursuing associate's degrees and implications for the workforce and the cost of higher education
- Disciplines in which community colleges have a comparative advantage
- Foreign-born STEM-educated individuals—countries of birth, immigration visas, etc.
- Stay rates of foreign students
- Trends in online learning and MOOCs

STEM Workforce/Talent

- Postdoctoral levels and trends in various STEM fields by country of birth and country of highest degree
- Number of postdoctorates in health, specific fields
- STEM employment
- Labor mobility and workforce migration
- Demographic composition of people who would enter specific occupations (e.g., clean energy, ICT, biotechnology, health services)
- Fraction of STEM degree holders that hold STEM jobs
- Earnings by degree type and occupation
- Feeder fields in agricultural science
- On-the-job training activities in S&T, manufacturing, and services
- STEM demand

- Employment in knowledge-intensive activities (manufacturing and services) as a share of total employment

Socioeconomic Impacts/Well-Being (The "Are We Better Off" Question)

- Economic growth
- Productivity
- Other measures of impact on GDP and jobs
- Agricultural preparedness
- Energy preparedness
- Return on investment on grants to universities by type of S&T
- National security/defense
- Environment
- Energy use
- Geographic hot spots

Linkages

Organizations/Institutions

- Public-private copublications per million population
- University-industry research collaborations
- Number and value of international collaborations
- Business structure dynamics
- Technology transfer between academic institutions and businesses, including mechanisms
- Technology transfer (including programs such as Manufacturing Extension Partnership Technology Transfer/Transition Pilot Initiative)
- Technology transfer from national laboratories
- Bilateral S&T agreements (including international)
- Collaboratories
- Industry clusters
- Incubators
- Consortia (Defense Advanced Research Projects Agency [DARPA], Advanced Research Projects Agency-Energy [ARPA-E], Technology Innovation Program at the National Institute for Standards and Technology)
- Intellectual property rights and policies
- Standards
- Market planning assistance (Department of Commerce [DoC], Bureau of Labor Statistics [BLS], Small Business Administration [SBA])
- Research and experimentation (R&E) tax credits (federal and state)
- Innovative SMEs collaborating with others as a share of SMEs
- Alumni contributions to R&D

Culture

- Public value of S&T
- Business climate
- Entrepreneurial activities
  - Mappings of entrepreneurial density
  - All establishments and firms with at least one employee, including start-ups, from 1976 to the present
  - All nonemployer firms and integrated-with-employer firms from 1994 to the present
  - All employer-employee matches and transitions (hires, separations, job creation, and job destruction) from 1990 to the present
  - Information on innovation policies and outcomes (contexts; national, regional, sectoral levels)
  - Data on existence of dealmakers and entrepreneurs and their connections in a given market
- Risk tolerance
- Social networks
  - Attendance at conferences (number of times per year)
  - Membership in associations (number of associations linked directly and indirectly to field of study or occupation)
  - Linkages to sources of capital
  - Linkages to sources of knowledge or ingenuity used in occupation

Subnational Indicators

- State, county, and metropolitan tables of data from the Business Research and Development and Innovation Survey (BRDIS) (covering R&D performance, workforce, and intellectual property)
- Academic R&D expenditures
- Federal R&D expenditures
- Industry support for R&D in universities
- Total R&D (from a resurrected nonprofit R&D survey)
- Degrees granted in STEM (production and migration)
- STEM jobs (Occupational Employment Statistics from BLS)
- STEM workforce migration (data on Local Employment Dynamics from the U.S. Census Bureau)
- Patent applications, grants, and citations (from the U.S. Patent and Trademark Office)
- STI equity investments (from various sources, including venture capital)
- STEM occupational projections (from BLS and the Employment and Training Administration [ETA])
- STEM occupation classification (from ETA)
- STEM graduate and workforce migration (National Center for Education Statistics, from the Census Bureau and BLS)

- Firm innovation processes (from the Economic Research Service [ERS] at the U.S. Department of Agriculture [USDA])
- Propensity-to-innovate ratings
- Mappings of entrepreneurial density
- Firm births, mergers and acquisitions, deaths ("business dynamics" as characterized by Haltiwanger at the panel's July 2011 workshop, including geography, industry, business size, business age)
- State and federal grants and loans (from Science and Technology for America's Reinvestment: Measuring the Effect of Research on Innovation, Competitiveness and Science [STAR METRICS])
- Initial public offerings
- New products (from Thomasnet.com)
- Drug and other approvals (from the Food and Drug Administration)
- Data on dealmakers and entrepreneurs, including number of connections
- Data on emerging industries, based on universities, government laboratories, firms, value chains, key occupations, and individuals

Appendix C

Workshop on Developing Science, Technology, and Innovation Indicators for the Future Agenda and Participants*

AGENDA

This workshop is part of a study by the Panel on Developing Science, Technology, and Innovation Indicators for the Future. During this two-day meeting, scholars and practitioners from around the world will discuss specific datasets, frameworks, methods, and tools for measuring science, technology and innovation (STI) activities at the national and subnational levels, and for developed and developing countries. Participants will discuss (1) metrics that have been shown to track changes in national economic growth, productivity, and other indicators of social development; (2) frameworks for gathering data on academic inputs to research, development, and translation processes toward commercialization of new scientific outputs, with specific regional outlooks; and (3) next-generation methods for gathering and disseminating data that give snapshot views of scientific research and innovation in sectors such as biotechnology and information and communication technology (ICT). Presentations and networked discussions will focus attention on the policy relevance of redesigned or new indicators.

DAY 1: Monday, July 11, 2011

8:00-9:00 AM REGISTRATION IN LOBBY *(breakfast available in main foyer)*

9:00-9:10 WELCOME AND OVERVIEW
Cochairs: **Robert Litan** (Kauffman Foundation) and **Andrew Wyckoff** (OECD)
- **Connie Citro** (Committee on National Statistics, National Research Council)
- **Steve Merrill** (Board on Science, Technology, and Economic Policy, National Research Council)

9:10-10:35 SESSION I: NEXT-GENERATION STI STATISTICS—FRAMEWORKS AND DATA
Objective: Identify what the federal statistical system can produce now regarding STI trends. Specific measures of business and university inputs and outputs and related outcomes and impacts are welcomed. What can be done now with existing surveys and administrative data? What do users want that requires *new methods* of gathering and disseminating data (types of data, linkages of agency surveys and periodicity)? *What should NCSES produce to meet demand?*
Chair: **Bob Litan** (Kauffman Foundation)
Discussant: **John Rolph** (University of Southern California)
Presenters:
- **John Haltiwanger** (University of Maryland)
- **Alicia Robb** (Kauffman Foundation)
- **Stefano Bertuzzi** (National Institutes of Health, STAR METRICS)
- **Matthieu Delescluse** (European Commission)

10:35-10:45 ***BREAK** (refreshments available in lobby)*

*All listed affiliations are as of July 2011.

10:45-12:45 PM **SESSION II: INTERNATIONAL STI INDICATORS—RECENT DEVELOPMENTS**

Objective: Identify recent developments in measuring STI and what is currently planned for the *future*. Discussion should reveal what has been successfully and unsuccessfully measured. What are critical bottlenecks and perceived opportunities? Policy relevance of indicators is key. *What global STI metrics and indicators should NCSES develop in the near and medium terms (the next 5-10 years)?*

Chair: **Michael Mandel** (University of Pennsylvania)

Discussant: **Andrew Wyckoff** (OECD)

Presenters:

- **Jonathan Haskel** (Imperial College Business School, U.K.)
- **Brian MacAulay** (National Endowment for Science, Technology and the Arts, U.K.)
- **Hugo Hollanders** (United Nations University-Maastricht Economic and Social Research Institute on Innovation and Technology [UNU-MERIT], Netherlands)
- **Shinichi Akaike** (Institute of Innovation Research, Hitotsubashi University, Japan)
- **Cheonsik Woo** (Korean Development Institute)

12:45-1:45 ***LUNCH** (in main foyer)*

1:45-3:45 **SESSION III: NEXT-GENERATION STI STATISTICS—FRONTIER METHODS**

Objective: Identify frameworks and tools beyond survey instruments that yield measurements of research and commercialization productivity. Details on how a statistical agency can utilize these tools are key. Education and workforce indicators are needed. *Which tools are ripe for application that NCSES should use to produce new STI indicators?*

Chair: **Geoff Davis** (Google)

Discussant: **Richard Freeman** (Harvard University)

Presenters:

- **Erik Brynjolfsson** (Massachusetts Institute of Technology)
- **Lee Giles** (Penn State University)
- **Carl Bergstrom** (University of Washington)
- **Richard Price** (Academia.edu)

3:45-4:00 ***BREAK** (refreshments available in lobby)*

4:00-5:00 **SESSION IV: ROUNDTABLE: INDUSTRY, ACADEMIC, AND GOVERNMENT PERSPECTIVES**

Objective: Identify what firms, universities, and statistical agencies can be expected to contribute to data inputs for STI indicators. Determine new uses for STI indicators at firms, particularly multinationals. Establish what policy makers and university sponsored research officers/technology transfer managers need to know in their respective decision-making processes. Where will the indicators be used and why, and why have they not already been developed? *What are the new data inputs and new statistical outputs that should be the laser focus for NCSES in the next 5-10 years?*

Chairs: **Barbara Fraumeni** (University of Southern Maine)

Discussants:

- **Nick Donofrio** (IBM)
- **Richard Freeman** (Harvard University)
- **David Goldston** (Natural Resources Defense Council)

5:00-5:10 **WRAP-UP**

- Cochairs: **Robert Litan** and **Andrew Wyckoff**
- Study Director: **Kaye Husbands Fealing** (National Academies/ Committee on National Statistics)

5:30-7:00 **RECEPTION** (*in main foyer*)

7:00 PM **ADJOURN**

DAY 2: Tuesday, July 12, 2011

8:00 -8:30 AM — REGISTRATION IN LOBBY *(breakfast available in main foyer)*

8:30 -10:30 — SESSION V: INTERNATIONAL STI INDICATORS—NEW REGIONS

Objective: Identify new foci for STI indicator initiatives. Includes presentations on emerging economies' measurement of STI diffusion and impacts. Discussion of service-sector measures and measures of design activities. *What indicators should NCSES develop to measure technological diffusion and design?*

Chair: **Carl Dahlman** (Georgetown University)

Discussant: **Fred Gault** (UNU-MERIT)

Presenters:

- **Howard Alper** (University of Ottawa, Canada)
- **Changlin Gao** (Chinese Academy of Science and Technology for Development)
- **Philippe Mawoko** (The New Partnership for Africa's Development [NEPAD], South Africa)
- **Gustavo Crespi** (Inter-American Development Bank, Uruguay)
- **Jayanta Chatterjee** (Indian Institute of Technology-Kanpur)

10:30-10:45 — ***BREAK*** *(refreshments available in lobby)*

10:45-12:30 PM — SESSION VI: SUBNATIONAL STI INDICATORS

Objective: Identify state and regional indicators of entrepreneurial activities and hot spots of innovation. *What indicators should NCSES develop to measure state and regional STI and diffusion activities?*

Chair: **Lee Wilkinson** (SYSTAT)

Discussant: **David Goldston** (Natural Resources Defense Council)

Presenters:

- **Andrew Reamer** (George Washington University)
- **Robert Atkinson** (Information Technology and Innovation Foundation)
- **Maryann Feldman** (University of North Carolina)
- **David Winwood** (University of Alabama-Birmingham Research Foundation)/Robert Samors (Association of Public and Land-grant Universities [APLU], by telephone)

12:30-12:40 — **Wrap-up**

Cochairs: **Robert Litan** and **Andrew Wyckoff**

12:40-2:00 — ***LUNCH*** *(in main foyer)*

2:00 PM — ADJOURN

PARTICIPANTS

Panel Members

Robert E. Litan *(Cochair),* The Ewing Marion Kauffman Foundation
Andrew W. Wyckoff *(Cochair),* OECD, Paris
Carl J. Dahlman, Georgetown University
Geoff Davis, Google, Inc.
Barbara M. Fraumeni, University of Southern Maine
Richard B. Freeman, Harvard University
Fred Gault, United Nations University-Maastricht Economic and Social Research Institute on Innovation and Technology, The Netherlands
David Goldston, Natural Resources Defense Council
Michael Mandel, University of Pennsylvania
John E. Rolph, University of Southern California
Leland Wilkinson, SYSTAT Software, Inc.

Presenters

Shinichi Akaike, Hitotsubashi University, Japan
Howard Alper, Canada's Science, Technology, and Innovation Council
Rob Atkinson, Information Technology and Innovation Foundation
Carl Bergstrom, University of Washington
Stefano Bertuzzi, National Institutes of Health
Eric Brynjolfsson, Massachusetts Institute of Technology
Jayanta Chatterjee, Indian Institute of Technology, Kanpur, India
Gustavo Crespi, Inter-American Development Bank
Matthieu Delescluse, European Commission, Brussels, Belgium
Nick Donofrio, IBM
Maryann Feldman, University of North Carolina
Changlin Gao, Chinese Academy of Science and Technology, Beijing, China
Lee Giles, Penn State University
John Haltiwanger, University of Maryland
Jonathan Haskel, Imperial College Business School, United Kingdom
Hugo Hollanders, Maastrict University, The Netherlands
Brian MacAulay, National Endowment for Science, Technology and the Arts, United Kingdom
Philippe Mawoko, The New Partnership for Africa's Development (NEPAD), Pretoria, South Africa
Richard Price, Academia.edu
Andrew Reamer, George Washington University
Alicia Robb, The Ewing Marion Kauffman Foundation
Robert Samors (by telephone), Association of Public and Land-grant Universities
David Winwood, Association of Public and Land-grant Universities

Guests

Ana Aizcorbe, U.S. Bureau of Economic Analysis
Jeff Alexander, SRI International
Gary Anderson, Jr., National Institute of Standards and Technology
Clara Asmail, National Institute of Standards and Technology
B.K. Atrostic, U.S. Census Bureau
David Ballard, GRA, Inc.
David Beede, U.S. Department of Commerce
Bob Bell, National Center for Science and Engineering Statistics, National Science Foundation
Brittany Bond, U.S. Department of Commerce
Patrice Bourdelais, Centre National de la Récherche Scientifique
Paul Bugg, U.S. Office of Management and Budget
Lynda Carlson, National Center for Science and Engineering Statistics, National Science Foundation
Carolyn Carroll, STAT TECH, Inc.
Arthur Cho, Japan Science and Technology Agency
Carol Corrado, Georgetown Center for Business and Public Policy
Louis Marc Ducharme, Statistics Canada
Louise Earl, Statistics Canada
Uchenna Egenti, East Tennessee State University
Jonathan Epstein, U.S. Senate
Paul Fakes, American Society for Mechanical Engineers
Chris Fall, U.S. Military
Jean Favero, Centre National de la Récherche Scientifique
Lauren Gilchrist, Center for Regional Economic Competitiveness
Stuart Graham, U.S. Patent and Trademark Office
Myron Gutmann, Directorate for the Social, Behavioral, and Economic Sciences, National Science Foundation
John Hall, PA Alliance for STEM Education
Kim Hamilton, U.S. Patent Board
Lee Herring, Office of Legislative and Public Affairs, National Science Foundation
Robert Hershey, Capital PC User Group
Chris Hill, George Mason University
Richard Hough, U.S. Census Bureau
Tommy Hudzik, Independent Consultant
Charles Hulten, University of Maryland
Elmer Iglesias, U.S. Department of Commerce
Takashi Inutsuka, Science Counselor, Embassy of Japan
John Jankowski, National Center for Science and Engineering Statistics, National Science Foundation
Ken Jarboe, Athena Alliance
Richard Johnson, Global Helix, LLC
David Kahaner, Asian Technology Information Program
Nimmi Kannankutty, National Center for Science and Engineering Statistics, National Science Foundation
Michael Kehoe, American Association for the Advancement of Science

Todd Kuiken, Woodrow Wilson International Center for Scholars
Karen Laney, U.S. International Trade Commission
Chuck Larson, Innovation Research International
Marc Legault, Science, Technology, and Innovation Council, Canada
Rolf Lehming, National Center for Science and Engineering Statistics, National Science Foundation
Wendy Li, U.S. Bureau of Economic Analysis
Ying Lowrey, U.S. Small Business Administration
Shelley Martinez, U.S. Office of Management and Budget
Tony Mazzaschi, Association of American Medical Colleges
Christine McDonald, U.S. Office of Management and Budget
Javier Miranda, U.S. Census Bureau
T.C. Moore, Independent Consultant
Francisco Moris, National Center for Science and Engineering Statistics, National Science Foundation
Paul Morris, National Center for Science and Engineering Statistics, National Science Foundation
Manuel Mota, Massachusetts Institute of Technology
Jeri Mulrow, National Center for Science and Engineering Statistics, National Science Foundation
Vinh Nguyen, Independent Consultant
Kimberly Noonan, National Center for Science and Engineering Statistics, National Science Foundation
Pamela O'Neil, National Center for Science and Engineering Statistics, National Science Foundation
Takashi Ohama, Japan Science and Technology Agency
Sumiya Okubo, Independent Consultant
Erik Pages, EntreWorks Consulting
Diane Palmintera, Innovation Associates
Sapun Parekh, National Science Foundation
Jongwon Park, SRI International
Joel Parriott, U.S. Office of Management and Budget
Sebastian Pfotenhauer, Massachusetts Institute of Technology
Pallivi Phartiyal, American Association for the Advancement of Science
Brian Reinhardt, Defense Threat Reduction Agency
Sally Rood, Science Policy Works International
Robert Shelton, World Technology Evaluation Center, Inc.
Stephanie Shipp, IDA Science and Technology Policy Institute
Debbie Stine, President's Council of Advisors on Science and Technology
Andrea Stith, International Higher Education and Science
Ezequiel Tacsir, Inter-American Development Bank
Greg Tassey, National Institute of Standards and Technology
Chris Thomas, Independent Consultant
Hua Tian, Arizona State University
James Tsang, Independent Consultant
Nick Vonortas, George Washington University
Philip Webre, Congressional Budget Office
Brittany Westlake, American Chemical Society
Jeremy Wise, U.S. International Trade Commission
Tim Wojan, U.S. Department of Agriculture
Rieko Yajima, American Association for the Advancement of Science
Arthur Yong Yeung Cho, Japan Science and Technology Agency
Pluvia Zuniga, United Nations University-Maastricht Economic and Social Research Institute on Innovation and Technology

NRC Staff

Connie Citro, Committee on National Statistics
Gail Greenfield, Policy and Global Affairs
Kaye Husbands Fealing, Committee on National Statistics
Anthony Mann, Committee on National Statistics
Steve Merrill, Board on Science, Technology, and Economic Policy
Miron Straf, Division of Behavioral and Social Sciences and Education

Appendix D

OECD-National Experts on Science and Technology Indicators (NESTI) Workshop and Attendees*

* All listed affiliations are as of June 2012.

For Official Use **DSTI/EAS/STP/NESTI/AH/A(2012)2**

Organisation de Coopération et de Développement Économiques
Organisation for Economic Co-operation and Development

English - Or. English

DIRECTORATE FOR SCIENCE, TECHNOLOGY AND INDUSTRY
COMMITTEE FOR SCIENTIFIC AND TECHNOLOGICAL POLICY

DSTI/EAS/STP/NESTI/AH/A(2012)2
For Official Use

Working Party of National Experts on Science and Technology Indicators

Developing Science, Technology, and Innovation Indicators for the Future - the Importance of International Comparability

NESTI Advisory Board workshop

6 June 2012

The NESTI Advisory Board will be holding a special workshop to discuss the process by which countries prioritise amongst STI indicators and the role that international comparability plays in that decision. The Workshop, co-organised with the US National Academies, is set out to assess and provide recommendations regarding the need for revised, refocused, and newly developed indicators designed to better reflect fundamental and rapid changes that are reshaping global science, technology and innovation (STI) systems.

Delegates wishing to participate are asked to register separately for this event using EMS or contacting directly Catherine.bignon@oecd.org and Chrystyna.harpluk@oecd.org.

CONTACT: Mr. Fernando GALINDO-RUEDA (OECD/STI/EAS); tel: +33 (0) 1 45 24 87 49;
e-mail: fernando.galindo-rueda@oecd.org

Complete document available on OLIS in its original format
This document and any map included herein are without prejudice to the status of or sovereignty over any territory, to the delimitation of international frontiers and boundaries and to the name of any territory, city or area.

English - Or. English

WORKING PARTY OF NATIONAL EXPERTS ON SCIENCE AND TECHNOLOGY INDICATORS (NESTI)

Developing Science, Technology, and Innovation Indicators for the Future—The Importance of International Comparability
NESTI Advisory Board workshop

DRAFT AGENDA

Paris, 6 June 2012

OECD Conference Centre
2 Rue André Pascal, 75016 Paris

Indicative timing	*No.*	*Item*
14:15	1.	**Welcome and introductions** Mr. Fred Gault, Workshop Chair, UNU-MERIT, Chair of NESTI Advisory Board
14:20	2.	**Background to the workshop** Mr. Andrew Wyckoff, Director DSTI, OECD and Cochair of the U.S. National Academies Panel on Developing Science, Technology and Innovation Indicators for the Future
14:30	3.	**Overview presentation of the National Academies panel and interim findings** Ms. Kaye Husbands Fealing, Professor, Hubert H. Humphrey School of Public Affairs, University of Minnesota and U.S. Committee on National Statistics, Board on Science, Technology and Economic Policy Link to interim report and related material: http://www8.nationalacademies.org/cp/CommitteeView.aspx?key=49353; http://sti-indicators.ning.com/ Participants and discussants are invited to examine the material in advance of the workshop and to react on the proposals in their interventions.
14:45	4.	**Discussion: The policy context and demands for developing and prioritizing STI indicators** Lead discussants: • Mr. Dominique Guellec, Head of Country Studies and Outlook Division, DSTI, OECD • Mr. Pierre Vigier, DG Research and Innovation, European Commission (TBC) • Mr. Philippe Mawoko, Interim Director, The African Observatory for STI (AOSTI), African Union Commission (TBC) • Mr. Ken Guy, Head of Science and Technology Policy Division, DSTI, OECD
15:30		*Coffee break*
15:45	5.	**Discussion: Implementing measurement priorities—Enabling international comparability** Lead discussants: • Ms. Alessandra Colecchia, Head of Economic Analysis and Statistics Division, DSTI, OECD • Mr. Leonid Gokhberg, Director of the Institute for Statistical Studies and Economics of Knowledge, National Research University—Higher School of Economics • Mr. Veijo Ritola, Head of Section, Science, Technology and Innovation Statistics, Eurostat • Mr. Joaquim Oliveira Martins, Head of Regional Development Policy Division, GOV, OECD

16:30	**6.**	**Updates from delegates on national priorities and general discussion**
17:00	**7.**	**Summary of workshops main points and conclusions** Mr. Fred Gault (Chair)
17:15	**8.**	**Workshop concludes**

Developing Science, Technology and Innovation Indicators for the Future—The Importance of International Comparability
NESTI Advisory Board Workshop
Paris, 6 June 2012, OECD Conference Centre, CC01, starting at 14:15

Participants

Member Countries and the European Union

Australia, Austria, Belgium, Chile, Czech Republic, Estonia, European Union, Germany, Hungary, Israel, Italy, Japan, Korea, Luxembourg, Netherlands, Norway, Poland, Portugal, Russia, Slovak Republic, Spain, Sweden, Switzerland, Turkey, United Kingdom, United States

Observers and International Organizations

African Union, Brazil, European Union, India, La Red Iberoamericana e Interamericana de Indicadores de Ciencia y Tecnología (RICYT), South Africa

Speakers

Fred Gault, UNU-MERIT, TUT-IERI and Chair of the NESTI Advisory Board
Andrew Wyckoff, Director DSTI, OECD and Cochair of the U.S. National Academies
Kaye Husbands Fealing, Senior Study Director, National Academy of Sciences
Dominique Guellec, Head Country Studies and Outlook Division, DSTI, OECD
Matthieu Delescluse, Policy Officer, Economic Analysis and Indicators Unit, DG Research and Innovation, European Commission
Almamy Konté, African Observatory for STI (AOSTI), African Union
Joaquim Oliveria Martins, Head of the Regional Development Policy Division, Public Governance and Territorial Development (GOV), OECD
Leonid Gokhberg, Director of the Institute for Statistical Studies and Economics of Knowledge, National Research University—Higher School of Economics
Veijo Ritola, Head of Section, Science, Technology and Innovation Statistics, Eurostat
Alessandra Colecchia, Head of Economic Analysis and Statistics Division, DSTI, OECD

Appendix E

National Center for Science and Engineering Statistics Research Abstracts 2012*

Linda Cohen, University of California, Irvine
"Doctoral Dissertation Research: The Price Elasticity of R&D: Evidence from State Tax Policies"

In tax year 2008, the federal research and development (R&D) tax credit paid out over $8 billion to businesses, which was 7% of total federal expenditures on research. The intent of the tax credit is to provide an incentive for firms to raise their private level of R&D funding. This project will investigate how effective tax incentives are at increasing R&D. This evaluation is difficult because, while we can observe R&D spending before and after a tax policy change, we can only speculate on what R&D would have been without the tax policy change. Because policy makers implement tax incentives in response to current and/or expected economic conditions, a simple comparison of R&D before and after a tax incentive is implemented will lead to inaccurate inferences about the effects of the tax incentive. For example, if R&D in a given year is low, then policy makers may respond with a tax incentive. While a rebound in the following year could be due to the tax incentive, it might also reflect R&D simply returning to its mean value. Alternatively, policy makers might foresee a decline in R&D and implement a tax incentive to prevent the decline. Subsequently observing no change in R&D after the tax incentive takes place would be evidence supporting the efficacy of the tax incentive. To correct for the endogeneity of tax incentives, we will use state-level tax variation driven by changes in the U.S. federal R&D tax credit. While state governments are attentive to state-level economic conditions when forming their idiosyncratic state-level tax policies, the federal government sets a uniform national tax policy and is less attentive to individual state economic conditions. In addition, changes in the federal R&D tax credit have differential impacts on state-level tax incentives across states due to the interaction of federal and state taxes. These two features imply a regression mode that can generate an unbiased estimate of the effect of tax incentives on R&D.

Broader Impacts: This study will make several important contributions, in addition to supporting the training of a doctoral candidate. First, the project will create a dataset on state corporate tax laws that will be more detailed than any existing dataset on state R&D tax incentives. These data will allow a descriptive analysis of how the overall tax burden for R&D has changed over time and across states/regions. Second, the project will generate an unbiased estimate of how tax incentives affect R&D. The final contribution will be an estimate of the endogeneity bias driven by self-selection of tax policies, which will help future economic research on tax incentives and uncover evidence on mechanisms behind the implementation of tax policies.

Frank Dobbin, Harvard University
"The Retention and Promotion of Women and Minority Faculty Members: Effects of Institutional Hiring, Promotion, Diversity and Work-Life Initiatives, 1993-2008"

U.S. colleges and universities have implemented a wide range of programs to promote diversity in the professoriate. Special recruitment programs, tenure extension policies for new parents, paid maternity leaves, mentoring programs targeting female and minority faculty members, dual-career hiring initiatives, and ethnic affinity networks for faculty are but a few of the initiatives. Which of these programs work? It is anyone's guess, and critics argue that many of the programs may have no effects, or even adverse effects. This goal of this project is to understand the role of university recruitment, promotion, diversity, and work-family programs in attracting and retaining female and minority professors. The project will address questions such as: Do tenure extension programs for new parents help female faculty members

*All listed affiliations are as of June 2012.

win tenure, or do they do more to help male faculty members, who more often have spouses with low-demand careers? Do networking programs help African American and Latino faculty members to succeed, or do they stigmatize and isolate those faculty members? Do formal promotion requirements help women and minorities to win promotion, or do they serve as window-dressing?

NCSES' Survey of Doctorate Recipients (SDR) data from 1993 to 2008, as well as new survey data, will be used to address these questions. The project will develop and pilot a questionnaire designed to obtain historical data on university recruitment, promotion, diversity, and work-family programs, develop a sample of colleges and universities and matched SDR respondents, and build methods for analyzing the data. The goal is to show the effects of the presence, and adoption, of different programs on the career progression of male and female, majority and minority Ph.D.'s. New methods will be developed for analyzing individual-level data from the SDR panels, using multinomial logit methods in hierarchical linear models, in which individuals are embedded in institutions. We will develop models that account for both left and right censoring in the data, but which make use of the multiple years of observation available for SDR respondents.

Broader Impacts: The result of this project will be to show which types of programs help schools to attract, retain, and promote women and minority faculty members, and will guide future administrators in making choices about program utilization and design. This project will also train 3 to 5 doctoral students to analyze data from the Survey of Doctorate Recipients, and to use advanced statistical techniques to examine factors shaping the careers of U.S. scientists and engineers.

Jeffery Gibeling, University of California, Davis "Doctoral Dissertation Research: Analysis of Institutional Characteristics That Contribute to Extended Time to Doctoral Degree"

The purpose of this research is to identify institutional factors that impact time to degree for the doctoral students who take the longest to complete their studies and graduate. Comparisons are made relative to their disciplinary peers, across nationally representative samples, without disclosing the identity of any institution or student. This study merges data from two nationally collected sources: (1) The Survey of Earned Doctorates (SED) and (2) The supplemental data (not rankings) from the National Research Council's (NRC) *A Data-Based Assessment of Research-Doctorate Programs in the United State*s. The SED and NRC data are merged to determine the patterns of time to degree and the point of extended time to degree within each discipline using the NRC taxonomy. The analysis then looks for interactions between the different levels of data—student qualities, socio-demographic factors, and institutional factors—to identify which factors influence extended time to degree. The driving force behind the research is a void in the current literature. We know that the length of time to doctoral degree varies widely within and across disciplines. While other research has evaluated interactions between various individual and program factors on time to doctoral degree, the impact of institutional factors on extended time to degree has not been specifically investigated. Furthermore, a statistical analysis has not previously been conducted using merged SED and NRC data to evaluate extended time to doctoral degree. This research seeks to fill that void and to add new information to the body of knowledge.

Broader Impacts: One significant outcome from this study will be the research-based identification of institutional factors associated with extended time to degree. Institutions, doctoral students, and researchers will all be able to identify different fields and populations impacted by the phenomenon of extended time to degree and thereby make more informed decisions about effective strategies to promote timely doctoral degree completion. This project will also train a new researcher in the use of multiple large-scale national datasets.

Alan Karr, National Institute of Statistical Sciences "Value-Added Postdoctoral Research on the Scientific Workforce"

This postdoctoral research program at the National Institute of Statistical Sciences (NISS) comprises performing innovative research and creating usable products that not only support the mission of the National Center for Science and Engineering Statistics (NCSES) but also address the needs of the nation. From a technical perspective, the research is framed by two statistical themes and two key societal issues. The first statistical theme is characterization of uncertainties arising from novel methods of integrating and analyzing data, addressing a critical need in an era of declining data collection budgets and decreasing participation in government surveys. The second theme centers on conducting experiments with real data, simulating phenomena of interest in order to evaluate, and in some cases enable, methodological advances. Key issues regarding surveys, such as how many times and by what means to contact nonrespondents, are too complex to be treated analytically, and infeasible to address with real world experiments; therefore simulation is effectively the only laboratory available. Specific research topics include data integration, prediction, model to design feedback, data-quality-aware statistical disclosure limitation and cost data quality tradeoffs. All federal statistical agencies stand to benefit from the research, which will produce innovative theory, novel, methodology and algorithmic implementations, together with datasets, analyses, software and insights that inform future data collections.

Broader Impacts: The societal issues are labor economics as it relates to the science, engineering, and health workforce

(SEHW). Understanding phenomena such as salaries, fringe benefits, mobility, and training/job relationships is crucial to maintaining the United States' competitiveness in a global economy, as well as to facing the challenges of difficult economic times. The second issue is aging, because other than the role of students born outside of the United States, aging is the most important phenomenon taking place in the SEHW (and, arguably, in society as a whole). For both issues, understanding the dramatically increasing richness of observed behaviors within the SEHW is a profound opportunity. New kinds of family structures, shared positions, and an array of forms of post-first-retirement employment are among the central social trends of our times. This project will generate new insights that inform both future research and sound policy.

Anne Marie Knott, Washington University
"The Impact of R&D Practices on R&D Effectiveness (RQ)"

In January 2011, President Obama signed into law the America COMPETES Reauthorization Act of 2010. The goal of the act was to invest in innovation through R&D and to improve the competitiveness of the United States. However, increasing investment in and of itself is unlikely to produce desired results. We need to understand who should increase spending and how. NCSES is well-positioned to provide that understanding through its data on firm innovative activities in the Survey of Industrial Research and Development (SIRD) (1987-2007), and its successor, the Business R&D and Innovation Survey (BRDIS) (2008-2011). The proposed study empirically examines the impact of U.S. firms' innovative activities on economic outcomes by matching a new measure of economic performance, firms' Research Quotient (RQ) to the SIRD and BRDIS data. This matching enables us to test major hypotheses within the economics of innovation literature that have been unresolved previously due to lack of reliable firm-level measures of innovative outcomes. These hypotheses pertain to the impact of firm size, market structure, firm heterogeneity, innovation type, innovation source, and appropriability on the incentives to conduct R&D as well as the effectiveness of that R&D.

Broader Impact: At the policy level, the study provides theoretically motivated and empirically rigorous insights for directing investment in innovation for the America COMPETES Reauthorization Act of 2010: characteristics of firms likely to generate the highest returns to that investment (who). Second, for practitioners, the study offers firms prescriptions for increasing their R&D effectiveness (how). Thus the study has the potential to increase the aggregate R&D productivity in the United States. Finally, for academics, the study will answer long-standing questions in the economics of innovation literature to support future theory development on the optimal conditions for innovation.

Peter Miller, Northwestern University
"Doctoral Dissertation Research: Testing Information and Communication Technology (ICT) Recall Aids for Surveys of Personal Networks"

This study seeks to develop recall aids for the name generator procedure from the General Social Survey and examine empirically whether these aids can improve the recall accuracy of the information about who comprises their personal networks from survey participants. It hypothesizes that researchers can obtain more comprehensive personal network data by encouraging survey respondents to consult the actual records that they keep in the contact directories provided by various ICTs (such as the phone book stored in a mobile phone and the address book functionality of email applications). Thus far, although the past literature has suggested a few techniques to reduce respondents' burden in the survey setting, there is little work addressing the issue of the recall accuracy for personal network data collection. This study employs a survey experiment; a Web survey will be administered to college students to gather information about their personal networks. College students consist of a homogeneous sample appropriate for this study, given the concern of internal validity in the expected findings. Students who agree to participate in the survey will be randomly assigned to three conditions. The control group will take the questionnaire without any recall aid, while the two experimental groups will take the survey with two different forms of prompts and probes respectively.

Broader Impacts: The study will provide an effective technique to collect personal network information from individuals such as scientists and engineers. The proposed technique can then be used in surveys to collect information to develop new social capital indicators for the science and engineering workforce. As a result, researchers can use these indicators to investigate how various dimensions (e.g. advice, support, etc.) of personal networks may explain the productivity and career outcomes of scientists and engineers. More generally, this project will advance the understanding about individuals' personal networks as well as the data collection technique for personal network research. It will also offer new insights into the understanding of the psychology of survey response.

Sharon Sassle, Cornell University
"Race and Gender Variation in STEM Employment and Retention: A Cohort Analysis Using SESTAT Data"

The continuing underrepresentation of women, Blacks, and Hispanics in science and engineering occupations impedes efforts to increase the size of the science, technology, engineering, and mathematics (STEM) workforce, a concern for policy makers who view science and scientists as critical to the future of the U.S. economy. Existing research shows that one factor contributing to this underrepresentation is that gains in women's, Blacks', and Hispanics' representation among STEM college majors do not necessarily translate into equal gains in STEM employment. Additionally, women, Blacks, and Hispanics remain far less likely than White or Asian men to be employed in most STEM occupations, particularly outside the life sciences. But little existing research studies trends over time in gender and race-ethnic differences in STEM employment or factors underlying these patterns. The proposed project will use seven waves of the National Science Foundation's (NSF's) Scientists and Engineers Statistical Data System (SESTAT) to study gender and race-ethnic differences in employment in STEM occupations among college graduates who hold a STEM degree.

Broader Impacts: Government spending to educate and train STEM workers is considerable, reaching nearly 900 million by NSF in 2011. These investments and the need to increase the numbers of women and underrepresented minority scientists to maintain the future health of the STEM workforce make retention of STEM workers in related occupations a critical policy issue. This project will enhance the ability of public- and private sector policy makers and program directors to develop and implement practices that encourage the retention of women and underrepresented minorities in STEM occupations.

Appendix F

Science, Technology, and Innovation Databases and Heat Map Analysis

Leland Wilkinson and Esha Sinha[1]

The panel assembled and analyzed underlying data on research and development (R&D), science and technology (S&T), human capital, and innovation to determine the following:

- What are the primary indicators that are necessary for the National Center for Science and Engineering Statistics (NCSES) to disseminate, and are they produced by traditional or frontier methods? To address this question, cluster analysis, primarily a heat map tool, was used together with knowledge gleaned from the literature on the performance of science, technology, and innovation (STI) indicators. Reference is made to the National Science Board's *Science and Engineering Indicators (SEI)* biennial publication when appropriate, but this analysis is not a full review of the *SEI* publication.
- What are the redundant indicators that NCSES does not need to produce going forward? These indicators might be low performers; highly correlated with other, more useful indicators; or measures that are gathered by other organizations. NCSES could target these indicators for efficiency gains while curating the statistics that are in demand but reliably produced elsewhere.

This appendix describes the main data on R&D, S&T, human capital, and innovation that the panel assembled and analyzed. It is divided into four sections. The first two sections contain descriptions of data sources from NCSES and other international statistical organizations. The third section presents the heat map analysis, citing the literature on methodological underpinnings of this technique. The final section gives observations based on this analysis. Not all of the data sources described were analyzed, because it was not feasible to investigate such a wide variety of data culled from various sources. Only databases of the five main STI data providers were analyzed: NCSES; OECD; Eurostat; the United Nations Educational, Scientific and Cultural Organization (UNESCO); Institute of Statistics (UIS); and Statistics Canada. Indicators published in the *SEI 2012 Digest* were also analyzed.

ASSEMBLED DATA

National Center for Science and Engineering Statistics

NCSES communicates its S&T data through various publications, ranging from *InfoBriefs* to Detailed Statistical Tables (DSTs) derived using table generation tools. The three table generation tools—the Integrated Science and Engineering Resource Data System (WebCASPAR), the Scientists and Engineers Statistical Data System (SESTAT), and the Survey of Earned Doctorates (SED) Tabulation Engine (National Center for Science and Engineering Statistics, 2013b)—are each supported by application-specific database systems. The Industrial Research and Development Information System (IRIS) is an additional searchable database of prepopulated tables.

- WebCASPAR hosts statistical data for science and engineering (S&E) at U.S. academic institutions (National Science Foundation, 2012e). This database is compiled from several surveys, including:

[1]Esha Sinha, CNSTAT staff, compiled the data used in the heat map analysis. Leland Wilkinson, panel member, initially ran the heat map program, based on an algorithm that he developed. Sinha then ran several versions of the program on different datasets and over several different time periods. She presented the results of the heat map analysis to the panel during its April 2012 panel meeting. She subsequently ran more sensitivity analyses to ensure the stability of the results. Panel member John Rolph reviewed the work, concluding that the statistical analysis was sound and potentially instructive as an indicators prioritization exercise that NCSES might perform in the future.

— SED[2]/Doctorate Records File;
— Survey of Federal Funds for Research and Development;
— Survey of Federal Science and Engineering Support to Universities, Colleges, and Nonprofit Institutions;
— Survey of Research and Development Expenditures at Universities and Colleges/Higher Education Research and Development Survey;
— Survey of Science and Engineering Research Facilities;
— National Science Foundation (NSF)/National Institutes of Health (NIH) Survey of Graduate Students and Postdoctorates in Science and Engineering; and
— National Center for Education Statistics (NCES) data sources—Integrated Postsecondary Education Data System (IPEDS):
  - IPEDS Completions Survey;
  - IPEDS Enrollment Surveys;
  - IPEDS Institutional Characteristics Survey (tuition data); and
  - IPEDS Salaries, Tenure, and Fringe Benefits Survey.

- SESTAT (National Science Foundation, 2013d) is a database of more than 100,000 scientists and engineers in the United States with at least a bachelor's degree. This is a comprehensive data collection on education, employment, work activities, and demographic characteristics, covering 1993 to 2008.[3] The SESTAT database includes data from:
  — the National Survey of College Graduates (NSCG);
  — the National Survey of Recent College Graduates (NSRCG);
  — the Survey of Doctorate Recipients (SDR); and
  — an integrated data file (SESTAT).
- IRIS (National Center for Science and Engineering Statistics, 2013a) is a database containing industrial R&D data published by NSF from 1953 through 2007. It comprises more than 2,500 statistical tables, which are constructed from the Survey of Industrial Research and Development (SIRD). It is, therefore, a databank of statistical tables rather than a database of microdata of firm-specific information. The data are classified by Standard Industrial Classification and North American Industrial Classification codes (as appropriate), and by firm size, character of work (basic, applied, development), and state. Employment and sales data for companies performing R&D are also included in IRIS.

The data outlined above focus on academic and industrial R&D expenditures and funding and on human capital in S&T. NCSES conducts five surveys to capture R&D support and performance figures for various sectors of the economy. The *National Patterns of Research and Development Resources* series of publications presents a national perspective on the country's R&D investment. R&D expenditure and performance data are available, as well as employment data on scientists and engineers. The *National Patterns* data are useful for international comparisons of R&D activities, and they also report total U.S. R&D expenditures by state. The data series span 1953 through 2011 and are a derived product of NCSES's above-referenced family of five active R&D expenditure and funding surveys:

1. Business Research and Development and Innovation Survey (BRDIS; for 2007 and earlier years, the industrial R&D data were collected by the SIRD);
2. Higher Education Research and Development Survey (HERD; for 2009 and earlier years, academic R&D data were collected by the Survey of Research and Development Expenditures at Universities and Colleges);
3. Survey of Federal Funds for Research and Development;
4. Survey of Research and Development Expenditures at Federally Funded R&D Centers (FFRDCs); and
5. Survey of State Government Research and Development.[4]

The *SEI* biennial volume is another notable contribution from NCSES, published by the National Science Board. It not only contains tables derived from the table generation tools described above but also amalgamates information from NCSES surveys, administrative records such as patent data from government patent offices, bibliometric data on publications in S&E journals, and immigration data from immigration services. For example, tables on the U.S. S&E

[2]SED data on race, ethnicity, citizenship, and gender for 2006 and beyond are available in the SED Tabulation Engine. All other SED variables are available in WebCASPAR except for baccalaureate institution. For more details on the WebCASPAR database, see https://webcaspar.nsf.gov/Help/dataMapHelpDisplay.jsp?subHeader=DataSourceBySubject&type=DS&abbr=DRF&noHeader=1.

[3]Data for 2010 were released in 2013.

[4]For details on each of these surveys, see http://nsf.gov/statistics/question.cfm#ResearchandDevelopmentFundingandExpenditures [November 2012]. A sixth survey, the Survey of Research and Development Funding and Performance by Nonprofit Organizations, was conducted in 1973 and for the years 1996 and 1997 combined. The final response rate for the 1996-1997 survey was 41 percent (see http://www.nsf.gov/statistics/nsf02303/sectc.htm). This lower-than-expected response rate limited the analytical possibilities for the data, and NSF did not publish state-level estimates. The nonprofit data cited in *National Patterns* reports either are taken from the Survey of Federal Funds for Research and Development or are estimates derived from the data collected in the 1996-1997 survey. See National Science Foundation (2013c, p. 2), which states: "Figures for R&D performed by other nonprofit organizations with funding from within the nonprofit sector and business sources are estimated, based on parameters from the Survey of R&D Funding and Performance by Nonprofit Organizations, 1996-97."

labor force are generated using data from the American Community Survey, the Current Population Survey (U.S. Census Bureau), SESTAT, and Occupational Employment Statistics (Bureau of Labor Statistics) (National Science Board, 2012a, Table 3-1, p. 3-8).

Along with information on U.S. R&D capacity and outputs, the *SEI Digest 2012* contains analysis of the data. The *SEI* indicators can be classified as follows (National Science Board, 2012b) (see Box F-1): (1) global R&D and innovation; (2) U.S. R&D funding and performance; (3) U.S. R&D federal portfolio; (4) science, technology, engineering, and mathematics (STEM) education; (5) U.S. S&E workforce trends and composition; (6) knowledge outputs; (7) geography of S&T; and (8) country characteristics.

BOX F-1
NCSES's STI Indicators

1 Global research and development (R&D) and innovation
- Worldwide R&D expenditure by regions and countries
- Average annual growth of R&D expenditure for the U.S., European Union (EU), and Asia-10 economies
- Annual R&D expenditure as share of economic output (R&D/gross domestic product [GDP])
- R&D testing by affiliation, region/country
- U.S. companies reporting innovation activities
- Exports and imports of high-tech goods

2 U.S. R&D funding and performance (including multinationals and affiliates)
- U.S. R&D expenditure by source of funds (including venture capital)
- Types of U.S. R&D performed
- Types of U.S. R&D performed by source of funds
- U.S. academic R&D expenditure by source of funds

3 U.S. R&D federal portfolio
- U.S. federal R&D expenditure by type of R&D
- U.S. federal support for science and engineering (S&E) fields
- U.S. federal R&D budget by national objectives
- U.S. federal R&D spending on R&D by performer
- Federal research and experimentation tax credit claims by North American Industrial Classification System (NAICS) industry
- Federal technology transfer activity indicators
- Small Business Innovation Research (SBIR) and Technology Innovation Program

4 Science, technology, engineering, and mathematics (STEM) education (most measures have demographic breakouts)
- Average mathematics and science scores of U.S. students (National Assessment of Educational Progress [NAEP] and Programme for International Student Assessment [PISA])
- Teacher participation, degrees, and professional development
- High school students taking college classes
- First university degrees in natural sciences and S&E fields by country/region
- S&E degrees, enrollments, and related expenditures—associate's, bachelor's, master's, doctoral
- Doctoral degrees in natural sciences and S&E fields by country/region
- Distance education classes

5 U.S. S&E workforce trends and composition
- Individuals in S&E occupations and as a percentage of the U.S. workforce
- S&E work-related training
- Unemployment rate for those in U.S. S&E occupations
- Change in employment from previous year for those in STEM and non-STEM U.S. jobs
- Women and underrepresented minorities in U.S. S&E occupations
- Foreign-born percentage of S&E degree holders in the United States by field and level of S&E degree

6 Knowledge outputs
- S&E journal articles by region/country
- Engineering journal articles as a share of total S&E journals by region/country
- Citations in the Asian research literature to U.S., EU, and Asian research articles
- Patents and citations of S&E articles in United States Patent and Trademark Office (USPTO) patents
- U.S. patents granted to non-U.S. inventors by region/country/economy
- Share of U.S. utility patents awarded to non-U.S. owners that cite S&E literature
- Value added of knowledge and technology

7 Geography of S&T
- Location of estimated worldwide R&D expenditure
- Average annual growth rates in number of researchers by country/economy
- Value added of high-tech manufacturing by region/country
- Exports of high-tech manufactured goods by region/country
- Cross-border flow of R&D funds among affiliates
- State S&T indicators

8 Country characteristics
- Macroeconomic variables
- Public attitudes toward and understanding of S&T

SOURCE: National Science Board (2012b).

The primary conclusions of the *SEI Digest* are drawn from the indicators outlined above and are supported by more detailed STI data collected by NCSES. The list of variables is presented in Table F-1. Because of space limitations, it was not possible to highlight in this table the fact that most of the information in the *SEI*—such as assessment scores, S&E degrees, individuals in S&E occupations, R&D expenditures and their various components, federal R&D obligations, patents, and venture capital—is available at the state level.

OECD

R&D statistics generated by OECD are based on three databases: Analytical Business Enterprise Research and Development (ANBERD), Research and Development Statistics (RDS), and Main Science and Technological Indicators (MSTI).

The ANBERD database presents annual data on industrial R&D expenditures. These data are broken down by 60 manufacturing and service sectors for OECD countries and selected nonmember economies. The reported data are expressed in national currencies as well as in purchasing power parity (PPP) U.S. dollars, at both current and constant prices. Estimates are drawn from the RDS database and other national sources. ANBERD is part of the Structural Analysis Database (STAN) family of industrial indicators produced by the Science, Technology, and Industry directorate at OECD.

The RDS database covers expenditures by source of funds, type of costs, and R&D personnel by occupation (in both head counts and full-time equivalents [FTEs]). This database is the main source of R&D statistics collected according to the guidelines set forth in OECD's *Frascati Manual* (OECD, 2002). It covers R&D expenditures by sector of performance, source of funds, type of costs, and estimates of R&D personnel and researchers by occupation (in both head counts and FTEs). It also includes data on government budget appropriations or outlays on R&D (GBAORD) (OECD, 2013b). Data are provided to OECD by member countries and observer economies through the joint OECD/Eurostat International Survey on the Resources Devoted to R&D. Series are available from 1987 to 2010 for 34 OECD countries and a number of nonmember economies. Information on sources and methods used by countries for collecting and reporting R&D statistics is provided in the Sources and Methods database.

OECD's MSTI publication provides indicators of S&T activities for OECD member countries and seven nonmember economies (Argentina, China, Romania, Russian Federation, Singapore, South Africa, and Chinese Taipei). Going back to 1981, MSTI includes indicators on financial and human resources in R&D, GBAORD, patents, technology balance of payments, and international trade in R&D-intensive industries (see http://www.oecd.org/sti/msti).

OECD Patent Database comprises information on patent applications from the European Patent Office (EPO) and the United States Patent and Trademark Office (USPTO), as well as patent applications filed under the Patent Cooperation Treaty (PCT) that designate the EPO and Triadic Patent Families.[5] The EPO's Worldwide Patent Statistical (PATSTAT) database is the primary source of these data. The following patent statistics are available on OECD's statistical portal: patents by country and technology fields (EPO, PCT, USPTO, Triadic Patent Families); patents by regions and selected technology fields (EPO, PCT); and indicators of international cooperation in patents (EPO, PCT, USPTO). OECD has developed four different sets of "raw" patent data for research and analytical purposes, which may be downloaded from its server. OECD also provides tables on biotechnology indicators (see http://www.oecd.org/innovation/innovationinsciencetechnologyandindustry/keybiotechnologyindicators.htm).

At present, no standard OECD database covers innovation statistics based on the *Oslo Manual*. The reason for this is the difficulty of comparing results based on different survey methodologies, particularly those used by countries that follow the Eurostat Community Innovation Survey (CIS) model questionnaire and those used by non-European Union (EU) countries that implement the same concepts and definitions in different ways. Ad hoc data collection on selected innovation indicators has been carried out in recent years, and the results have been published in the *STI Scoreboard* and other related publications.

UNESCO Institute of Statistics

UIS collects its STI data from approximately 150 countries and territories. It has also partnered with three organizations to acquire additional data: on 25 Latin American countries, from the Network on Science and Technology Indicators—Ibero-American and Inter-American (RICYT); on 40 OECD member states and associated countries, from OECD; and on 7 European countries, from Eurostat. UIS conducts a biennial R&D survey, which is administered to the office responsible for national S&T policy or statistics of United Nations (UN) member nations. Even though the survey is administered every 2 years, the questionnaire items request annual information for the previous 5 years. Therefore, the data series is available for 1996 to 2010. A major accomplishment of UIS is that it adapted survey instruments and methodologies and developed other key indicators that are suited to the needs of developing countries. The aim was to enable those countries to apply concepts of the *Frascati Manual* that would in turn produce comparable S&T statistics across nations. The UIS S&T survey not only collects data on R&D expenditures but also elicits information on researchers involved in R&D. The survey uses a standardized occupational classification of researchers: "professionals

[5]See http://www.oecd.org/sti/inno/oecdpatentdatabases.htm for more details and links to sources of these data.

engaged in the conception or creation of new knowledge, products, processes, methods, and systems and also in the management of the projects concerned" (OECD, 2002, p. 93). The classification includes Ph.D. students who are involved in R&D activities.

In 2011, UIS conducted a pilot survey on innovation in the manufacturing sector. Countries surveyed were Brazil, China, Colombia, Egypt, Ghana, Indonesia, Israel, Malaysia, the Philippines, the Russian Federation, South Africa, and Uruguay. The survey included both technological and nontechnological innovation. Survey items were (1) firms involved in innovation, (2) cooperative arrangements, and (3) factors hampering innovation.

Eurostat

Eurostat is the European Commission's statistical office. Its main function is to provide statistical information on European nations to the European Commission. The main themes of Eurostat's statistical portfolio are policy indicators; general and regional statistics; economy and finance; population and social conditions; industry, trade, and services; agriculture and fisheries; external trade, transport, environment, and energy; and STI. Within the STI theme, there are five domains:

1. Research and development—Data are collected from national R&D surveys using definitions from the *Frascati Manual* (OECD, 2002).
2. CIS—Data originate from the national CIS on innovation activity in enterprises that are based on the *Oslo Manual* (OECD-Eurostat, 2005).
3. High-tech industry and knowledge-intensive services—Various origins and methodologies are used; statistics are compiled at Eurostat.
4. Patents—Data originate from the patent database PATSTAT, hosted by EPO. PATSTAT gathers data on applications from the EPO and from about 70 national patent offices around the world (mainly USPTO and the Japan Patent Office); statistics are compiled at Eurostat.
5. Human resources in S&T—Data are derived at Eurostat from the EU Labour Force Survey (LFS) and the Data Collection on Education Systems (UOE) according to the guidelines in the *Canberra Manual* (OECD, 1995).

Statistics Canada

Statistics Canada is the Canadian federal statistical agency with a mandate under the Statistics Act:

(a) to collect, compile, analyze, abstract and publish statistical information relating to the commercial, industrial, financial, social, economic and general activities and condition of the people;
(b) to collaborate with departments of government in the collection, compilation and publication of statistical information, including statistics derived from the activities of those departments;
(c) to take the census of population of Canada and the census of agriculture of Canada as provided in this Act;
(d) to promote the avoidance of duplication in the information collected by departments of government; and
(e) generally, to promote and develop integrated social and economic statistics pertaining to the whole of Canada and to each of the provinces thereof and to coordinate plans for the integration of those statistics.[6]

The Canadian Socio-economic Information Management System (CANSIM) is a socioeconomic database of Statistics Canada and contains data tables from censuses and 350 active surveys. Data are provided under various subjects, such as the system of national accounts, labor, manufacturing, construction, trade, agriculture, and finance.

There are four areas within S&T:

1. R&D—Statistics on R&D expenditures and funding are collected by six surveys focused on various performing and funding sectors:
   a. Research and Development in Canadian Industry
   b. Research and Development of Canadian Private Non-Profit Organizations
   c. Provincial Research Organizations
   d. Provincial Government Activities in the Natural Sciences
   e. Provincial Government Activities in the Social Sciences
   f. Federal Science Expenditures and Personnel, Activities in the Social Sciences and Natural Sciences
2. Human resources in S&T—Data on personnel engaged in R&D are derived from the Federal Science Expenditures and Personnel, Activities in the Social Sciences, and Natural Sciences surveys.
3. Biotechnology—Currently inactive, the 2005 Biotechnology Use and Development Survey provided information on innovation by biotechnology companies.
4. Innovation—CANSIM includes tables from the 2003 and 2005 survey cycles of the Survey of Innovation. Jointly with Industry Canada and Foreign Affairs and International Trade Canada, Statistics Canada conducted the first Survey of Innovation and Business

[6]Available: http://www.statcan.gc.ca/edu/power-pouvoir/about-apropos/5214850-eng.htm [July 2013].

TABLE F-1 Subtopics of Science, Technology, and Innovation Data Produced by Agencies/Organizations, Showing Level of Detail and Unique Variables (boldface text)

Agency/Organization	Total R&D	Industrial R&D	Academic R&D	Federal R&D	GBAORD	Nonprofit R&D
NCSES **(NB: Statistics on R&D expenditure and SEH degrees available by state)**	Total R&D by performer and funder, character of work	Industrial R&D by funder, character of work, NAICS classification, company size	Academic R&D by funder, character of work; entities and subrecipients of academic R&D	Federal R&D by funder, character of work	**R&D obligations and outlays by character of work and performing sector;** reported in *Science and Engineering Indicators* only	Nonprofit R&D by funder, character of work, S&E field, extramural entity, type of nonprofit organization
NCSES and NSB	Total R&D by performer and funder, character of work, country/ economy	Industrial R&D by funder, NAICS classification, company size; R&D performed by multinational companies, foreign affiliates	Academic R&D in S&E and non-S&E fields	Federal R&D by major socioeconomic objectives, country/region	Federal obligations for R&D and R&D plant by agency, performer, character of work	Domestic R&D performed by private nonprofit sector, domestic R&D funded by private nonprofit sector
Statistics Canada	GERD by performer and funder	BERD by funding sector	HERD	GOVERD by socioeconomic objectives, type of science, components		Private nonprofit R&D by funder
OECD	GERD by performer, funder, field of science, socioeconomic objectives	BERD by funding sector, type of cost, size class, field of science, performing industry	HERD, HERD financed by industry	GOVERD, GOVERD financed by industry	GBAORD by socioeconomic objectives	GERD performed by private nonprofit sector
Eurostat **(NB: Almost all data available at regional level)**	GERD by funding source, sector of performance, type of cost, socioeconomic objectives, field of science	BERD by funding source, type of cost, size class, economic activity	HE intramural expenditure by funding source, type of cost, field of science, socioeconomic objectives	Government intramural expenditure by funding source, sector of performance, type of cost, socioeconomic objectives, field of science	GBAORD by socioeconomic objectives	GERD performed by private nonprofit sector; GERD funded by private nonprofit sector
UNESCO	GERD by performing sector and funding sector, field of science, character of work	GERD performed by business enterprise sector; GERD funded by business enterprise sector	GERD performed by higher education sector; GERD funded by higher education sector			GERD performed by private nonprofit sector; GERD funded by private nonprofit sector

Agency/ Organization	Technology BOP and International Trade in R&D-Intensive Industries	Patents and Venture Capital	R&D Personnel, Scientists and Engineers	Science, Engineering, and Health Degrees; Assessment Scores	Innovation	Public Attitudes Toward Science and Technology
NCSES **(NB: Statistics on R&D expenditure, SEH degrees available by state)**			Scientists and engineers by gender, age, race/ ethnicity, level of highest degree, occupation, labor force status, employment sector, primary/ secondary work activity, median annual salaries	Graduate students, doctorate holders, postdoctorates, nonfaculty research staff by gender, race/ethnicity, citizenship, academic field, Carnegie classification; SEH doctorates by gender, age, race/ethnicity, occupation, labor force status, employment sector, primary/secondary work activity, postdoctoral appointments, median annual salaries	Product and process innovation by NAICS classification	
NCSES and NSB	U.S. trade balance in research, development, and testing services by affiliation; exports of high-technology and manufactured products by technology level, product, region/ country/ economy; global value added by type of industry; ICT infrastructure index; U.S. high-technology microbusinesses	USPTO patents granted by selected technology area, region/country/ economy; patenting activity in clean energy and pollution control technologies; patents granted by BRIC nations by share of resident and nonresident inventors; patent citations to S&E articles by patent technology area, article field; patenting activity of employed U.S.-trained SEH doctorate holders; U.S. venture capital investment by financing stage and industry/technology; venture capital disbursed per $1,000 of GDP; venture capital deals as a percentage of high-technology business establishments; venture capital disbursed per venture capital deal by state	Workers in S&E and STEM occupations by MSA, occupation category, educational background, R&D work activities, age, ethnicity/ race; scientists and engineers reporting international engagement by demographic characteristics, education, employment sector, occupation, salary, work-related training; foreign-born workers in S&E occupations by education level	SEH doctorate holders by gender, race/ethnicity, field of doctorate, sector of employment, academic appointment, salary, unemployment rate; S&E doctorate recipients and full-time S&E graduate students by source, primary mechanism of support, Carnegie classification; foreign recipients of U.S. S&E doctorates by field, country/economy of origin; field switching among postsecondary students; time taken to receive an S&E doctorate; community college attendance among recent recipients of S&E degrees by sex, race/ ethnicity, degree level, degree year, citizenship status; NAEP assessment scores in mathematics and science; advanced placement exam taking by public school students	Small Business Innovation Research funding per $1 million of GDP by state	Media coverage, news stories by topic area; correct answers to S&T and S&T-related questions by gender and country/ region; public perceptions of various occupational groups' contribution to society and public policy-making process; public assessment/ opinion of stem cell research and environmental problems; source of information for S&T issues

continued

TABLE F-1 Continued

Agency/Organization	Total R&D	Industrial R&D	Academic R&D	Federal R&D	GBAORD	Nonprofit R&D
Statistics Canada		**Intellectual property commercialization by higher education sector; intellectual property management by federal department and agency**	Researchers, support staff, technicians by R&D performing sector and type of science; federal personnel engaged in S&T by activity, type of science, S&T component	University degrees, diplomas, and certificates granted by program level and Classification of Instructional Programs, gender, immigration status	Innovation activities; product and process innovation; degree of novelty; hampering factors of and obstacles to innovation; important sources of information; cooperation arrangements; innovation impacts; methods of protection; geomatic activities	
OECD	BOP—payments and receipts; trade—imports and exports by R&D-intensive industries	Patent applications, Triadic Patent Families, patents in selected technologies by region, international cooperation	R&D personnel by field of science, sector of performance, qualification; researchers by sector of performance and gender	Graduates by field and level of education; PISA scores in science and mathematics		
Eurostat **(NB: Almost all data available at subnational level)**	Trade in high-tech industries and knowledge-intensive sectors within EU and ROW; employment in these sectors by gender, occupation, educational qualification, mean earnings	**Patent applications at USPTO and EPO by priority year and sector, ownership of patents, patent citations;** European and international copatenting, Triadic Patent Families	Human resources in S&T, R&D personnel, and researchers by gender, field of science, sector of performance, qualification; citizenship of researchers in government and higher education sector	Doctorate holders by gender, activity status; employed doctorate holders by gender, sector, occupation, field of science, **job mobility**	Innovation activities; product and process innovation; degree of novelty; hampering factors of and public funding for innovation; important sources of information; cooperation arrangements; environmental innovation; objectives of innovation; impacts of innovation; methods of protection; employees in innovation sector	

UNESCO	R&D personnel and researchers (FTE and head count) by gender, performing sector, educational qualification, field of science; technicians and other supporting staff by performing sector	Innovation in manufacturing sector—firms involved in innovation; cooperation arrangements; hampering factors of innovation; available for 12 nations

NOTES: BERD = business enterprise expenditure on research and development; BOP = balance of payments; BRIC = Brazil, Russia, India, and China; EPO = European Patent Office; EU = European Union; FTE = full-time equivalent; GBAORD = government budget appropriations or outlays for research and development; GDP = gross domestic product; GERD = gross domestic expenditure on research and development; GOVERD = government intramural expenditure on research and development; HE = higher education; HERD = higher education expenditure on research and development; ICT = information and communication technology; MSA = metropolitan statistical area; NAEP = National Assessment of Educational Progress; NAICS = North American Industry Classification System; NSB = National Science Board; NCSES = National Center for Science and Engineering Statistics; PISA = Programme for International Student Assessment; R&D = research and development; ROW = rest of the world; S&E = science and engineering; SEH = science, engineering, and health; STEM = science, technology, engineering, and mathematics; UNESCO = United Nations Educational, Scientific and Cultural Organization; USPTO = United States Patent and Trademark Office.

SOURCES: Adapted from BRDIS, see http://www.nsf.gov/statistics/industry/ [November 2012]. Federal Funds, see http://www.nsf.gov/statistics/fedfunds/ [November 2012]. R&D Expenditure at FFRDCs, see http://www.nsf.gov/statistics/ffrdc/ [November 2012]. HERD, see http://www.nsf.gov/statistics/herd/ [November 2012]. Science and Engineering State Profiles, see http://www.nsf.gov/statistics/pubseri.cfm?seri_id=18 [November 2012]. S&E I 2012, see http://www.nsf.gov/statistics/seind12/tables.htm [November 2012]. UNESCO, see http://www.uis.unesco.org/ScienceTechnology/Pages/default.aspx [November 2012]. OECD, see http://stats.oecd.org/Index.aspx?DataSetCode=MSTI_PUB [November 2012]. Eurostat, see http://epp.eurostat.ec.europa.eu/portal/page/portal/science_technology_innovation/data/database European Union, 1995-2013 [November 2012]. Statistics Canada, CANSIM, see http://www5.statcan.gc.ca/cansim/a33?lang=eng&spMode=master&themeID=193&RT=TABLE [November 2012].

Strategy (SIBS) in 2010. Information was collected from enterprises for the period 2007-2009, and survey estimates were published in 2011. SIBS data are not available in CANSIM.

World Intellectual Property Organization (WIPO)

WIPO is a specialized UN agency focused on intellectual property—patents, trademarks, copyrights, and design. It collects data by sending questionnaires to the intellectual property offices of 185 member states. It produces annual statistics on patents, utility models, trademarks, industrial designs, and plant varieties, thus creating a comprehensive national database of intellectual property. Some of the data series for particular nations go back to 1885.

Innovation Data from Other U.S. Agencies

Apart from NCSES, other U.S. agencies collect innovation statistics, ranging from patents and trademarks to grants and federal awards (see Table F-2):

- USPTO—Three datasets are available from USPTO—the Patent Assignments Dataset, Trademark Casefile Dataset, and Trademark Assignments Dataset. As their names suggest, these datasets comprise ownership and changes in ownership for USPTO-granted patents and trademarks.
- Economic Research Service (ERS), U.S. Department of Agriculture (USDA)—In 2013, ERS began fielding its first Rural Establishment Innovation Survey, which is aimed at business establishments funded through USDA's Rural Development Mission Area. The purpose of the survey is threefold: to collect information on the adoption of innovative practices and their contribution to firm productivity; to discover how participation in federal, state, and local programs aids the growth of rural business units; and to determine usage of available local and regional assets, such as workforce education and local business associations, by rural business units.
- NIH, NSF, and the White House Office of Science and Technology Policy—Science and Technology for America's Reinvestment: Measuring the Effect of Research on Innovation, Competitiveness, and Science (STAR METRICS) is a multiagency venture that relies on the voluntary participation of science agencies and research institutions to document the outcomes of science investments for the public. Currently, more than 90 institutions are participating in the program. The STAR METRICS data infrastructure contains recipient-based data that include information on contract, grant, and loan awards made under the American Recovery and Reinvestment Act of 2009.
- NSF—The U.S. government's research spending and results webpage provides information on active NSF and National Aeronautics and Space Administration (NASA) awards, such as awardees, funds obligated, and principal investigator.
- National Institute of Standards and Technology (NIST), Department of Commerce (Anderson, 2011)—NIST has been responsible for preparing the Department of Commerce's report on technology transfer utilization. The *Federal Laboratory (Interagency) Technology Transfer Summary Reports* cover federal laboratories and FFRDCs. They contain data tables on patent applications, invention licenses, cooperative R&D agreements, and R&D obligations, both extramural and intramural.
- Small Business Administration (SBA)—The Small Business Innovation Research (SBIR) program and the Small Business Technology Transfer (STTR) program fall under the administration of the SBA's Technology Program Office. These programs award more than $2 billion each year to small high-tech businesses.[7] The SBA-Tech.net website includes a searchable database on federal R&D funds/awards by agency, category, and state.
- Department of Energy (DOE)—DOE's visual patent search tool allows users to collect information on issued U.S. patents and published patent applications that result from DOE funding.

Data collected by federal statistical agencies, either through surveys or from administrative databases, contain rich information on various economic and social issues. Most of this information is used by private corporations and educational institutions (sometimes the agencies themselves) that either present the data in a comprehensive fashion or develop tools for dissemination and analysis. Some of those efforts are outlined below:

- Google—On its website, Google hosts a bulk download tool that allows users to download data tables on patents and trademarks issued by USPTO.
- NIH, NSF, and the White House Office of Science and Technology Policy—Applications of the STAR METRICS data platform include the Portfolio Explorer Project, a tool for examining public research award information by topic, region, institution, and researcher. STAR METRICS currently uses four tools to view scientific portfolios (Feldman, 2013b):
  - —The Portfolio Viewer provides information on proposals, awards, researchers, and institutions by program level and scientific topic.

[7]For details, see http://www.sba.gov/about-sba-services/7050 [July 2013].

TABLE F-2 Innovation Data from U.S. Agencies Other Than NCSES

Agency	Database/Survey/Data Collection Mechanism	Indicator/Data Items	Time Period
United States Patent and Trademark Office (USPTO)	Patent Assignments Dataset	Patent assignments and change of ownership of patents that are granted by USPTO	2010 onward
	Trademark Casefile Dataset	Trademarks granted by USPTO	1884-2010
	Trademark Assignments Dataset	Change of ownership of trademarks granted by USPTO	2010 onward
Economic Research Service, U.S. Department of Agriculture	Rural Establishment Innovation Survey	Inventory of innovation activities; use of technology by labor force, establishment, and community characteristics; factors hampering innovation; funding source for innovation; applications for intellectual property and trademarks; sources of information on new opportunities	First survey cycle was conducted in 2013
National Institutes of Health, National Science Foundation, and White House Office of Science and Technology Policy	STAR METRICS	Recipient-based data containing information on contract, grant, and loan awards made under the American Recovery and Reinvestment Act of 2009	2009-2012
National Science Foundation	Research Spending and Results	Recipient-based data containing information on awards made by the National Science Foundation and the National Aeronautics and Space Administration	2007 onward
National Institute of Standards and Technology, U.S. Department of Commerce	Federal Laboratory (Interagency) Technology Transfer Summary Reports	Patent applications, invention licenses, cooperative R&D agreements, R&D obligations—extramural and intramural	1987-2009
Small Business Administration	Small Business Innovation Research (SBIR) and Small Business Technology Transfer (STTR) programs	Federal R&D funds/ awards by agency, category, and state	1983-2012
U.S. Department of Energy **NB: Information available at the state, county, and municipal levels, as well as from utilities and nonprofits**	Energy Innovation Portal—Visual Patent Search Tool	Issued U.S. patents and published patent applications that are created using Department of Energy funding	1979 onward
	Advanced Manufacturing Office—State Incentives and Resource Database	Energy-saving incentives and resources available for commercial and industrial plant managers	

SOURCES: USPTO databases, see http://www.gwu.edu/~gwipp/Stuart%20Graham%20020712.pdf. Rural Development, USDA, see http://www.gpo.gov/fdsys/pkg/FR-2011-06-22/html/2011-15474.htm. STAR METRICS, see https://www.starmetrics.nih.gov/Star/Participate#about. Research Spending and Results, see https://www.research.gov/research-portal/appmanager/base/desktop?_nfpb=true&_eventName=viewQuickSearchFormEvent_so_rsr. Federal Laboratory (Interagency) Technology Transfer Summary Reports, see http://www.nist.gov/tpo/publications/federal-laboratory-techtransfer-reports.cfm. SBIR and STTR, see http://www.sbir.gov/. Energy Innovation Portal, see http://techportal.eere.energy.gov/. Advanced Manufacturing Office, see http://www1.eere.energy.gov/manufacturing/.

— The Expertise Locator provides information on proposals and coprincipal investigators related to different topic areas to make it possible to locate researchers working on that topic.
— The Patent Viewer provides data on patents from NSF grantees.
— The Map Viewer offers a geographic tool for viewing NSF investments by institution and topic.

- DOE—DOE's Green Energy Data Service contains bibliographic data for patents relating to various forms of green energy (e.g., solar, wind, tidal, bioenergy) resulting from research sponsored by DOE and predecessor agencies.
- Indiana University—Innovation in American Regions is a project funded in part by the U.S. Commerce Department's Economic Development Administration. The research is conducted by the Purdue Center for Regional Development, Indiana University, Kelley School of Business. The web tools available to users are the Innovation Index, Cluster Analysis, and Investment Analysis. The Innovation Index is a weighted index of indicators based on four components—human capital, economic dynamics, productivity and employment, and economic well-being. Cluster Analysis depicts occupation and industry clusters for any state, metro area, micro area, district, or county in the nation. Investment Analysis provides various kinds of information to aid regional investors.
- Innovation Ecologies Inc.—The Regional Innovation Index is a single data platform consisting of a host of indicators from various sources. The indicators measure venture capital, labor inputs, personal income, education and training, globalization, Internet usage, R&D inputs, universities, quality of life, knowledge, employment outcomes in firms and establishments, social and government impacts, and innovation processes.

Innovation Data from Private Sources

A number of educational institutions and corporate organizations collect and disseminate innovation data (see Table F-3):

- University of California, Los Angeles (UCLA)—Zucker and Darby (2011) developed the COMETS (Connecting Outcome Measures in Entrepreneurship Technology and Science) database. COMETS is an integrated database of principal investigators, dissertation writers and advisers, inventors, and employees at private-sector firms. COMETS data can be used to trace government expenditures on R&D from the initial grant through knowledge creation, translation, diffusion, and in some cases commercialization.
- Association of University Technology Managers (AUTM)—AUTM has been conducting licensing surveys on U.S. and Canadian universities, hospitals, and research institutions since 1991. Twenty years of data from participating institutions are placed in Statistics Access for Tech Transfer (STATT), a searchable and exportable database. It contains information on income from, funding source of, staff size devoted to, and legal fees incurred for licensing; start-ups that institutions created; resultant patent applications filed; and royalties earned.
- Association of Public and Land-grant Universities (APLU), Commission on Innovation, Competitiveness, and Economic Prosperity (CICEP)—APLU is involved in creating new metrics with which to measure the economic impact of universities at the regional and national levels. APLU's CICEP has been working to identify and investigate the efficacy of potential metrics in the areas of human capital and knowledge capital. Indicators being investigated range from unfunded agreements between universities and industry (e.g., material transfer agreements, nondisclosure agreements), to student engagement in economic activities, to the impacts of technical assistance provided by universities to various actors in the region's economy.
- Harvard University—Patent Network Dataverse (Feldman, 2013a) is an online database created and hosted by the Institute for Quantitative Social Science at Harvard University. This is a "virtual web archive" that has, among other things, matched patents and publication data. Researchers use Dataverse to publish, share, reference, extract, and analyze data.
- PricewaterhouseCoopers and the National Venture Capital Association—*The MoneyTree Report* is published quarterly and is based on data provided by Thomson Reuters. The report contains data on venture capital financing, including the companies that supply and receive the financing.
- Venture capital database—CB Insights, Venture Deal, Grow Think Research, and Dow Jones VentureSource have venture capital databases that profile venture capital firms and venture capital-financed firms.

TYPES OF INFORMATION CAPTURED BY VARIOUS STI DATABASES

STI data can be broadly categorized into three distinct subtopics:

1. R&D expenditure—Total R&D activity in a nation can be further broken down into:
   — Total R&D expenditure or gross domestic expenditure on R&D (GERD),

TABLE F-3 Innovation Product Data from Private Sources

Agency	Database/Survey/Data Collection Mechanism	Indicator/Data Items	Time Period
University of California, Los Angeles	Connecting Outcome Measures in Entrepreneurship, Technology, and Science (COMETS) database	Integrates data on government grants, dissertations, patents, and publicly available firm data; currently contains information on patents granted by the U.S. Patent and Trademark Office (USPTO) and on National Science Foundation (NSF) and National Institutes of Health (NIH) grants	2007-2012
Association of University Technology Managers	Statistics Access for Tech Transfer (STATT)	Academic licensing data from participating academic institutions: licensing activity and income, start-ups, funding, staff size, legal fees, patent applications filed, royalties earned	1991-2010
Association of Public and Land-grant Universities—Commission on Innovation, Competitiveness, and Economic Prosperity	New Metrics to Measure Economic Impact of Universities	Relationship with industry: agreements, clinical trials, sponsored research, external clients Developing the regional and national workforce: student employment, student economic engagement, student entrepreneurship, alumni in workforce Knowledge incubation and acceleration programs: success in knowledge incubation and acceleration programs, ability to attract investments, relationships between clients/program participants and host university	Pilot conducted in spring 2012 with 35 participating institutions
Harvard University	Patent Network Dataverse: U.S. Patent Inventor Database	Patent coauthorship network	1975 onward
PricewaterhouseCoopers and National Venture Capital Association	*Money Tree Report*	Venture capital firms and firms receiving financing: quarterly and yearly investment amounts, number of deals by industry, stage of development, first-time financings, clean technology, and Internet-specific financings	Quarterly data, 1st quarter 1995 onward
CB Insights, Venture Deal, Grow Think Research, Dow Jones VentureSource	Venture Capital Database	Profile of venture capital firms and venture capital-financed firms	

SOURCES: COMETS Database, see http://scienceofsciencepolicy.net/?q=node/3265. STATT database, see http://www.autm.net/source/STATT/index.cfm?section=STATT. APLU Economic Impact, see http://www.aplu.org/page.aspx?pid=2693. Patent Network Dataverse, see http://thedata.harvard.edu/dvn/dv/patent. Money Tree Report, see https://www.pwcmoneytree.com/MTPublic/ns/index.jsp. Venture Capital databases, see http://www.cbinsights.com/; http://www.venturedeal.com/; http://www.growthinkresearch.com/; https://www.venturesource.com/login/index.cfm?CFID=2959139&CFTOKEN=53e4cab-1e600d5d-9089-411f-a010-949554ae0978.

— Business R&D expenditure or business enterprise expenditure on R&D (BERD),
— Academic R&D expenditure or higher education expenditure on R&D (HERD),
— Federal R&D expenditure,
— Government intramural expenditure on R&D (GOVERD),
— Government budget appropriations or outlays on R&D (GBAORD), and
— R&D performed and/or funded by nonprofit organizations.

2. Human capital/human resources in S&T—It comprises human capital in S&T, including individuals in S&T occupations and those with degrees in S&T fields. Most of the above-mentioned agencies/organizations produce statistics on both subgroups. The variables reported are:
   — total R&D personnel;
   — researchers;
   — technicians;
   — other supporting staff;
   — scientists and engineers;
   — number of degrees in science, engineering, and health (SEH) fields; and
   — number of graduates in S&E fields.
3. Innovation—Statistics on business innovation are being collected by NCSES through BRDIS. NCSES has released two *InfoBriefs* (NSF 11-300 and NSF

12-307) that provide information on technologically innovative firms and usage of methods for protecting intellectual property, both by North American Industrial Classification System (NAICS) classification. Data in both *InfoBriefs* were gathered from the 2008 BRDIS. BRDIS focuses on technological innovation (product and process) and was inspired by Eurostat's CIS, which looks at both technological and nontechnological innovation. Similarly, Statistics Canada's SIBS contains many elements borrowed from the CIS. To some extent, this ensures that questions across the three surveys align, and may be helpful for international data comparisons. As the survey results become available, it will be possible to answer the question of whether the data across all three surveys are truly comparable; for now, however, it is too early to say. The subtopics within innovation statistics stem from sections/questions in survey questionnaires and can divided into nine categories:[8]

1. type of innovation activity—product, process, organizational, marketing;
2. innovation activity and expenditure;
3. turnover due to innovative products;
4. objectives of innovation;
5. sources of information on innovation;
6. cooperation in innovation activity;
7. factors hampering innovation activity;
8. government support/public funding for innovation; and
9. innovation with environmental benefits.

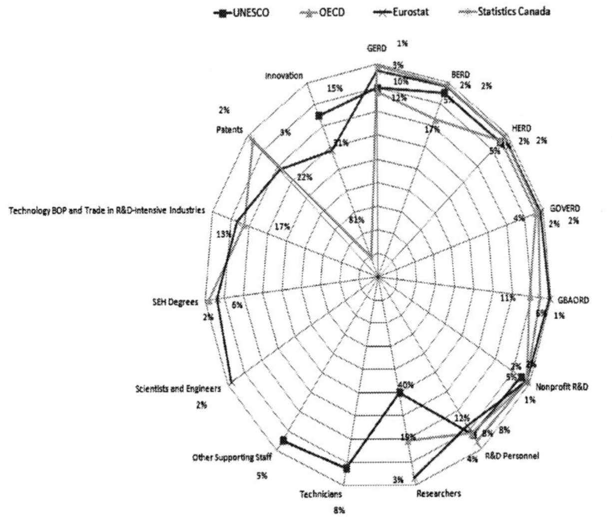

FIGURE F-1 Subtopics of science, technology, and innovation data produced by agencies/organizations other than the National Center for Science and Engineering Statistics.
NOTES: The scale is in reverse order. As one moves closer to the epicenter, the value increases. BERD = business enterprise expenditure on R&D; BOP = balance of payments; GBAORD = government budget appropriations or outlays for research and development; GERD = gross domestic expenditure on research and development; GOVERD = government intramural expenditure on R&D; HERD = higher education expenditure on research and development; R&D = research and development; SEH = science, engineering, and health; UNESCO = United Nations Educational, Scientific and Cultural Organization.
SOURCES: Adapted from UNESCO, see http://www.uis.unesco.org/ScienceTechnology/Pages/default.aspx [November 2012]. OECD, see http://stats.oecd.org/Index.aspx?DataSetCode=MSTI_PUB [November 2012.] Eurostat, see http://epp.eurostat.ec.europa.eu/portal/page/portal/science_technology_innovation/data/database European Union, 1995-2013 [November 2012]. Statistics Canada, CANSIM, see http://www5.statcan.gc.ca/cansim/a33?lang=eng&spMode=master&themeID=193&RT=TABLE [November 2012].

Table F-1, presented earlier, outlines the level of detail available in the STI data produced by NCSES, Statistics Canada, OECD, Eurostat, and UNESCO. Unique variables produced by these agencies—those not available from other organizations—are highlighted in the table.

Even though agencies and other organizations try to produce STI statistics covering the subtopics, some of them clearly have an advantage over others in certain areas. Staff of the Committee on National Statistics looked at the concentration of agencies and other organizations in various subtopics (see Figures F-1 and F-2). The metric used in these figures is the percentage of tables produced on a particular subtopic relative to the total tables generated by the STI database. Using Eurostat's statistics database as an example, it has 330 tables on various STI subtopics (see Table F-1). Of those, 9 tables show GERD values of member nations disaggregated by economic activity, costs, and so on. Similarly, there are 40 tables on R&D personnel and their various attributes, which brings the percentage of tables on the R&D personnel subtopic to 12 percent (40 divided by 330). A separate figure was created for NCSES to avoid confusion. As seen in Figures F-1 to F-3, STI data produced by NCSES are oriented toward scientists and engineers and SEH degrees; Statistics Canada and Eurostat focus more on innovation topics, and OECD and UNESCO on researchers.

METHODOLOGY

The panel used cluster analysis, which includes multidimensional scaling (MDS), and a heat map tool to understand the redundancy in the main S&T indicators produced by the above-mentioned organizations/agencies. Although MDS and the heat map are not exclusive approaches to analyzing STI data, they are among many possible paths to understanding the issue of redundancy in the multitude of variables published by various agencies and organizations. Both methods

[8]Subtopics in the CIS.

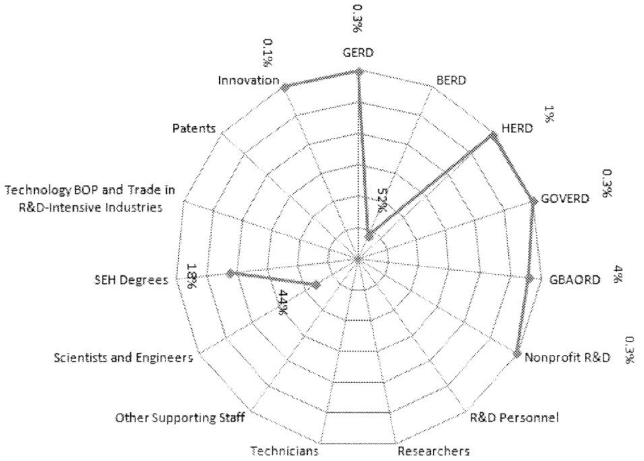

FIGURE F-2 Subtopics of science, technology, and innovation data produced by the National Center for Science and Engineering Statistics.
NOTES: The scale is in reverse order. As one moves closer to the epicenter, the value increases. BERD = business enterprise expenditure on R&D; BOP = balance of payments; GBAORD = government budget appropriations or outlays for research and development; GERD = gross domestic expenditure on research and development; GOVERD = government intramural expenditure on R&D; HERD = higher education expenditure on research and development; R&D = research and development; SEH = science, engineering, and health.
SOURCES: Adapted from BRDIS, see http://www.nsf.gov/statistics/industry/ [November 2012]. Federal Funds, see http://www.nsf.gov/statistics/fedfunds/ [November 2012]. R&D Expenditure at FFRDCs, see http://www.nsf.gov/statistics/ffrdc/ [November 2012]. HERD, see http://www.nsf.gov/statistics/herd/ [November 2012]. Science and Engineering State Profiles, see http://www.nsf.gov/statistics/pubseri.cfm?seri_id=18 [November 2012].

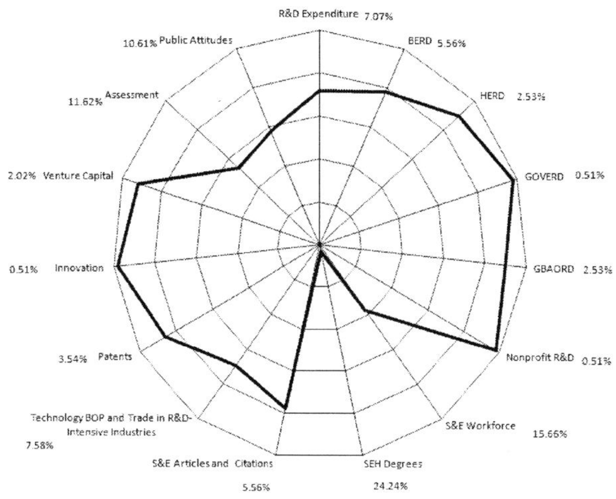

FIGURE F-3 Subtopics of science, technology, and innovation indicators published in *Science and Engineering Indicators 2012 Digest.*
NOTES: The scale is in reverse order. As one moves closer to the epicenter, the value increases. BERD = business enterprise expenditure on R&D; BOP = balance of payments; GBAORD = government budget appropriations or outlays for research and development; GERD = gross domestic expenditure on research and development; GOVERD = government intramural expenditure on R&D; HERD = higher education expenditure on research and development; R&D = research and development; S&E = science and engineering; SEH = science, engineering, and health.
SOURCE: Adapted from *Science and Engineering Indicators 2012*, see http://www.nsf.gov/statistics/seind12/tables.htm.

offer wide-ranging applications in various fields and have helped researchers gain some amount of understanding of the dataset on which they are working.

Generally, cluster analysis[9] is a collection of methods for finding distinct or overlapping clusters in data. It is an analytic procedure for grouping sets of objects into subsets that are relatively similar among themselves. In a broad sense, there are two methods of clustering—hierarchical and partitioning. With hierarchical methods, small clusters are formed that merge sequentially into larger and larger clusters until only one remains, resulting in a tree of clusters. Partitioning methods split a dataset into a set of discrete clusters that are nonhierarchical in nature because they do not fit into a tree or hierarchy. Different numbers of clusters on the same dataset can result in different partitioning that may overlap. To produce clusters, there must be some measure of dissimilarity or distance among objects. Similar objects should appear in the same cluster and dissimilar objects in different clusters. Different measures of similarity produce different hierarchical clusterings. If there are two vectors consisting of values on p features of two objects, popular distance measures are:[10]

- Euclidean—the square root of the sum of squared elementwise differences between the two vectors;
- City Block—the sum of absolute differences between the two vectors;
- cosine—the inner product of the two vectors divided by the product of their lengths (norms);
- Pearson 1—the Pearson correlation between two vectors; and
- Jaccard—the sum of the mismatches between the elements of one vector and the elements of the other.

[9] For examples of the broad analytical capabilities of cluster analysis, see Feser and Bergman (2000) on industrial clusters, Myers and Fouts (2004) on K-12 classroom environments for learning science, and Newby and Tucker (2008) in the area of medical research.

[10] This explanation of distance measures and linkage methods is based on the Data Analysis output of *AdviseStat* (see http://www.skytree.net/products-services/adviser-beta/ [December 2012]).

Euclidean distances are sensitive to differences between values because the differences are squared, and larger differences carry more weight than small. City Block distances are similarly sensitive to differences between values, but these differences are not squared. In contrast with Euclidean or City Block distances, cosine-based distances are invariant to scaling—multiplying all values by a constant will not change cosine distance. Pearson-based distances are invariant to scaling and location—adding a constant to the values will not change these distances. Jaccard distances are based only on mismatches of values and are invariant under one-to-one recodings of unique values. All of these measures are real (metric) distances; they obey the metric axioms.[11]

Multidimensional Scaling

The panel used MDS models to discover which sets of indicators are more similar to each other and to help in arriving at a set of primary and derivative indicators. Table F-1 and Figure F1, presented earlier, show the great number of variables capturing numerous pieces of STI information. From the viewpoint of an agency or organization, it is important to understand which indicators are necessary for addressing key policy questions and in turn make the production of STI variables more efficient. Various applications of MDS are documented by Young and Hamer (1987). MDS is frequently applied to political science data, such as voting preferences. For example, Minh-Tam and colleagues (2012) used MDS to embed the network of capital cities of European nations based on their pairwise distances (Minh-Tam et al., 2012; Nishimura et al., 2009).

The original motivation for MDS was to fit a matrix of dissimilarities or similarities to a metric space. Since its origins, however, MDS has had many other applications. A popular use is to compute a distance matrix on the columns of a rectangular matrix using a metric distance function (Euclidean, cosine, Jaccard, power, etc.). The result is that MDS projects the original variables into a low-dimensional (usually 2-dimensional) space. This approach is an alternative to principal components analysis. If the projection is intrinsically nonlinear for a given dataset, MDS can provide a better view than principal components.

Young (2013, p. 1) describes the process as follows:

> Multidimensional scaling (MDS) is a set of data analysis techniques that display the structure of distance-like data as a geometrical picture. It is an extension of the procedure discussed in scaling. . . . MDS pictures the structure of a set of objects from data that approximate the distances between pairs of the objects. . . . Each object or event is represented by a point in a multidimensional space. The points are arranged in this space so that the distances between pairs of points have the strongest possible relation to the similarities among the pairs of objects. That is, two similar objects are represented by two points that are close together, and two dissimilar objects are represented by two points that are far apart.

Given a configuration of points in a metric space, one can compute a symmetric matrix of pairwise distances on all pairs of points. By definition, the diagonal of this matrix is zero, and the off-diagonal elements are positive. Now suppose a condition is inverted. One has an input matrix **X** and wants to compute a distance matrix **Y** containing the coordinates of points in the metric space using the distance formula provided by the metric. A general formula for a distance metric is:

$$d_{ij}^{p} = \Sigma_{a}^{r} \mid X_{ia} - X_{ja} \mid^{P}, (p \geq 1), X_{i} \neq X_{j}$$

where there are r dimensions, where X_{ia} is the coordinate of point *i* on dimension a, and where X_i is an r-element row vector from the ith row of the n by r matrix **X** containing the coordinates X_{ia} of all n points on all r dimensions. For d_{ij} to satisfy all of the properties of a metric, d_{ij} must be positive. Therefore, only the positive root of d_{ij} must be used in determining d_{ij}. This is known as a Minkowski model. Three special cases of the Minkowski model are of primary interest. One of these is the Euclidean model, which is obtained when the Minkowski exponent (p) is 2. The second is the city block or taxicab model; when p = 1, d_{ij} is simply the sum of absolute difference in the coordinates of the two points. When p is infinitely large, the Dominance model is obtained.

The MDS analysis in this report uses the Euclidean model, as described earlier in this chapter. For the application to STI indicators, the input matrix **X** needed to be symmetric, which refers to $X_{ia} = X_{ai}$. Since the input matrix was not symmetric initially, a matrix of correlation coefficients of the variables in the analysis was used.[12]

Heat Map

Another notable method is cluster heat maps. In certain fields, the analyst wants to cluster rows and columns of a matrix simultaneously. The popular display is called the cluster heat map.[13] Wilkinson and Friendly (2009) describe heat map analysis as follows:

> The cluster heat map is a rectangular tiling of a data matrix with cluster trees appended to its margins. Within a relatively compact display area, it facilitates inspection of row, column,

[11]Metric axioms are: Identity, where distance (A, A) = 0; Symmetry, where distance (A, B) = distance (B, A); and Triangle Inequality, where distance (A, C) ≤ distance (A, B) + distance (B, C). See http://www.pigeon.psy.tufts.edu/avc/dblough/metric_axioms.htm [July 2013].

[12]For more details on the derivation of the Euclidean model from the general formula for a distance metric, see Young and Hamer (1987, p. 87).

[13]The whole explanation of clustering methods is taken from "A Second Opinion on Cluster Analysis," *Whitepaper on a Second Opinion*, downloaded from the *AdviseAnalystics* website (https://adviseanalytics.com/advisestat [December 2012]).

> and joint cluster structure. Moderately large data matrices (several thousand rows/columns) can be displayed effectively on a higher-resolution color monitor, and even larger matrices can be handled in print or in megapixel displays.

The heat map also orders the variables such that similar variables are closer to each other. Cluster heat maps are built from two separate hierarchical clusterings on rows and columns, and as a consequence, each rests on the same foundation that one-way hierarchical clustering involves.

ANALYSIS

STI databases consist of variables, some of which measure R&D expenditures, some the numbers of scientists and engineers, others the amount of trade taking place, and so on. As there is no uniform scale for all of these variables (dollar figures and actual counts), the panel decided to use Pearson distance, which is 1 – the Pearson correlation between two vectors. Therefore, if a correlation is negative, the distance will be greater than 1. For standardized variables (z-scores), 1 – Pearson is equivalent to Euclidean distance. One can also conclude that hierarchical clustering depends on linkage methods. These methods determine how the distance between two clusters is calculated. They are:

- single—distance between the closest pair of objects, one object from each group;
- complete—distance between the farthest pair of objects, one object from each group;
- average—average of distances between all pairs of objects, one object from each group;
- centroid—distance between the centroids of the clusters;
- median—distance between the centroids of the clusters weighted by the size of the clusters; and
- Ward—increase in the within-cluster sum of squares as a result of joining two clusters.

Once distances between clusters have been computed, the closest two are merged. Single linkage tends to produce long, stringy clusters, whereas complete linkage tends to produce compact clusters. Ward clustering usually produces the best hierarchical trees when the clusters are relatively convex and separated. Since the panel believes Ward linkage is the best all-around method, it was used for this analysis.

To analyze STI data, the panel used hierarchical clustering, cluster heat map, and MDS methods. For purposes of analysis, we used the statistics program AdviseStat, an expert system for statistics akin to an intelligent analytics advisor. In the cluster analysis, we used Pearson correlations as the similarity measure, and the hierarchical clustering used Ward linkage. Variables were standardized before similarities were computed. Standardizing puts measurements on a common scale and prevents one variable from influencing the clustering because it has a larger mean and/or variance. In general, standardizing makes overall level and variation comparable across measurements. As mentioned above, comprehensive evaluation of STI variables leads to scale issues as variables capture different types of information. Hence, it is necessary to put the variables in a common scale.

The analysis consisted of three segments, depending on the type of data. OECD, UNESCO, and Eurostat collect STI information from member nations; NCSES and Statistics Canada collect similar information from a single nation. Analyzing variables from all five databases would be intractable. Therefore, it was necessary to separate the analysis into two groups—many nations and single nation. In the many nations analysis, we concentrated on (1) variables from OECD, UNESCO, and Eurostat and (2) indicators from the *SEI 2012 Digest*. The single nation analysis has two components: (1) U.S. R&D expenditures and funding as published by NCSES, OECD, UNESCO, and Eurostat and (2) U.S. R&D human capital as published by NCSES, OECD, UNESCO, and Eurostat. The third segment focused on innovation data published by NCSES, Eurostat, and Statistics Canada. The conclusions and observations from the analysis are summarized below.

Many Nations Analysis

This analysis is restricted to OECD, UNESCO, and Eurostat as their databases contain data on more than one nation. As mentioned earlier, the data series of OECD and Eurostat go back to 1981, while that of UNESCO begins in 1996. Here again, the data were divided into two parts. The first part of the analysis focused on main S&T variables from the three databases for 1996 to 2011 (see Figures F-4, F-5, and F-6). The second part of the analysis was based on a subset of those variables, for which information was available going back to 1981, and hence was restricted to OECD and Eurostat (see Figures F-7, F-8, and F-9). The selected variables are listed in Table F-4. It should be noted that this is not an exhaustive list of all the variables available in the three databases. Early in the analysis, the panel took a "kitchen sink" approach whereby the analysis was run on all variables. We ran into multiple clusters, as many of the variables are tabs of a main variable. For example, "number of foreign citizen female researchers in government sector" is a subset of "number of female researchers." As can be seen in Table F-1, the number of variables that can be gleaned from a single subtopic is large; for example, Eurostat's STI database contains at least 180 variables on human resources in S&T. The aim of the cluster analysis and MDS is to understand what redundancy exists in main S&T variables, and our analysis was therefore restricted to the selected variables shown in Table F-4 that address various subtopics. The variable names start with EURO, OECD, or UN, denoting the S&T database to which the variables belong. It should not be assumed that the excluded variables are unimportant—Chapter 3 of this report

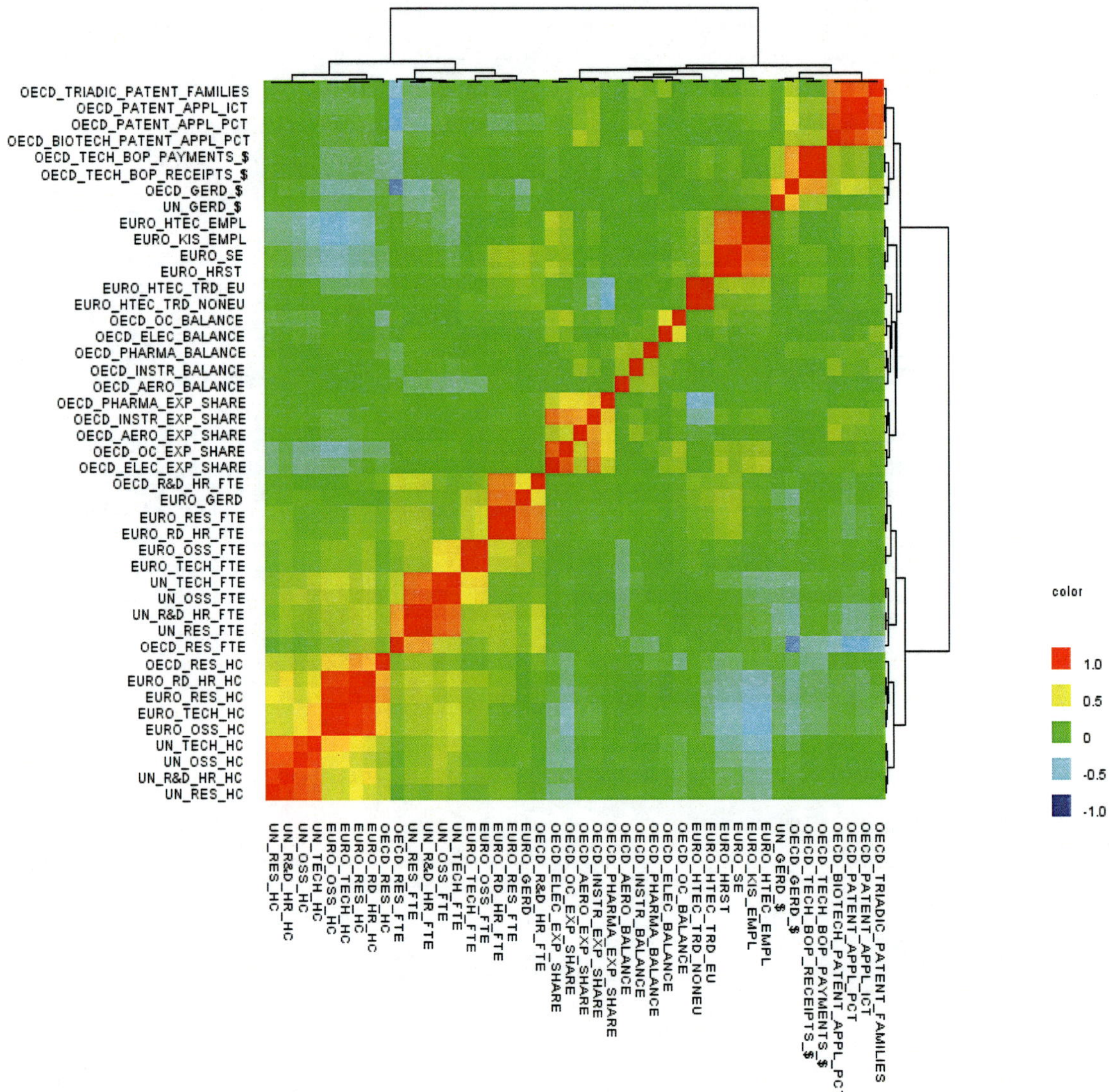

FIGURE F-4 Heat map of main science and technology variables from OECD, UNESCO, and Eurostat databases.

NOTES: AERO = aerospace industry; BOP = balance of payments; ELEC = electronic industry; EMPL = employment; EU = European Union; EURO = Eurostat; EXP = export industry; FTE = full-time equivalent; GERD = gross domestic expenditure on research and development; HC = head count; HR = human resources; HRST = human resources in science and technology; HTECH = high technology; ICT = information and communication technology; INSTR = instrument industry; KIS = knowledge-intensive services; NONEU = non-European Union; OC = office machinery and computer; OSS = other supporting staff; PCT = Patent Cooperation Treaty; PHARMA = pharmaceutical industry; R&D = research and development; RD = R&D; RES = researchers; SE = science and engineering; TRD = Trade; UN = United Nations; UNESCO = United Nations Educational, Scientific and Cultural Organization.

SOURCES: Panel analysis and UNESCO, see http://www.uis.unesco.org/ScienceTechnology/Pages/default.aspx [November 2012]. OECD, see http://stats.oecd.org/Index.aspx?DataSetCode=MSTI_PUB [November 2012]. Eurostat, see http://epp.eurostat.ec.europa.eu/portal/page/portal/science_technology_innovation/data/database European Union, 1995-2013 [November 2012].

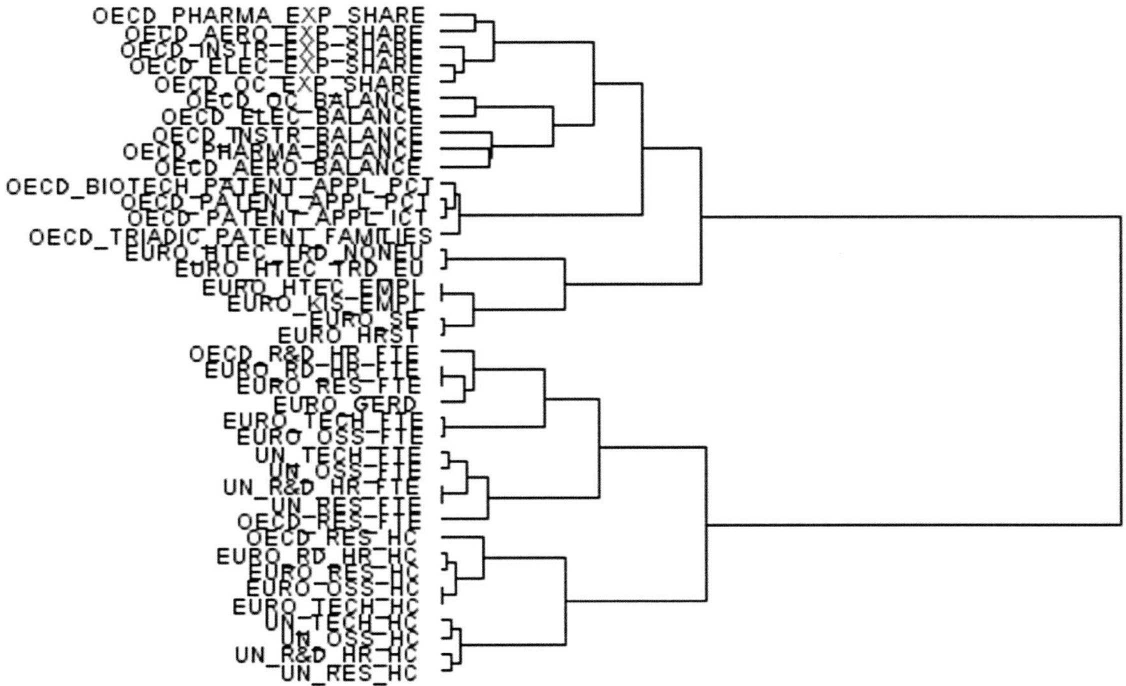

FIGURE F-5 Hierarchical cluster of main science and technology variables from OECD, UNESCO, and Eurostat databases.
NOTES: AERO = aerospace industry; ELEC = electronic industry; EMPL = employment; EU = European Union; EURO = Eurostat; EXP = export industry; FTE = full-time equivalent; GERD = gross domestic expenditure on research and development; HC = head count; HR = human resources; HRST = human resources in science and technology; HTECH = high technology; ICT = information and communication technology; INSTR = instrument industry; KIS = knowledge-intensive services; NONEU = non-European Union; OC = office machinery and computer; OSS = other supporting staff; PCT = Patent Cooperation Treaty; PHARMA = pharmaceutical industry; R&D = research and development; RD = R&D; RES = researchers; SE = science and engineering; TRD = trade; UN = United Nations; UNESCO = United Nations Educational, Scientific and Cultural Organization.
SOURCES: Panel analysis and UNESCO, see http://www.uis.unesco.org/ScienceTechnology/Pages/default.aspx [November 2012]. OECD, see http://stats.oecd.org/Index.aspx?DataSetCode=MSTI_PUB [November 2012]. Eurostat, see http://epp.eurostat.ec.europa.eu/portal/page/portal/science_technology_innovation/data/database European Union, 1995-2013 [November 2012].

points out that they have a crucial role in answering policy questions. We also encountered the problem of missing data, especially in the UNESCO dataset. As described earlier, the UNESCO database consists of 216 nations, but the set is reduced to 80 if one excludes those countries for which the number of time series is limited. For the second part of the analysis, the number of observations was reduced still further to 52 nations. It is important to understand that we attempted to make our analysis as comprehensive as possible by merging information from three databases, but that effort was hampered by the unavailability of data in certain cases.

In addition to reviewing redundancy in statistics produced by UNESCO, OECD, and Eurostat, the panel expanded the analysis to include data taken from *SEI 2012*. The online version of *SEI 2012* is available at http://www.nsf.gov/statistics/seind12/start.htm. The site provides access to tables and figures used in the digest. These tables and figures provide information on the United States, the EU, Japan, China, other selected Asian economies (the Asia-8: India, Indonesia, Malaysia, the Philippines, Singapore, South Korea, Taiwan, and Thailand), and the rest of the world. A more detailed breakdown is available in the appendix tables.[14] Because the digest is intended to inform the reader of broad trends, the source data for the figures and tables do not show a continuous time series.[15] Note that this is not a review of everything that is in *SEI 2012*, because many tables and figures are used to highlight findings and conclusions. As pointed out in the introduction to *SEI 2012*:

> The indicators included in *Science and Engineering Indicators 2012* derive from a variety of national, international, public, and private sources and may not be strictly comparable in a statistical sense. As noted in the text, some data are weak, and the metrics and models relating them to each other and to economic and social outcomes need further development. Thus, the emphasis is on broad trends; individual data points and findings should be interpreted with care.

[14]Detailed appendix tables are available at National Science Foundation (2012c).

[15]For example, see Table 6-6 on p. 6-41 in the *S&E 2012 Digest* (National Science Board, 2012b).

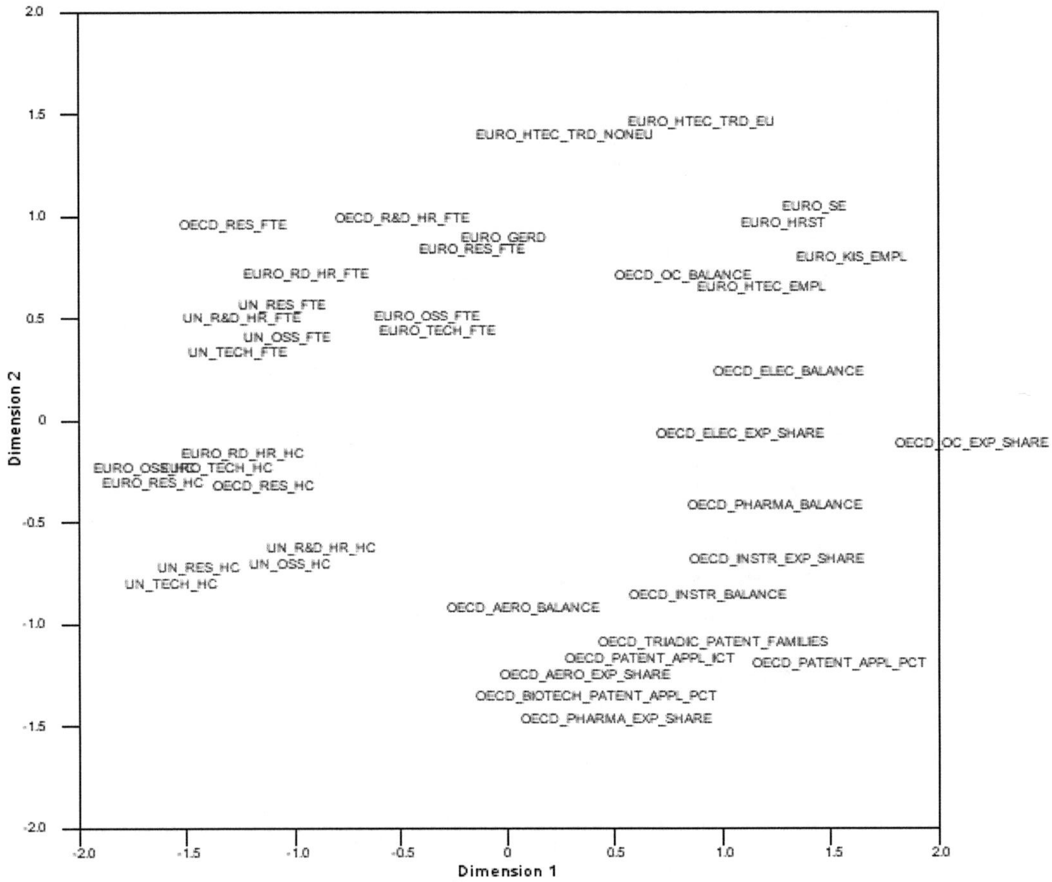

FIGURE F-6 Multidimensional scaling of main science and technology variables from OECD, UNESCO, and Eurostat databases.
NOTES: EMPL = employment; EU = European Union; EURO = Eurostat; EXP = export; FTE = full-time equivalent; GERD = gross domestic expenditure on research and development; HC = head count; HR = human resources; HRST = human resource in science and technology; HTECH = high technology; ICT = information and communication technology; KIS = knowledge-intensive services; NONEU = non-European Union; OC = office machinery and computer; OSS = other supporting staff; PHARMA = pharmaceutical industry; PCT = Patent Cooperation Treaty; R&D = research and development; RD= R&D; RES = researchers; SE = science and engineering; TRD = trade; UN = United Nations; UNESCO = United Nations Educational, Scientific and Cultural Organization.
SOURCES: Panel analysis and UNESCO, see http://www.uis.unesco.org/ScienceTechnology/Pages/default.aspx [November 2012]. OECD, see http://stats.oecd.org/Index.aspx?DataSetCode=MSTI_PUB [November 2012]. Eurostat, see http://epp.eurostat.ec.europa.eu/portal/page/portal/science_technology_innovation/data/database European Union, 1995-2013 [November 2012].

Table F-5 shows a list of indicators used in the panel's analysis. Every effort was made to be as congruent as possible with the list outlined earlier. Because this is a many nations analysis, not all the indicators are included in the table, because many were specific to the United States.

Single Nation Analysis: International Comparability and Human Capital in Science and Engineering

International Comparability

Complying with the *Frascati Manual*, NCSES reports R&D expenditures by performer and funder (see Table F-6). For comparability purposes, NCSES reports GERD for the United States in *National Patterns*, the *SEI*, and various *InfoBriefs*. Categorization of R&D expenditures by government priorities provides a broad picture of the distribution of R&D activities and a basis for international comparisons (National Science Foundation, 2010b). The standards for collecting data on socioeconomic objectives were introduced in the third edition of the *Frascati Manual* (OECD, 2002). Godin (2008) points out that the third edition of the manual expanded the scope of the previous edition to include research on the social sciences and humanities and place greater emphasis on "functional" classifications, notably the distribution of R&D by "objectives." The *Frascati Manual*

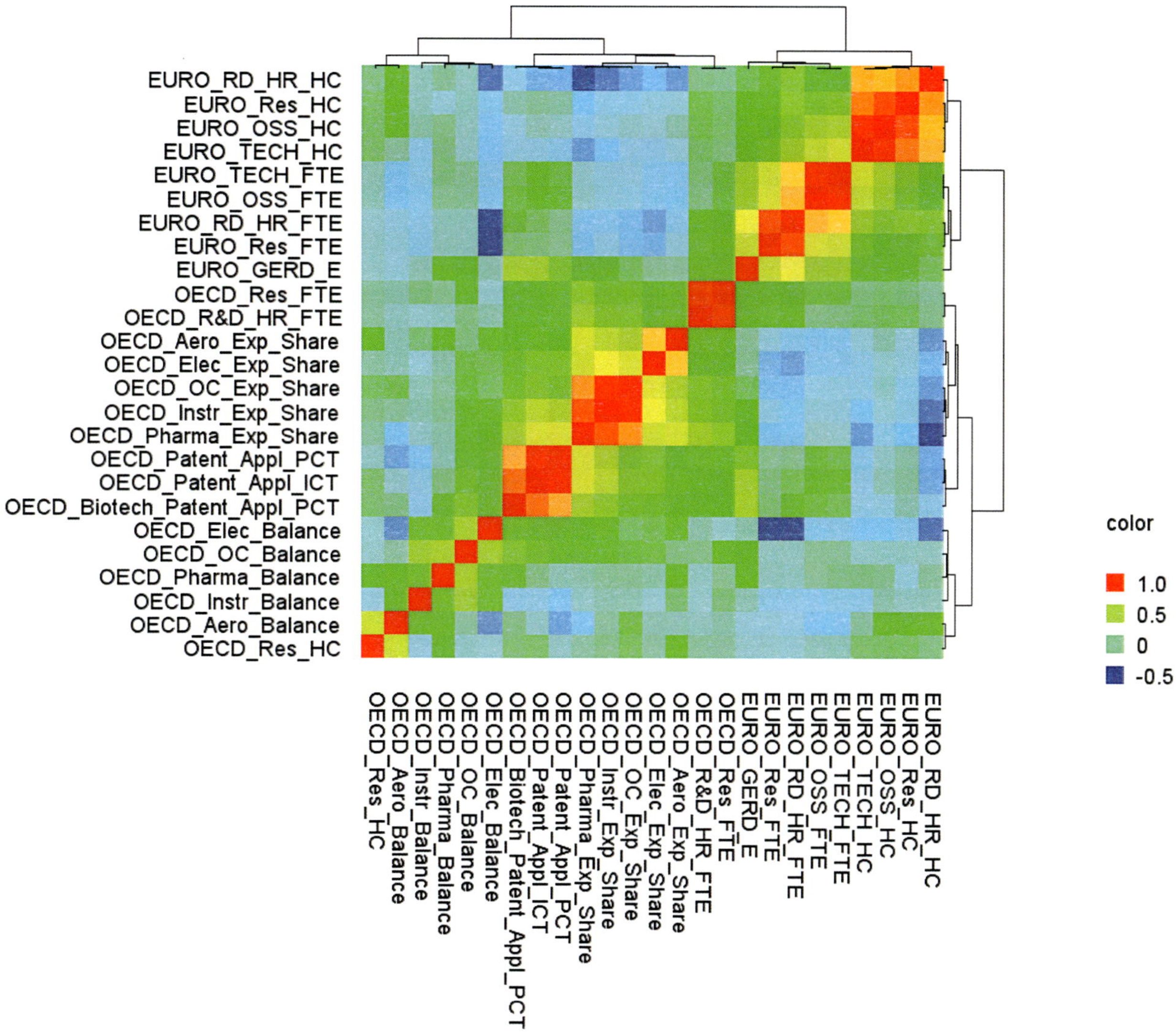

FIGURE F-7 Heat map of main science and technology variables from OECD and Eurostat databases.
NOTES: E = European currency unit; EURO = Eurostat; EXP = export; FTE = full-time equivalent; GERD = gross domestic expenditure on research and development; HC = head count; HR = human resources; ICT = information and communication technology; OC = office machinery and computer; OSS = other supporting staff; PCT = Patent Cooperation Treaty; R&D = research and development; RD = R&D.
SOURCES: Panel Analysis and UNESCO, see http://www.uis.unesco.org/ScienceTechnology/Pages/default.aspx [November 2012]. OECD, see http://stats.oecd.org/Index.aspx?DataSetCode=MSTI_PUB [November 2012]. Eurostat, see http://epp.eurostat.ec.europa.eu/portal/page/portal/science_technology_innovation/data/database European Union, 1995-2013 [November 2012].

recommends collecting performer-reported data in all sectors for two priorities: (1) defense and (2) control and care of the environment. In Table 4-23 of the *SEI 2012 Digest*, U.S. GBAORD values are reported by socioeconomic objectives for 1981, 1990, 2000, and 2009. The agency publishes those figures using special tabulations because aggregate R&D funding data from federal agencies are already allocated to various socioeconomic objective categories, but performer-based R&D totals are not. BRDIS does intermittently survey companies to report their R&D performance for defense purposes and for environmental protection applications, even though the latter category is not fully compliant with the *Frascati Manual*. Along with GBAORD, Eurostat and OECD report GERD by socioeconomic objectives. The *Frascati Manual* also recommends that major fields of S&T be adopted as the functional fields of a science classification system. This classification should be used for R&D expenditures of the governmental, higher education, and private non-profit sectors; if possible for the business enterprise sector; and for personnel data in all sectors (OECD, 2007). OECD,

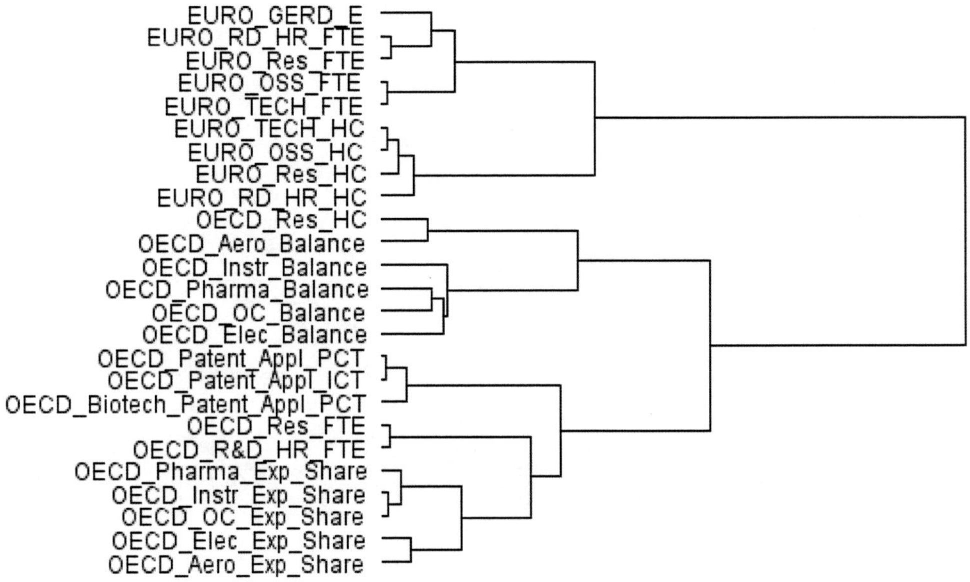

FIGURE F-8 Hierarchical cluster of main science and technology variables from OECD and Eurostat databases.
NOTES: E = European currency unit; EURO = Eurostat; EXP = export; FTE = full-time equivalent; GERD = gross domestic expenditure on research and development; HC = head count; HR = human resources; ICT = information and communication technology; OC = office machinery and computer; OSS = other supporting staff; PCT = Patent Cooperation Treaty; R&D = research and development; RD = R&D.
SOURCES: Panel analysis and UNESCO, see http://www.uis.unesco.org/ScienceTechnology/Pages/default.aspx [November 2012]. OECD, see http://stats.oecd.org/Index.aspx?DataSetCode=MSTI_PUB [November 2012]. Eurostat, see http://epp.eurostat.ec.europa.eu/portal/page/portal/science_technology_innovation/data/database European Union, 1995-2013 [November 2012].

Eurostat, and UNESCO publish GERD figures by fields. NCSES publishes academic expenditures by S&E subfields. The original targets for categorization of R&D expenditures by socioeconomic objectives and fields of science were GBAORD and HERD, respectively. Revisions of the *Frascati Manual* have expanded the scope of the categorization to include all kinds of R&D expenditures.

Table F-6 shows total R&D expenditure and its components for the United States as published by NCSES and other international organizations. Figures F-10 and F-11 show the results of the cluster analysis performed on the data in Table F-6.

Human Capital in Science and Engineering

NCSES produces a multitude of STI human capital variables, as seen in Table F-7:[16]

- *Scientists and engineers*—Scientists and engineers are individuals who satisfy one of the following criteria: (1) have ever received a U.S. bachelor's or higher degree in an S&E or S&E-related field, (2) hold a non-S&E bachelor's or higher degree and are employed in an S&E or S&E-related occupation, and (3) hold a non-U.S. S&E degree and reside in the United States.
- *Doctoral scientists and engineers*—Doctoral scientists and engineers are scientists and engineers who have earned doctoral degrees from U.S. universities and colleges.
- *Bachelor's and master's degrees in S&E*—Estimates of recent college graduates in S&E fields were generated from the biennial NSRCG, which provides information about individuals who recently obtained bachelor's or master's degrees in an SEH field. As the NSRCG was a biennial survey, it collects information for two academic years; therefore, the estimates of S&E bachelor's and master's degrees produced from the NSRCG are for two consecutive academic years. NCSES also requests special tabulations from NCES on S&E bachelor's, master's, and doctoral degrees, which are published in *Women, Minorities and Persons with Disabilities in Science and Engineering*.
- *Doctorate recipients*—Doctorate recipients are individuals who received a doctoral degree from a U.S. institution in an SEH field.

[16]Definitions of these terms are found on NCSES's website at http://www.nsf.gov/statistics.

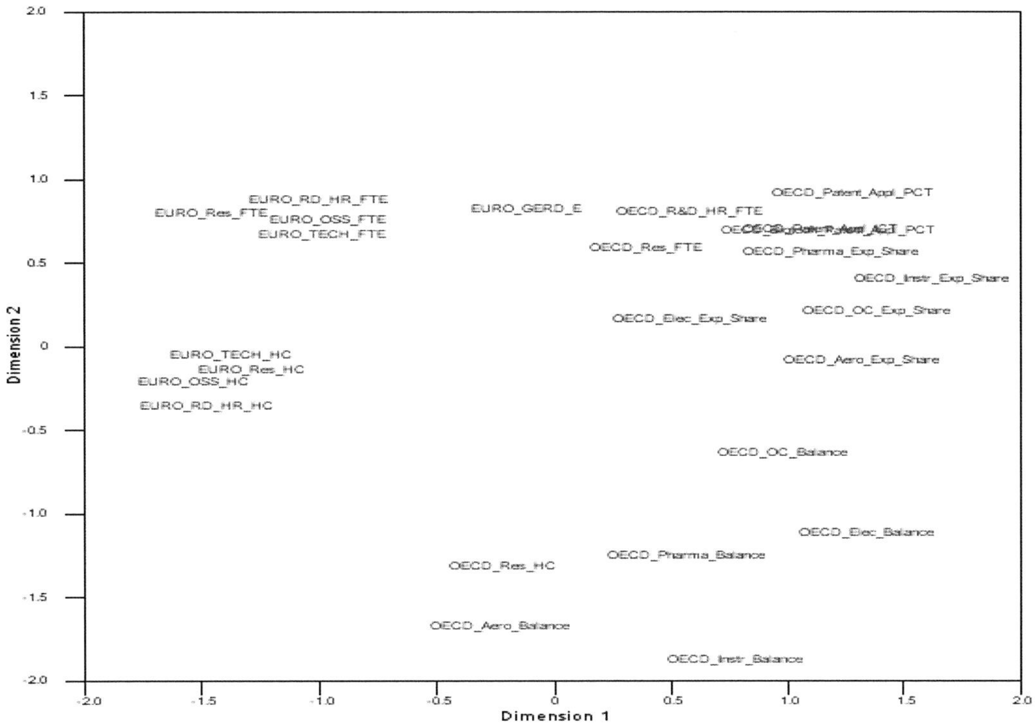

FIGURE F-9 Multidimensional scaling of main science and technology variables from OECD and Eurostat databases.
NOTES: EMPL = employment; EU = European Union; EURO = Eurostat; EXP = export; FTE = full-time equivalent; GERD = gross domestic expenditure on research and development; HC = head count; HR = human resources; HRST = human resources in science and technology; HTECH = high technology; ICT = information and communication technology; KIS = knowledge-intensive services; NONEU = non-European Union; OC = office machinery and computer; OSS = other supporting staff; PCT = Patent Cooperation Treaty; PHARMA = pharmaceutical industry; R&D = research and development; RD = R&D; RES = researchers; SE = science and engineering; TRD = trade; UN = United Nations; UNESCO = United Nations Educational, Scientific and Cultural Organization.
SOURCES: Panel analysis and UNESCO, see http://www.uis.unesco.org/ScienceTechnology/Pages/default.aspx [November 2012]. OECD, see http://stats.oecd.org/Index.aspx?DataSetCode=MSTI_PUB [November 2012]. Eurostat, see http://epp.eurostat.ec.europa.eu/portal/page/portal/science_technology_innovation/data/database European Union, 1995-2013 [November 2012].

- *Graduate students in S&E*—Graduate students in S&E are those who have been enrolled for credit in any SEH master's or doctorate program in the fall of the survey cycle year.
- *Postdoctorates in S&E*—Postdoctorates are defined as individuals who (1) hold a recent doctorate or equivalent, a first-professional degree in a medical or related field, or a foreign equivalent to a U.S. doctoral degree and (2) have a limited-term appointment primarily for training in research or scholarship under the supervision of a senior scholar in a unit affiliated with a GSS[17] institution.
- *Nonfaculty researchers*—Doctorate-holding, nonfaculty researchers are defined as individuals involved principally in research activities who are not postdoctorates or members of a faculty.

Relative to NCSES, other organizations/agencies do not publish a large set of human capital variables, but they capture information on certain R&D occupations (see Table F-1) that were missing from NCSES's surveys very recently. OECD produces statistics on R&D personnel and researchers. In accordance with the *Frascati Manual*, R&D personnel include all persons employed directly in R&D activities, as well as those providing direct services, such as R&D managers, administrators, and clerical staff, while researchers are professionals engaged in the conception or creation of new knowledge, products, processes, methods, and systems and in the management of the projects concerned. Eurostat defines human resources in S&T as people who fulfill one of the following conditions:

[17]NSF/NIH Survey of Graduate Students and Postdoctorates in Science and Engineering at http://www.nsf.gov/statistics/srvygradpostdoc/ [December 2012].

TABLE F-4 Selected Science and Technology Variables from OECD, UNESCO, and Eurostat Database Used in the Panel's Analysis

Variable Label	Variable
EURO_GERD	Gross domestic expenditure on R&D (millions of current PPP dollars)
EURO_HRST	Human resources in science and technology
EURO_HTEC_EMPL	Employment in high-tech sectors
EURO_HTEC_TRD_EU	Trade with EU partners in high-tech sectors
EURO_HTEC_TRD_NONEU	Trade with non-EU partners in high-tech sectors
EURO_KIS_EMPL	Employment in high-tech knowledge-intensive services
EURO_OSS_FTE	Other supporting staff—full-time equivalent
EURO_OSS_HC	Other supporting staff—head count
EURO_RD_HR_FTE	R&D personnel—full-time equivalent
EURO_RD_HR_HC	R&D personnel—head count
EURO_RES_FTE	Researchers—full-time equivalent
EURO_RES_HC	Researchers—head count
EURO_SE	Scientists and engineers
EURO_TECH_FTE	Technicians—full-time equivalent
EURO_TECH_HC	Technicians—head count
OECD_AERO_BALANCE	Trade balance: aerospace industry (millions of current dollars)
OECD_AERO_EXP_SHARE	Export market share: aerospace industry
OECD_BIOTECH_PATENT_APPL_PCT	Number of patents in the biotechnology sector—applications filed under the PCT (priority year)
OECD_ELEC_BALANCE	Trade balance: electronic industry (millions of current dollars)
OECD_ELEC_EXP_SHARE	Export market share: electronic industry
OECD_GERD_$	Gross domestic expenditure on R&D (millions of current PPP dollars)
OECD_INSTR_BALANCE	Trade balance: instruments industry (millions of current dollars)
OECD_INSTR_EXP_SHARE	Export market share: instruments industry
OECD_OC_BALANCE	Trade balance: office machinery and computer industry (millions of current dollars)

(1) successfully completed education at the third level in an S&T field of study; and
(2) were not formally qualified as above, but are employed in S&T occupations in which the above qualifications are normally required.[18]

Eurostat refers to scientists and engineers as persons who use or create scientific knowledge and engineering and technological principles, i.e., persons with scientific or technological training who are engaged in professional work on S&T activities and high-level administrators and personnel who direct the execution of S&T activities. UNESCO publishes information on researchers, technical professionals, and other supporting staff. OECD, Eurostat, and UNESCO produce human capital statistics by head count and FTEs.

Table F-7 shows various human capital variables for the United States that are published by NCSES and other international organizations. Figures F-12 and F-13 show results of the cluster analysis performed on the data in Table F-7. Doctoral scientists and engineers is the only NCSES variable that is closely related to the variables reported by Eurostat and OECD.

Single Nation Analysis: Innovation Statistics—Levels versus Percentages

Table F-8 provides a comparative view of innovation data by industry classification that are available from the three surveys on innovation—the CIS, BRDIS, and Canada's Survey of Innovation. SIBS 2009 has more recent data on the status of innovation activity in Canada, but the data are not available by industry classification; hence the 2003 Survey of Innovation data are presented here. NCSES data cover the period 2006-2008, because companies were asked to report on innovation activity for those years. The EU innovation data are taken from CIS 2006 and 2008. In Tables 1 and 2 of *InfoBrief* NSF 11-300, data on firms producing innovative products and processes are presented as percentages—for example, the percentage of innovative firms reporting that they produced a new/significantly improved product. This is also the case with innovation data produced by Statistics Canada, while data from the CIS are available in both level and percentage form. Staff of the Committee on National

[18]See http://epp.eurostat.ec.europa.eu/cache/ITY_SDDS/Annexes/hrst_st_esms_an1.pdf, page 1 [December 2012].

TABLE F-4 Continued

Variable Label	Variable
OECD_OC_EXP_SHARE	Export market share: office machinery and computer industry
OECD_PATENT_APPL_ICT	Number of patents in the ICT sector—applications filed under the PCT (priority year)
OECD_PATENT_APPL_PCT	Number of patent applications filed under the PCT (priority year)
OECD_PHARMA_BALANCE	Trade balance: pharmaceutical industry (millions of current dollars)
OECD_PHARMA_EXP_SHARE	Export market share: pharmaceutical industry
OECD_R&D_HR_FTE	R&D personnel—full-time equivalent
OECD_RES_FTE	Researchers—full-time equivalent
OECD_RES_HC	Researchers—head count
OECD_TECH_BOP_PAYMENTS_$	Technology balance of payments: payments (millions of current dollars)
OECD_TECH_BOP_RECEIPTS_$	Technology balance of payments: receipts (millions of current dollars)
OECD_TRIADIC_PATENT_FAMILIES	Number of Triadic Patent Families (priority year)
UN_GERD_$	Gross domestic expenditure on R&D (millions of current PPP dollars)
UN_OSS_FTE	Other supporting staff—full-time equivalent
UN_OSS_HC	Other supporting staff—head count
UN_R&D_HR_FTE	R&D personnel—full-time equivalent
UN_R&D_HR_HC	R&D personnel—head count
UN_RES_FTE	Researchers—full-time equivalent
UN_RES_HC	Researchers—head count
UN_TECH_FTE	Technicians—full-time equivalent
UN_TECH_HC	Technicians—head count

NOTES: AERO = aerospace industry; APPL = application; BOP = balance of payments; ELEC = electronic industry; EMPL = employment; EU = European Union; EURO = Eurostat; EXP = export market share; FTE = full-time equivalent; GERD = gross domestic expenditure on research and development; HC = head count; HR = human resources; HRST = human resources in science and technology; HTECH = high technology; ICT = information and communication technology; INSTR = instruments industry; KIS = knowledge-intensive services; NONEU = non-European Union; OC = office machinery and computer; OSS = other supporting staff; PCT = Patent Cooperation Treaty; PHARMA = pharmaceutical industry; PPP = purchasing power parity; R&D = research and development; RD = R&D; RES = researchers; SE = science and engineering; TECH = technicians; TRD = trade; UN = United Nations; UNESCO = United Nations Educational, Scientific and Cultural Organization.

Statistics manipulated the available data in Table 1 of NSF 11-300 and converted percentage figures into levels; the results are shown in Table F-8. Survey results from Canada's 2003 Survey of Innovation, which are available in CANSIM, are problematic to interpret as it is often difficult to understand what the denominator is. In some data tables, it is clear that the denominator is innovative firms, while for other tables the user must guess. One can calculate the total number of innovative firms receiving tax credits or total number of innovative firms reporting customers as an important source of innovation information if information on total innovative firms is available. Hence, staff of the Committee on National Statistics could not convert percentage figures into levels in the case of innovation data from CANSIM. It would be useful if more information on the surveyed population, such as total population, sample size, and response rate, were readily available. This information needs to be published by industry classification, as is evident from Tables F-8 and F-9.[19]

[19]For further information, see Lonmo (2005, Table 1).

OBSERVATIONS

Many Nations Analysis

Figures F-4 to F-6 show a cluster heat map, a hierarchical cluster tree, and the multidimensional scaling of a Pearson correlation matrix, respectively. The input matrix consists of the main S&T variables from the OECD, UNESCO, and Eurostat databases. The red and orange squares along the diagonal of the heat map in Figure F-4 show that those variables are very closely related to each other, and either they could be merged, or the most well-behaved and consistent variables among them could be selected. Figures F-5 and F-8 show clusters of variables. Broadly speaking, human resource variables form one category and trade variables another. Figures F-6 and F-9 show sets of variables that are either similar or dissimilar. In these two figures, the dimensions have no interpretation, and one is looking for clusters of variables that would indicate they belong together. Strong correlation patterns are observed in the variables on researchers, technicians, and other supporting staff. These variables are closely grouped together. Moreover, within these variables, those produced by the same organization

TABLE F-5 Science and Engineering Indicators from *SEI 2012 Digest* Used in the Panel's Analysis

Indicator Label	Indicator
RD_%_GDP	R&D expenditures as a share of economic output = R&D as percentage of GDP
Deg_NatSci	First university degrees in natural sciences
Deg_Eng	First university degrees in engineering
Doct_NatSci	Doctoral degrees in natural sciences
Doct_Eng	Doctoral degrees in engineering
S&E_Art	S&E journal articles produced
Eng_Share_S&E_Art	Engineering journal articles as a share of total S&E journal articles
Res_Art_Int_CoAuthor	Percentage of research articles with international coauthors
Share_Citation_Int_Lit	Share of region's/country's citations in international literature
Global_HighValue_Patents	Global high-value patents
Export_Comm_KIS	Exports of commercial knowledge-intensive services
HighTech_Exports	High-technology exports
Trade_Balance_KIS_IntAsset	Trade balance in knowledge-intensive services and intangible assets
VA_HighTech_Manu	Value added of high-technology manufacturing industries
VA_Health_SS	Global value added of health and social services
VA_Educ	Global value added of education services
VA_Whole_Retail	Global value added of wholesale and retail services
VA_Real_Estate	Global value added of real estate services
VA_Transport_Storage	Global value added of transport and storage services
VA_Rest_Hotel	Global value added of restaurant and hotel services

NOTES: GDP = gross domestic product; KIS = knowledge-intensive services; R&D = research and development; RD = R&D; S&E = science and engineering.
SOURCE: Panel analysis and *Science and Engineering Indicators 2012*, see http://www.nsf.gov/statistics/seind12/tables.htm [November 2012].

are more similar. Clusters of subtopics are also observed. Expenditure variables, trade variables, and patent variables are more similar to variables within their group. This shows that variables on a subtopic relay similar information; i.e., they are proxy variables. For example, if an analyst is looking at predictor variables for a regression model and is unable to obtain data on technical staff, then researchers can substitute. In some ways, this relieves the burden on statistical agencies/offices trying to follow the *Frascati Manual*'s recommendations. Even if they fall short in collecting certain variables, similar information can be gleaned from other variables on the same topic.

Single indicators highlighted for each subtopic as primary indicators are not shown here, as that would lead to conjecture. Nations should decide which variables to collect depending on ease of collection and budgetary constraints. The panel is not asserting that statistical offices around the world should stop collecting detailed S&T data, as the utility of variables is not limited to the ability to feed them into a regression model. National statistical offices collect detailed STI information through surveys and/or by using administrative records to answer specific policy questions, such as the mobility of highly skilled labor, the gender wage gap in S&T occupations, and the amount of investment moving into certain S&T fields. It can be said that the S&T community is interested in understanding the progress of nations in attracting the best talent, or the broad careers pursued by Ph.D. holders in particular fields, or the R&D investment in environmental projects. The main concern faced by the panel was the unavailability of detailed data as main variables undergo disaggregation. Apart from OECD and Eurostat member countries, the rest of the world has yet to keep pace in terms of capturing STI information in accordance with recommendations of the *Frascati* and *Oslo Manuals.* OECD and Eurostat have been frontrunners in pursuing valuable information, and they should be commended for their efforts. At the same time, the panel is not critical of non-OECD and non-Eurostat nations, as both data collection agencies and respondents must undergo a learning process to provide such fine data in a consistent fashion.

Figures F-14 to F-16 show a cluster heat map, a hierarchical cluster tree, and the multidimensional scaling of a Pearson correlation matrix, respectively. The input matrix consists of S&E indicators from the *SEI 2012 Digest*. The red and orange squares along the diagonal of the heat map show that those variables are very closely related to each other, and either they could be merged, or the most well-behaved and consistent variables among them could be selected. Figure F-15 shows clusters of indicators; Figure F-16 shows sets of indicators that are either similar or dissimilar. In these two figures, the dimensions have no interpretation, and one is looking for clusters of variables that would indicate they belong together. Indicators representing the service sector are observed to be highly correlated with each other. Indicators denoting first university degrees are closely grouped together. The same conclusion can be drawn for indicators on generation of S&E knowledge (articles and citations). Therefore, clusters of subtopics are observed, similar to those observed for STI variables from the OECD, Eurostat, and UNESCO databases. Certain indicators, such as R&D as

TABLE F-6 Statistics on U.S. R&D Expenditure Produced by NCSES, UNESCO, OECD, and Eurostat (in millions of current dollars)

National Center for Science and Engineering Statistics—National Patterns of R&D Resources								
Indicator	Total R&D Expenditure	Industry-Performed R&D	Industry FFRDC-Performed R&D	Federally-Performed R&D	Universities and Colleges-Performed R&D	University and College FFRDC-Performed R&D	Nonprofit-Performed R&D	Nonprofit FFRDC-Performed R&D
Variable Name Year	NCSES_RD	NCSES_RD_IND	NCSES_RD_IND_FFRDCS	NCSES_RD_FED	NCSES_RD_UC	NCSES_RD_UC_FFRDCS	NCSES_RD_NP	NCSES_RD_NP_FFRDCS
1981	72292	50425	1385	8605	7085	2484	1784	524
1982	80748	57166	1484	9501	7603	2544	1915	536
1983	89950	63683	1585	10830	8251	2840	2176	585
1984	102244	73061	1739	11916	9154	3243	2511	620
1985	114671	82376	1863	13093	10308	3616	2761	655
1986	120249	85932	1891	13504	11540	3973	2867	541
1987	126360	90160	1995	13588	12807	4287	3013	509
1988	133881	94893	2122	14342	14221	4581	3213	510
1989	141891	99860	2195	15231	15634	4756	3669	547
1990	151993	107404	2323	15671	16939	4894	4126	636
1991	160876	114675	2277	15249	18206	5120	4652	696
1992	165350	116757	2353	15853	19388	5259	4993	748
1993	165730	115435	1965	16531	20495	5289	5267	749
1994	169207	117392	2202	16355	21607	5294	5599	758
1995	183625	129830	2273	16904	22617	5367	5827	808
1996	197346	142371	2297	16585	23718	5395	6209	772
1997	212152	155409	2130	16819	24884	5463	6626	821
1998	226457	167102	2078	17362	26181	5559	7332	843
1999	245007	182090	2039	17851	28176	5652	8207	993
2000	267983	199961	2001	18374	30705	5742	9734	1465
2001	279755	202017	2020	22374	33743	6225	11182	2192
2002	278744	193868	2263	23798	37215	7102	12179	2319
2003	291239	200724	2458	24982	40484	7301	12796	2494
2004	302503	208301	2485	24898	43122	7659	13394	2644
2005	324993	226159	2601	26322	45190	7817	14077	2828
2006	350162	247669	3122	28240	46955	7306	13928	2943
2007	376960	269267	5165	29859	49010	5567	14777	3316
2008	403040	290681	6346	29839	51650	4766	16035	3724
2009	400458	282393	6446	30901	54382	4968	17531	3835

continued

TABLE F-6 Continued

	UNESCO				
Indicator	GERD	GERD Performed by Business Enterprise Sector	GERD Performed by Government Sector	GERD Performed by Higher Education Sector	GERD Performed by Private Nonprofit Sector
Variable Name Year	UN_GERD	UN_GERD_BEP	UN_GERD_GOVP	UN_GERD_HEP	UN_GERD_PNPP
1996	197792	142371	25504	2378	6209
1997	212709	155409	25801	24873	6626
1998	226934	167102	26320	26171	7341
1999	245548	182090	27041	28165	8252
2000	268121	199961	27685	30693	9782
2001	278239	202017	31358	33731	11133
2002	277066	193868	33647	37202	12349
2003	289736	200724	35703	40470	12839
2004	300293	208301	36567	43128	12297
2005	323047	226159	38526	45197	13164
2006	347809	247669	39573	46983	13584
2007	373185	269267	40472	49021	14425
2008	398194	289105	42225	51163	15701

TABLE F-6 Continued

	OECD					
Indicator	GERD	BERD	GBAORD	GOVERD	HERD	GERD Performed by Nonprofit Sector
Variable Name Year	OECD_GERD	OECD_BERD	OECD_GBAORD	OECD_GOVERD	OECD_HERD	OECD_GERD_PNPP
1981	72750	50425	33735	13455	7085	1784
1982	81166	57166	36115	14482	7603	1915
1983	90403	63683	38768	16294	8251	2175
1984	102874	73061	44214	18149	9154	2511
1985	115219	82376	49887	19775	10308	2761
1986	120562	85932	53249	20222	11540	2867
1987	126667	90160	57069	20686	12807	3013
1988	134202	94893	59106	21877	14220	3213
1989	142226	99860	62115	23065	15632	3669
1990	152389	107404	63781	23923	16936	4126
1991	161388	114675	65897	23858	18203	4652
1992	165835	116757	68398	24700	19385	4993
1993	166147	115435	69884	24956	20489	5267
1994	169613	117392	68331	25024	21598	5599
1995	184077	129830	68791	25813	22608	5827
1996	197792	142371	69049	25504	23708	6209
1997	212709	155409	71653	25801	24873	6626
1998	226934	167102	73569	26320	26171	7341
1999	245548	182090	77637	27041	28165	8252
2000	268121	199961	83613	27685	30693	9782
2001	278239	202017	91505	31358	33731	11133
2002	277066	193868	103057	33647	37202	12349
2003	289736	200724	114866	35703	40470	12839
2004	300293	208301	126271	36567	43128	12297
1005	325936	226159	131259	40378	45190	14209
1006	350923	247669	136019	42256	46955	14043
2007	377594	269267	141890	44474	49010	14843
2008	403668	290681	144391	45246	51650	16091
2009	401576	282393	164292	47118	54382	17683
2010			148448			

continued

TABLE F-6 Continued

Eurostat						
Indicator	GERD	BERD	GBAORD	GOVERD	HERD	GERD Performed by Nonprofit Sector
Variable Name Year	EURO_GERD	EURO_BERD	EURO_GBAORD	EURO_GERD_ GOVP	EURO_GERD_ HEP	EURO_GERD_ PNPP
1982	72750	50425	33735	13455	7085	1784
1082	81166	57166	36115	14482	7603	1915
1983	90403	63683	38768	16294	8251	2175
1984	102874	73061	44214	18149	9154	2511
1985	115219	82376	49887	19775	10308	2761
1986	120562	85932	53249	20222	11540	2867
1987	126667	90160	57069	20686	12807	3013
1988	134202	94893	59106	21877	14220	3213
1989	142226	99860	62115	23065	15632	3669
1990	152389	107404	63781	23923	16936	4126
1991	161388	114675	65897	23858	18203	4652
1992	165835	116757	68398	24700	19385	4993
1993	166147	115435	69884	24956	20489	5267
1994	169613	117392	68331	25024	21598	5599
1995	184077	129830	68791	25813	22608	5827
1996	197792	142371	69049	25504	23708	6209
1997	212709	155409	71653	25801	24873	6626
1998	226934	167102	73569	26320	26171	7341
1999	245548	182090	77637	27041	28165	8252
2000	268121	199961	83613	27685	30693	9782
2001	278239	202017	91505	31358	33731	11133
2002	277066	193868	103057	33647	37202	12349
2003	289736	200724	114866	35703	40470	12839
2004	300293	208301	126271	36567	43128	12297
2005	323047	226159	131259	38526	45197	13164
2006	347809	247669	136019	39573	46983	13584
2007	373185	269267	141890	40472	49021	14425
2008	398194	289105	144391	42225	51163	15701
2009			164292			
2010			148448			

NOTES: BERD = business enterprise expenditure on research and development; EURO = Eurostat; FFRDC = federally funded research and development center; GBAORD = government budget appropriations or outlays for research and development; GDP = gross domestic product; GERD = gross domestic expenditure on research and development; GOVERD = government intramural expenditure on research and development; HERD = higher education expenditure on research and development; IND = industry; NCSES = National Center for Science and Engineering Statistics; NP = nonprofit; PNPP = private nonprofit performed; R&D = research and development; RD = R&D; UC = universities and colleges; UNESCO = United Nations Educational, Scientific and Cultural Organization.
SOURCES: National Science Foundation (2012). National Patterns of R&D Resources: 2009 Data Update. NSF 12-321. National Center for Science and Engineering Statistics. Available: http://www.nsf.gov/statistics/nsf12321/ [November 2012]. UNESCO, see http://www.uis.unesco.org/ScienceTechnology/Pages/default.aspx [November 2012]. OECD, see http://stats.oecd.org/Index.aspx?DataSetCode=MSTI_PUB [November 2012]. Eurostat, see http://epp.eurostat.ec.europa.eu/portal/page/portal/science_technology_innovation/data/database [November 2012].

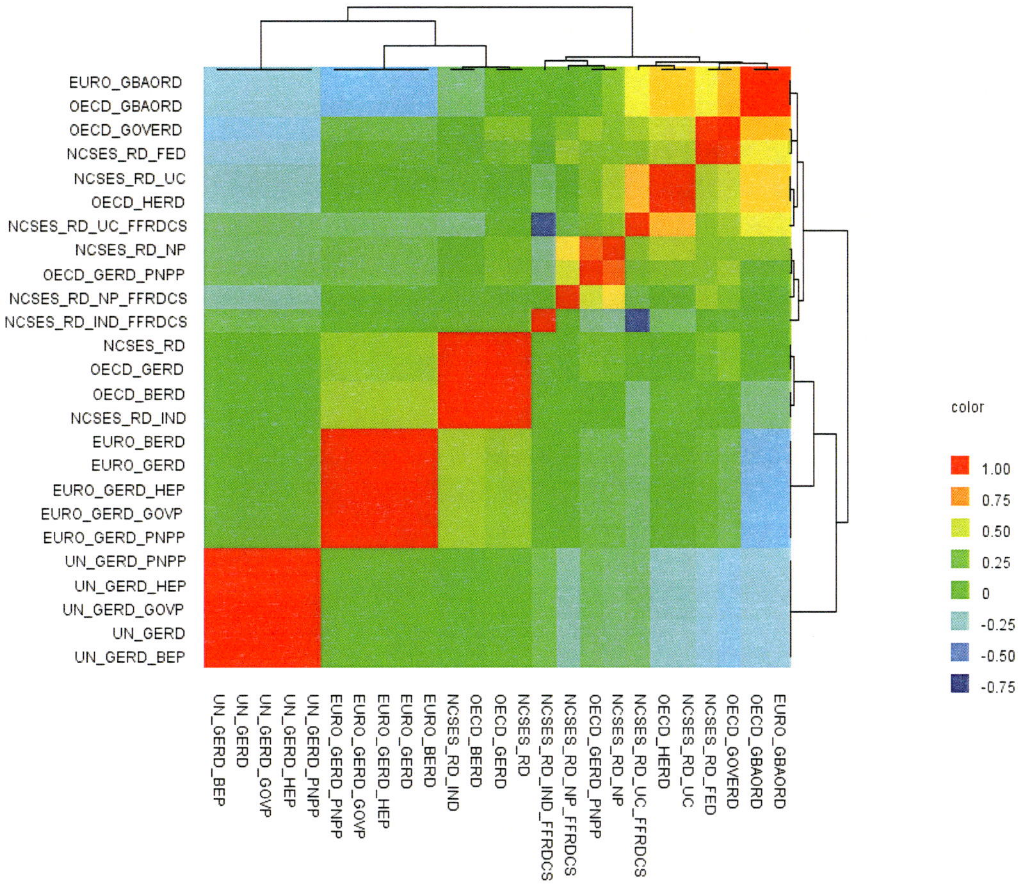

FIGURE F-10 Heat map of U.S. R&D expenditure variables from various STI databases.
NOTES: BEP = business enterprise performed; BERD = business enterprise expenditure on research and development; EURO = Eurostat; FFRDCs = federally funded R&D centers; GBAORD = government budget appropriations or outlays for research and development; GERD = gross domestic expenditure on research and development; GOVERD = government intramural expenditure on research and development; GOVP = government sector performed; HEP = higher education sector performed; HERD = higher education expenditure on research and development; IND = industry; NCSES = National Center for Science and Engineering Statistics; NP = nonprofits; PNPP = private nonprofit performed; RD = research and development; UC = universities and colleges.
SOURCES: Panel analysis and National Science Foundation. (2012). National Patterns of R&D Resources: 2009 Data Update. NSF 12-321. National Center for Science and Engineering Statistics, available: http://www.nsf.gov/statistics/nsf12321/ [November 2012]. UNESCO, see http://www.uis.unesco.org/ScienceTechnology/Pages/default.aspx [November 2012]. OECD, see http://stats.oecd.org/Index.aspx?DataSetCode=MSTI_PUB [November 2012]. Eurostat, see http://epp.eurostat.ec.europa.eu/portal/page/portal/science_technology_innovation/data/database [November 2012].

a share of GDP, global high-value patents, doctoral degrees in engineering, and doctoral degrees in natural science are not strongly correlated with other indicators. Hence within the set of indicators analyzed, these four indicators stand apart. The reader should not assume that these indicators are unique, because the list of indicators analyzed here is small. The uniqueness might not hold if more indicators were included in the input matrix.

Single Nation Analysis

The clusters shown in Figures F-10 and F-11 are not surprising, as sector-specific expenditure variables are clustered together; i.e., business R&D expenditure figures are similar to each other irrespective of the data source. The same conclusion can be drawn for figures on expenditures on federal R&D, nonprofit R&D, and academic R&D.

Eurostat, OECD, and UNESCO report numbers of FTE researchers for the United States, but it is not clear how that number is calculated. NCSES and NCES report head counts of S&E human resources. Therefore, a disparity is seen in

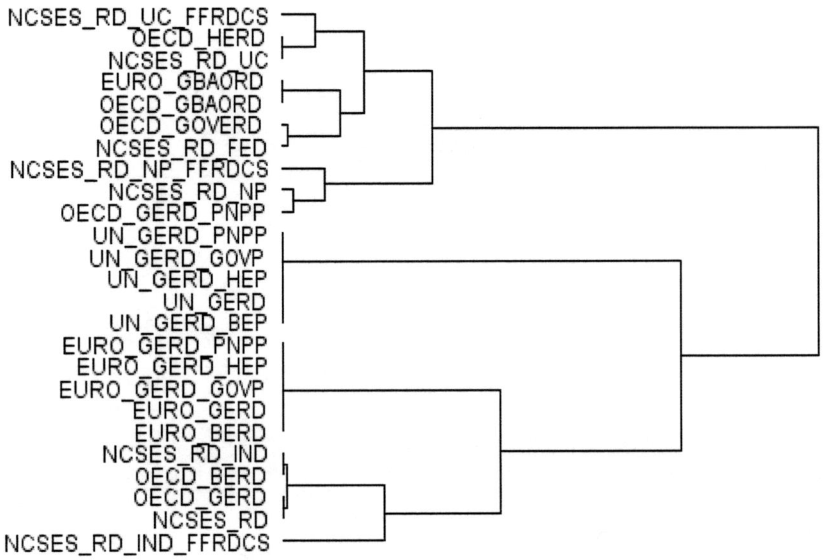

FIGURE F-11 Cluster map of U.S. R&D expenditure variables from various STI databases.
NOTES: BEP = business enterprise performed; BERD = business enterprise expenditure on research and development; EURO = Eurostat; FFRDCs = federally funded R&D centers; GBAORD = government budget appropriations or outlays for research and development; GERD = gross domestic expenditure on research and development; GOVERD = government intramural expenditure on research and development; GOVP = government sector performed; HEP = higher education sector performed; HERD = higher education expenditure on research and development; IND = industry; NCSES = National Center for Science and Engineering Statistics; NP = nonprofit; PNPP = private nonprofit performed; RD = research and development; UC = universities and colleges.
SOURCES: Panel analysis and National Science Foundation (2012). National Patterns of R&D Resources: 2009 Data Update. NSF 12-321. National Center for Science and Engineering Statistics, available: http://www.nsf.gov/statistics/nsf12321/ [November 2012]. UNESCO, see http://www.uis.unesco.org/ScienceTechnology/Pages/default.aspx [November 2012]. OECD, see http://stats.oecd.org/Index.aspx?DataSetCode=MSTI_PUB [November 2012]. Eurostat, see http://epp.eurostat.ec.europa.eu/portal/page/portal/science_technology_innovation/data/database [November 2012].

the metric that is reported, as U.S. head counts represent the supply of human capital not necessarily involved in R&D. FTE researchers show the contribution of labor hours to the R&D process; hence it is important for a researcher or an S&T policy maker to understand that different data sources may appear to report the same thing, but this usually is not the case. Figures F-12 and F-13 show that variables representing head counts are not strongly correlated with FTE researchers. One advantage of having so many variables is that as a user, one can select among them depending on the question being addressed. Table F-5 shows that the whole set of variables produced by NCSES is an attempt at capturing different segments of the S&E population, which range from scientists to medical researchers. Figures F-12 and F-13 show that variables from NCSES and NCES are clustered together, with variables reporting the same indicator being more strongly correlated (see the cluster of doctorate recipients, graduate students [NCSES], and doctoral degrees [NCES]). This suggests the possibility that NCSES may be overproducing some of the S&E human capital variables. As previously mentioned, however, agencies produce variables to answer particular policy questions. The end result is a trade-off between efficiency and addressing user needs. It is commendable that NCSES has been able to satisfy academicians and policy analysts alike, but a more resourceful approach is required under current budgetary conditions.

The panel would also like to highlight the efforts of NCSES to comply more closely with the recommendations of the *Frascati Manual*. The Survey of Industrial Research and Development (SIRD) (the old industrial survey) questionnaire contained items on FTE R&D scientists and engineers only. NCSES decided to resolve this data gap by including questions on researchers (FTE) and R&D personnel (head count) by gender; occupation (scientists and engineers, technicians, support staff); and location, including foreign locations. With the new data, it is possible to generate tabs, for example, on female technicians working in Belgium. The Survey of Research and Development Expenditures at Universities and Colleges (the old academic survey) contained a serious data gap in terms of information on R&D personnel in the academic sector. In 2010, NCSES began using the HERD survey to collect researcher and R&D personnel head counts. The HERD redesign investigation process indicated that collecting FTE data would be highly problematic, whereas collecting principal investigator data appeared to be rather reasonable. Therefore, obtaining information on FTE researchers or R&D personnel in the academic sector still is not possible, but one can obtain head counts of researchers, principal investigators, and R&D personnel.

One point that came to the panel's attention is that NCSES does not publish its main STI indicators on a single webpage. For national R&D expenditures, a user accesses *National Patterns*, while for human capital in S&E, one must generate tables from SESTAT. For further detail on academic R&D expenditures, WebCASPAR serves as a more useful tool. IRIS contains historical data tables on industrial R&D expenditures. SESTAT data feed into various NCSES publications, including (1) *Characteristics of Scientists and Engineers in the U.S.*; (2) *Characteristics of Doctoral Scientists and Engineers in the U.S.*; (3) *Doctoral Scientists & Engineers Profile;* (4) *Characteristics of Recent College Grads*; (5) *Women, Minorities, and Persons with Disabilities in Science and Engineering*; and (6) various *InfoBriefs*. It is difficult to find summary tables that combine information across all five publications. WebCASPAR contains detail on SEH degrees that is not available in SESTAT. When staff of the Committee on National Statistics downloaded STI databases of other agencies/organizations, they had an easier task because all variables were available on a single webpage.

TABLE F-7 Statistics on U.S. Science and Engineering Human Resources Produced by NCSES, NCES, UNESCO, OECD, and Eurostat

	National Center for Science and Engineering Statistics							
	SESTAT				WEBCASPAR			
Indicator	Scientists and Engineers	Doctoral Scientists and Engineers	S&E Bachelor's Degree Recipients	S&E Master's Degree Recipients	S&E Doctorate Recipients (Includes Medical and Other Life Sciences)	SEH Graduate Students	SEH Post-doctorates	SEH Nonfaculty Research Staff
Variable Name Year	SE	DSE	RCG_BACH	RCG_MAST	DOCREP_SE	GRADSTUD	POSTDOC	NONFACULTY_ RES_STAFF
1990					23823	452113	29565	5255
1991			308500	57000	25060	471212	30865	5478
1992					25785	493522	32747	5482
1993	11615200	513460	348900	73200	26640	504304	34322	6001
1994					27500	504399	36377	6209
1995	12036200	542540	354450	74750	27864	499640	35926	6534
1996			354450	74750	28564	494079	37107	6604
1997	12530700	582080	371500	78500	28650	487208	38481	6722
1998			371500	78500	28773	485627	40086	7100
1999	13050800	626700	379150	80050	27338	493256	40800	7573
2000			379150	80050	27557	493311	43115	7879
2001		656550	468850	123350	27,069	509607	43311	7531
2002			468850	123350	26263	540404	45034	7906
2003	21647000	685300	521833	138967	26916	567121	46728	8473
2004			521833	138967	27993	574463	47240	9075
2005			521833	138967	29768	582226	48555	9527
2006	22630000	711800	467000	102000	31774	597643	49343	10814
2007			467000	102000	33974	619499	50840	10752
2008	10204000	752000			34926	631489	54164	13747
2009					35562	631645	57805	14059
2010					35253	632652	63415	

TABLE F-7 Continued

	National Center for Science and Engineering Statistics			International Organizations			
	Women, Minorities and Persons with Disabilities in Science and Engineering			UNESCO	OECD	Eurostat	
Indicator	S&E Bachelor's Degrees	S&E Master's Degrees	S&E Doctoral Degrees	Researchers (FTE)	Researchers (FTE)	Researchers in Business Enterprise Sector (FTE)	Researchers (FTE)
Variable Name Year	BACH_SE	MAST_SE	DOC_SE	UN_RES_FTE	OECD_RES_FTE	EURO_RES_ BEMPOCC_FTE	EURO_RES_ PERSOCC_FTE
1990	329094	77788	22868			758500	
1991	337675	78368	24023		981659	776400	981659
1992	355265	81107	24675			772000	
1993	366035	86425	25443		1013772	766600	1013772
1994	373261	91411	26205			757300	
1995	378148	94309	26536		1035995	789400	1035995
1996	384674	95313	27243			859300	
1997	388482	93485	27232	1159908	1159908	918600	1159908
1998	390618	93918	27278			997700	
1999			25933	1260920	1260920	1033700	1260920
2000	398622	95683	25966	1293582	1293582	1041300	1293582
2001	400435	99,528	25453	1320305	1320305	1060000	1320096
2002	415983	99650	24254	1342454	1342454	1075300	1342454
2003	442755	108355	25425	1430551	1430551	1156000	1430551
2004	458658	119296	26573	1384536	1384536	1111300	1384536
2005	470214	120870	28561	1375304	1375304	1097700	1375304
2006	478858	120999	30452	1414341	1414341	1135500	1414341
2007	485772	120278	32588	1412639	1412639	1130500	1412639
2008	496168	126404	33359				
2009	505435	134517	33284				

NOTES: BACH = bachelor's degrees; BEMPOCC = business enterprise sector; DOCREP = doctorate recipients; DSE = doctoral scientists and engineers; EURO = Eurostat; FTE = full-time equivalent; GRADSTUD = graduate students; MAST = master's degrees; PERSOCC = researchers FTE; POSTDOC = postdoctorates; RCG = recent college graduates; RES = researchers; S&E = science and engineering; SEH = science, engineering, and health; SESTAT = Scientists and Engineers Statistical Data System; UN = United Nations; UNESCO = United Nations Educational, Scientific and Cultural Organization.
SOURCES: WebCASPAR, see https://webcaspar.nsf.gov [November 2012]. UNESCO, see http://www.uis.unesco.org/ScienceTechnology/Pages/default.aspx [November 2012]. OECD, see http://stats.oecd.org/Index.aspx?DataSetCode=MSTI_PUB [November 2012]. Eurostat, see http://epp.eurostat.ec.europa.eu/portal/page/portal/science_technology_innovation/data/database [November 2012].

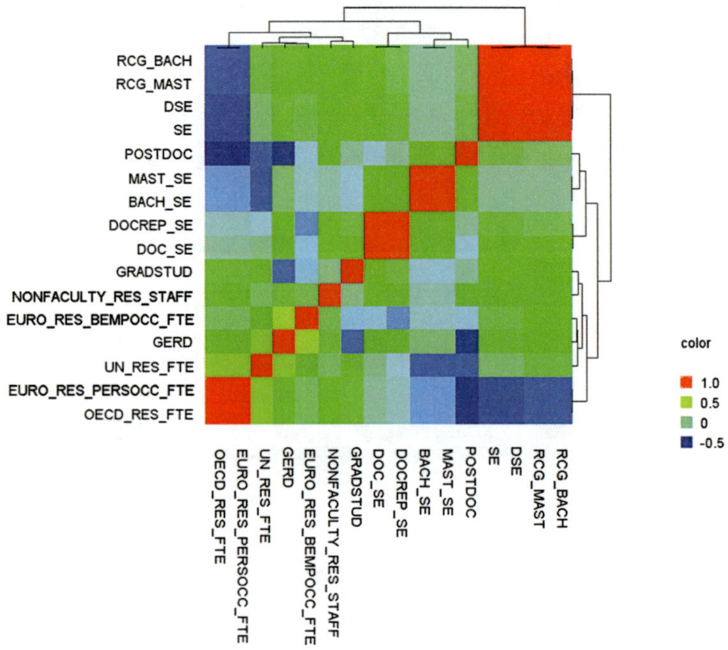

FIGURE F-12 Heat map of U.S. human capital variables from various STI databases.

NOTES: BACH = bachelor's degrees; BEMPOCC = business enterprise sector; DOC = doctorate; DOCREP = doctorate recipients; DSE = doctoral scientists and engineers; EURO = Eurostat; FTE = full-time equivalent; GERD = gross domestic expenditure on research and development; GRADSTUD = graduate students; MAST = master's degrees; PERSOCC = researchers FTE; RCG = recent college graduates; RES = researchers; SE = science and engineering; UN = United Nations.

SOURCES: Panel analysis and WebCASPAR, see https://webcaspar.nsf.gov [November 2012]. UNESCO, see http://www.uis.unesco.org/ScienceTechnology/Pages/default.aspx [November 2012]. OECD, see http://stats.oecd.org/Index.aspx?DataSetCode=MSTI_PUB November 2012]. Eurostat, see http://epp.eurostat.ec.europa.eu/portal/page/portal/science_technology_innovation/data/database [November 2012].

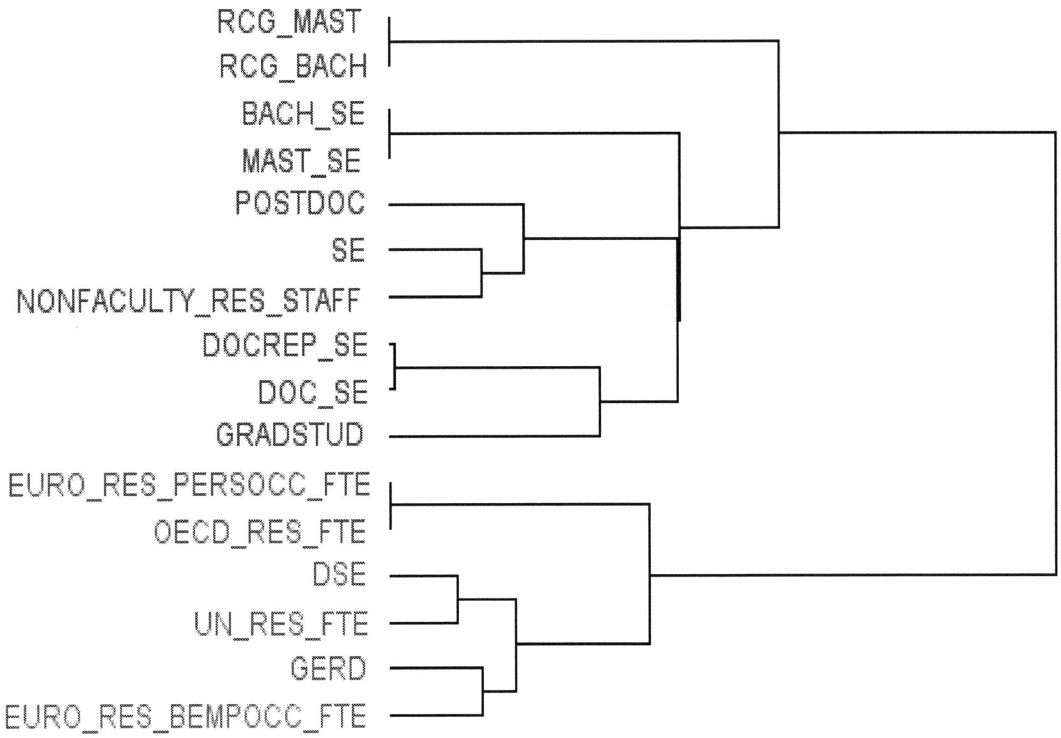

FIGURE F-13 Cluster map of U.S. human capital variables from various STI databases.
NOTES: BACH = bachelor's degrees; BEMPOCC = business enterprise sector; DOC = doctorate; DOCREP = doctorate recipients; DSE = doctoral scientists and engineers; EURO = Eurostat; FTE = full-time equivalent; GERD = gross domestic expenditure on research and development; GRADSTUD = graduate students; MAST = master's degrees; PERSOCC = researchers FTE; RCG = recent college graduates; RES = researchers; SE = science and engineering; UN = United Nations.
SOURCES: Panel analysis and WebCASPAR, see https://webcaspar.nsf.gov [November 2012]. UNESCO, see http://www.uis.unesco.org/ScienceTechnology/Pages/default.aspx [November 2012]. OECD, see http://stats.oecd.org/Index.aspx?DataSetCode=MSTI_PUB November 2012]. Eurostat, see http://epp.eurostat.ec.europa.eu/portal/page/portal/science_technology_innovation/data/database [November 2012].

TABLE F-8 Innovation Statistics from NCSES (2006-2008) and Statistics Canada (2003)

	United States: National Center for Science and Engineering Statistics						
	2006-2008: Number of Companies Who Answered Yes to Innovating						
Industry Classification	Any Good/ Service	Goods	Services	Any Process	Mfgr./ Production Methods	Logistic/ Delivery/ Distribution Methods	Support Activities
Manufacturing industries, 31–33	27962	22878	12710	27962	22878	8897	16523
Food, 311	1547	1456	455	1547	1183	546	637
Beverage/tobacco products, 312	204	156	72	180	108	72	96
Textile/apparel/leather and allied products, 313–16	1159	915	549	1098	793	427	732
Wood products, 321	549	366	305	976	793	183	366
Chemicals, 325	2419	1947	1062	2006	1298	1003	1180
Pharmaceuticals/medicines, 3254	675	360	405	630	270	405	480
Other, 325	1760	1584	660	1364	1012	616	704
Plastics/rubber products, 326	1464	1281	671	1708	1464	488	915
Nonmetallic mineral products, 327	702	594	270	756	594	216	432
Primary metals, 331	357	273	231	399	357	84	231
Fabricated metal products, 332	4176	2871	2349	5742	4959	1305	3132
Machinery, 333	3120	2760	1320	2880	2520	600	1800
Computer/electronic products, 334	3150	3010	1260	2310	1820	770	1260
Computers/peripheral equipment, 3341	336	282	156	276	186	114	126
Communications equipment, 3342	408	408	168	264	200	56	136
Semiconductor/other electronic components, 3344	675	625	250	625	475	175	350
Navigational/measuring/electro-medical/control instruments, 3345	1534	1508	624	1040	936	416	572
Other, 334	185	175	65	70	50	15	50
Electrical equipment/ appliance/ components, 335	1036	1008	308	784	672	336	588
Transportation equipment, 336	1512	1350	594	1242	972	270	810
Motor vehicles/trailers/parts, 3361–63	792	726	264	726	627	99	462
Aerospace products/parts, 3364	288	261	171	225	144	63	180
Other, 336	455	403	156	325	195	104	208
Furniture/related products, 337	1092	1014	468	1482	936	390	936
Manufacturing nec, other 31–33	5302	4097	2892	5543	4097	1687	3374
Nonmanufacturing industries, 21–23, 42–81	113432	42537	99253	113432	28358	42537	85074
Information, 51	6930	3696	5775	4620	1617	2310	3696
Software publishers, 5112	3080	2320	2240	2120	760	880	1720
Telecommunications/Internet service providers/ Web search portals/data processing services, 517–18	2331	945	2142	1386	441	693	1260
Other, 51	1548	516	1419	1290	516	903	774
Finance/insurance, 52	4472	559	4472	4472	559	1118	3913
Real estate/rental/leasing, 53	3430	2940	2450	2940	980	980	2940
Professional/scientific/technical services, 54	22568	10416	20832	20832	6944	8680	17360

TABLE F-8 Continued

Computer systems design/related services, 5415	6790	3880	5820	4850	1552	1358	4462
Scientific R&D services, 5417	1023	744	682	806	465	186	434
Other, 54	15110	6044	13599	15110	6044	6044	12088
Health care services, 621–23	18350	5505	16515	14680	5505	5505	12845
Nonmanufacturing nec, other 21–23, 42–81	55968	27984	46640	65296	18656	18656	46640

	Canada: Statistics Canada					
	2003: Percentage of Business Units					
Industry Classification	Both Product and Process Innovators	Innovators	Process Innovators	Process Innovators Only	Product Innovators	Product Innovators Only
Air transportation [481]	22.4	36.7	32.7	10.2	26.5	4.1
Airport operations [48811]	17.1	46.3	41.5	24.4	22.0	4.9
Cable and other program distribution [5175]	42.8	66.5	42.8	0	66.5	23.7
Computer and communications equipment and supplier wholesaler-distributors [4173]	28.2	65.1	37.3	9.1	56.0	27.8
Computer systems design and related services [54151]	35.4	87.2	42.0	6.6	80.6	45.2
Contract drilling (except oil and gas) [213117]	14.3	32.1	17.9	3.6	28.6	14.3
Data processing, hosting, and related services [5182]	50.0	72.4	63.8	13.8	58.6	8.6
Electronic and precision equipment repair and maintenance [8112]	18.4	53.3	33.4	15.1	38.3	19.9
Engineering services [54133]	21.1	55.3	32.0	10.9	44.5	23.4
Environmental consulting services [54162]	32.8	67.3	45.9	13.1	54.2	21.4
Geophysical surveying and mapping services [54136]	14.6	57.8	41.4	26.8	31.0	16.4
Industrial design services [54142]	27.6	53.9	31.3	3.7	50.2	22.5
Information and communication technology (ICT) service industries	37.2	78.2	44.1	6.9	71.3	34.1
Internet service providers [518111]	58.2	75.4	61.2	3	72.4	14.2
Interurban and rural bus transportation [4852]	18.8	43.8	25	6.3	37.5	18.8
Management consulting services [54161]	26.5	44.1	35	8.5	35.7	9.1
Management, scientific and technical consulting services [5416]	26.6	47.1	35.9	9.2	37.8	11.2
Office and store machinery and equipment wholesalers-distributors [41791]	37.2	61.8	42.7	5.5	56.3	19.1

TABLE F-8 Continued

	Canada: Statistics Canada					
	2003: Percentage of Business Units					
Industry Classification	Both Product and Process Innovators	Innovators	Process Innovators	Process Innovators Only	Product Innovators	Product Innovators Only
Office machinery and equipment rental and leasing [53242]	30.0	52.6	37.7	7.7	44.9	14.9
Other machinery, equipment and supplies wholesaler-distributors [4179]	26.0	63.8	33.7	7.7	56.1	30.1
Other scientific and technical consulting services [54169]	23.8	52.2	35.1	11.3	40.9	17.2
Other support activities for mining [213119]	20.0	34.5	29.1	9.1	25.5	5.5
Other telecommunications [5179]						
Port and harbour operations [48831]	20.7	41.4	41.4	20.7	20.7	0
Rail transportation [482]	0	53.3	33.3	33.3	20	20
Research and development in the physical, engineering and life sciences [54171]	32.2	68.1	44.3	12.1	56.1	23.9
Research and development in the social sciences and humanities [54172]	21.2	60.1	50.5	29.3	30.8	9.6
Satellite telecommunications [5174]	62.7	100	73.7	11.1	88.9	26.3
Scientific research and development services [5417]	30.1	66.6	45.5	15.4	51.3	21.1
Software publishers [5112]	53.1	94.3	59.3	6.2	88.1	35.0
Support activities for forestry [1153]	10.3	28.7	25.0	14.7	13.9	3.6
Surveying and mapping (except geophysical) services [54137]	23.6	51.2	48.2	24.6	26.6	3.0
Telecommunications resellers [5173]	29.4	74.5	29.4	0	74.5	45.0
Testing laboratories [54138]	20.0	51.9	33.5	13.5	38.4	18.4
Truck transportation [484]	10.9	25.7	20.8	9.9	15.8	5.0
Water transportation [483]	8.3	20.8	16.7	8.3	12.5	4.2
Web search portals [518112]						
Wired telecommunications carriers [5171]	57.8	75.4	60.5	2.6	72.8	15.0
Wireless telecommunications carriers (except satellite) [5172]	43.5	60.0	49.1	5.6	54.4	10.9

SOURCES: Adapted from National Science Foundation (2010). NSF Releases New Statistics on Business Innovation. NSF 11-300. National Center for Science and Engineering Statistics, available: http://www.nsf.gov/statistics/infbrief/nsf11300/ [November 2012]. Statistics Canada, Adapted from CANSIM Table 358-00321, 2 Survey of Innovation, selected service industries, percentage of innovative business units [November 2012].

TABLE F-9 Innovation Statistics from Eurostat, 2006 and 2008

European Union: Eurostat

Industry Classification	2006: Number of Enterprises with Type of Innovation Technological Innovation	Novel Innovators, Product Only	Novel Innovators, Process Only	Introduced Organizational/ Marketing Innovation
Agriculture, forestry and fishing				
Mining and quarrying	1402	239	559	652
Manufacturing	158629	35531	42018	76297
Electricity, gas, steam and air conditioning supply	2340	228	1191	1233
Water supply; sewerage, waste management and remediation activities				
Construction	19202	6528	7809	1751
Wholesale and retail trade; repair of motor vehicles and motorcycles	42233	7413	18292	15994
Transportation and storage				
Accommodation and food service activities				
Information and communication				
Financial and insurance activities				
Real estate activities				
Professional, scientific and technical activities				
Administrative and support service activities				
Hotels and restaurants	5422	1306	2999	333
Transport, storage and communication	24702	4304	7301	13065
Financial intermediation	8792	1416	2200	4847
Real estate, renting and business activities	22748	4849	6395	6442

Industry Classification	2008: Number of Enterprises with Type of Innovation Innovation Activity	Technological Innovation Only	Innovation Activity	Novel Innovators, Product Only	Novel Innovators, Process Only
Agriculture, forestry and fishing	2799	1223	579	524	862
Mining and quarrying	4072	595	523	181	648
Manufacturing	446126	50774	39981	36255	44047
Electricity, gas, steam and air conditioning supply	4919	541	792	208	682
Water supply; sewerage, waste management and remediation activities	13891	1785	1719	702	1784
Construction	42042	9470	17805	3716	11230
Wholesale and retail trade; repair of motor vehicles and motorcycles	75102	12742	30184	7422	14042
Transportation and storage	72156	6310	12699	2822	8501
Accommodation and food service activities	15062	2492	6628	939	2048
Information and communication	27343	5300	4337	6748	2874
Financial and insurance activities	28580	1655	3065	1948	2483
Real estate activities	2631	361	1198	327	588
Professional, scientific and technical activities	19809	4521	5565	2978	3870
Administrative and support service activities	7909	1563	3557	910	1947
Hotels and restaurants					
Transport, storage and communication					
Financial intermediation					
Real estate, renting and business activities					

SOURCES: Eurostat, see http://epp.eurostat.ec.europa.eu/portal/page/portal/science_technology_innovation/data/database[November 2012].

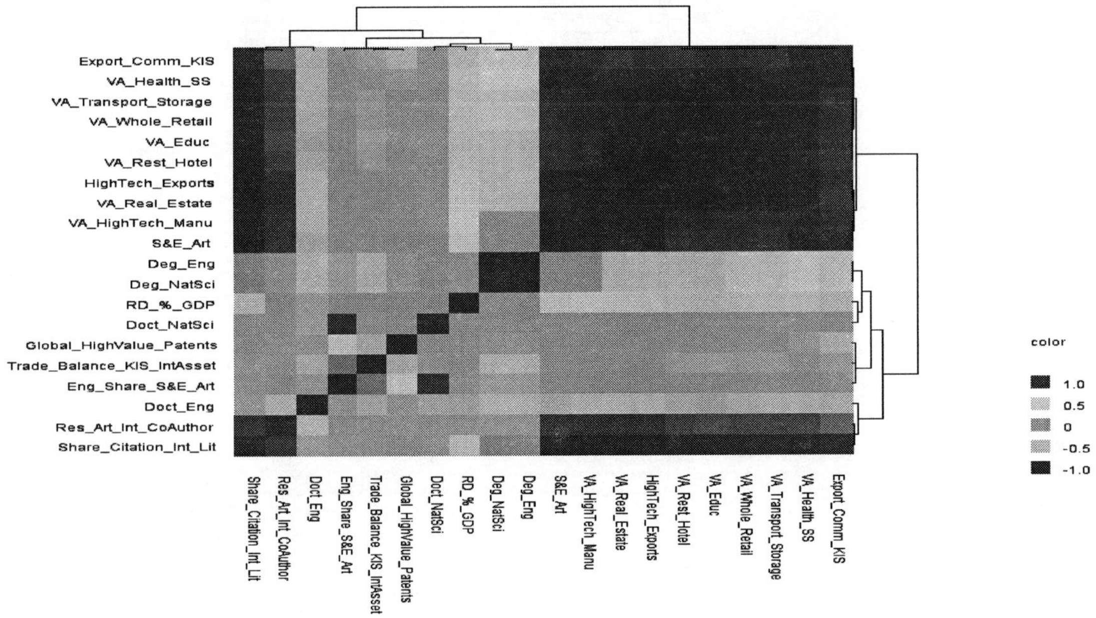

FIGURE F-14 Heat map of science and engineering indicators from *SEI 2012 Digest*.
NOTE: GDP = gross domestic product; KIS = knowledge-intensive services; RD = research and development; S&E = science and engineering; SS = social services; VA = global value added. SOURCE: Panel analysis and *Science and Engineering Indicators 2012*, see http://www.nsf.gov/statistics/seind12/tables.htm [November 2012].

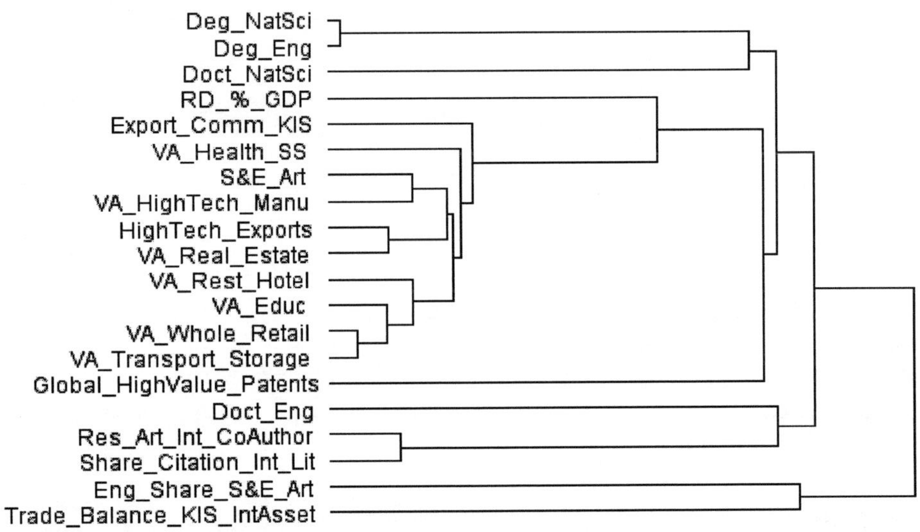

FIGURE F-15 Hierarchical cluster of science and engineering indicators from *SEI 2012 Digest*.
NOTES: GDP = gross domestic product; KIS = knowledge-intensive services; RD = research and development; S&E = science and engineering; SS = social services; VA = global value added.
SOURCE: Panel analysis and *Science and Engineering Indicators 2012*, see http://www.nsf.gov/statistics/seind12/tables.htm [November 2012].

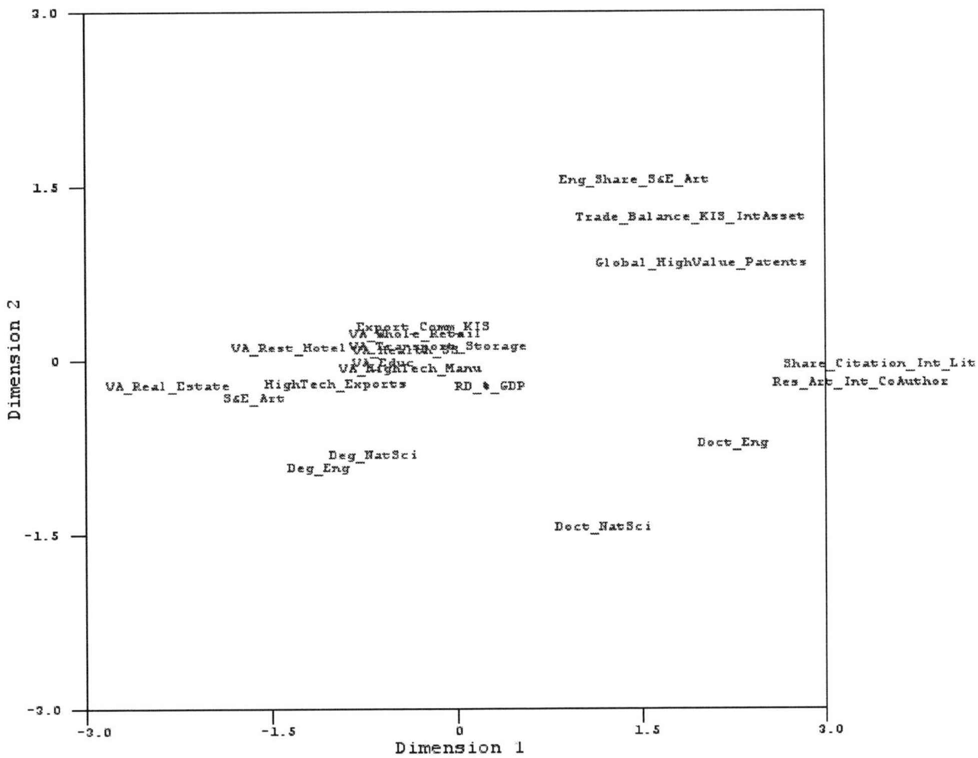

FIGURE F-16 Multidimensional scaling of science and engineering indicators from *SEI 2012 Digest*.
NOTES: GDP = gross domestic product; KIS = knowledge-intensive services; RD = research and development; S&E = science and engineering; SS = social services; VA = global value added.
SOURCE: Panel analysis and *Science and Engineering Indicators 2012*, see http://www.nsf.gov/statistics/seind12/tables.htm [November 2012].

Appendix G

2011 BRDIS Results

11011012

U.S. DEPARTMENT OF COMMERCE
Economics and Statistics Administration
U.S. CENSUS BUREAU

FORM
BRDI-1 (02-16-2012)

2011 BUSINESS R&D AND INNOVATION SURVEY

OMB No. 0607-0912: Approval Expires 01/31/2015

DUE DATE:

Report electronically:
econhelp.census.gov/brdis
Username:

Password:

Reporting electronically allows you to save your work as you go through the form and could save you time.

Report by mail:
If you report online, please do ***not*** mail in the paper form.

U.S. CENSUS BUREAU
1201 East 10th Street
Jeffersonville, IN
47132-0001

For information or assistance:
- econhelp.census.gov/brdis
- Call 1-800-772-7851, option "5" (8 a.m.– 5 p.m. EST, M-F)
- Write to the address above. Include your 11-digit ID printed on the mailing label.

Table 1 - 2011 Totals

See last page for Undistributed Lines

For questions with equivalent data collected in both 2010 and 2011, unless "no item imputation" is specified, a ratio of a company's prior year data was multiplied by that company's current year data to impute an estimate. This technique is called company trend.

Unless otherwise noted, data for estimates provided in this PDF are for companies who report R&D, either funded by the company or funded by others. This includes estimates for sales, employment, and patents.

(Please correct any errors in this mailing address)

Your Response is Required by Law

Title 13, United States Code, requires businesses and other organizations that receive this questionnaire to answer the questions and return the report to the Census Bureau.

Respondents are not required to respond to any information collection unless it displays a valid approval number from the Office of Management and Budget (OMB). The OMB number appears at the top of this page.

Your Response is Confidential by Law

Title 13, United States Code, requires that your response be seen only by persons sworn to uphold the confidentiality of Census Bureau information and may be used only for statistical purposes. The law also provides that copies retained in your company's files are immune from legal process.

About the Business R&D and Innovation Survey

The Business R&D and Innovation Survey is a national survey that collects critical information about research and development (R&D) and innovation at businesses operating in the U.S. Policy makers and data users in industry and academia make use of this information for short- and long-term planning.

Better information about R&D and innovation in the U.S. business sector will help leaders make better decisions to strengthen American competitiveness.

Thank You – Your Response is Important

Accurate and timely statistical information could not be produced without your continued cooperation and goodwill. Thank you.

~ This survey is jointly conducted by the U.S. Census Bureau and the National Science Foundation ~

11011020

Table of Contents

Survey Overview and Table of Contents

The survey is divided into six sections. Each section asks questions about different aspects of R&D or innovation at your company. Due to the specialized nature of each section, it may be necessary to collaborate with colleagues in different departments to answer the questions. The sections are color coded and cover the following topical areas:

Changes from the 2010 survey

On the basis of extensive conversations with many of the 2010 survey respondents, the 2011 survey has been improved. Sections and questions have been modified to make the concepts presented easier to understand. For a list of specific changes, go to econhelp.census.gov/brdis.

Filing electronically

You may submit your survey online via a secure website. Online submission allows you to save the data on secure Census Bureau servers as you go, so you can save, exit, and resume later without losing any of your data. It also allows you to save a paper or electronic copy of your completed survey. To submit online, follow the instructions at econhelp.census.gov/brdis.

Electronic materials

Electronic versions of the questionnaire and related documents are available to print or share with others in your company via e-mail. They can be found at econhelp.census.gov/brdis.
You can:

- Print and download copies of the questionnaire in PDF format
- Download Excel worksheets for each section
- Get question-by-question instructions, definitions, and examples that provide clarification
- Get answers to frequently asked questions, including how the data will be used

11011038

What is Research and Development (R&D)?

R&D is planned, creative work aimed at discovering new knowledge or developing new or significantly improved goods and services. This includes a) activities aimed at acquiring new knowledge or understanding without specific immediate commercial applications or uses (basic research); b) activities aimed at solving a specific problem or meeting a specific commercial objective (applied research); and c) systematic use of research and practical experience to produce new or significantly improved goods, services, or processes (development).

The term R&D does NOT include expenditures for:

- Costs for routine product testing, quality control, and technical services unless they are an integral part of an R&D project
- Market research
- Efficiency surveys or management studies
- Literary, artistic, or historical projects, such as films, music, or books and other publications
- Prospecting or exploration for natural resources

Does R&D include development of software and Internet applications?

Only development of software and Internet applications that include an element of uncertainty and that are intended to close gaps and meet scientific and technological needs should be reported as R&D on this survey.

R&D activity in software includes:

- Development of new software
- Significant improvement of existing software
- Construction of new theories and algorithms in the field of computer science

R&D activity in software does NOT include:

- Creation of new software based on known methods and applications
- Support for existing systems
- Conversion or translation of existing software and software languages
- Adaptation of a product to a specific client, unless knowledge that significantly improved the base program was added in that process
- Routine debugging of existing systems and software

Reporting unit

The reporting unit is your company, including all subsidiaries and divisions. Include subsidiary companies where there is more than 50 percent ownership.

Reporting period

Report data for the calendar year 2011, if possible, or for your company's fiscal year ending between April 2011 and March 2012.

Estimates are acceptable

Please report all items to the best of your ability.

To speak with a survey specialist, call 1-800-772-7851, option '1' for English, then option '5'.

Survey specialists are available 8 a.m. to 5 p.m. EST, M-F to help with any questions you may have.

11011046

SECTION 1

SECTION 1
Company Information

1-1 **Did another company other than a holding company own more than 50 percent of your company during 2011?**

☐ Yes → **Please provide the following information:**

Name of owner

(MM) (DD) (YYYY)

Date owner purchased your company

☐ No → **Skip to Question 1-3**

1-2 **If you answered "Yes" to Question 1-1, is your company's owner a foreign company?**

☐ Yes (foreign owner) → **Please follow these instructions:**

REPORTING INSTRUCTIONS FOR FOREIGN-OWNED COMPANIES:
If your company is foreign-owned, the reporting unit for the survey is your U.S.-based company, including all its majority-owned subsidiaries and divisions regardless of location. For reporting purposes, your foreign owner and any other foreign affiliates your company does not own should be treated the same as any other customer or business partner you do not own.

If you pay your foreign owner for R&D services, those costs should be included in your responses in Section 2 as "costs for purchased R&D services."

If your foreign owner pays or reimburses your company for R&D services, the costs for this R&D should be included in your responses in Section 3 as "costs funded, paid for, or reimbursed by others."

Report your survey data using U.S. generally accepted accounting principles (U.S. GAAP) as recognized by the Financial Accounting Standard Board (FASB). If your company follows International Financial Reporting Standards (IFRS), we request that you estimate any adjustments that would be required to conform to U.S. GAAP.

☐ No (U.S. owner) → **Please provide the following information:**

EIN of owner (9 digits)

REPORTING INSTRUCTIONS FOR U.S.-OWNED COMPANIES:
If your company was purchased between April 1, 2011 and December 31, 2011, report only for the period January 1, 2011 to the date of purchase. If your company was purchased before April 1, 2011, complete Question 1-6 and return this form to the Census Bureau – you are not required to complete the rest of this survey.
Disregard these instructions if your owner instructs you to complete this survey.

11011053

SECTION 1

1-3 **Did your company own more than 50 percent of any company operations or subsidiaries outside the 50 United States and D.C. during 2011?**

☐ Yes → **Include data for these operations/subsidiaries in your survey responses**

☐ No

1-4 **Has your company ceased operations?**

☐ Yes → **Please provide the following information:**

Date your company ceased operations (MM) (DD) (YYYY)

REPORTING INSTRUCTIONS:
If your company ceased operations between April 1, 2011 and December 31, 2011, report only for the period January 1, 2011 to the date your company ceased operations. If your company ceased operations before April 1, 2011, complete Question 1-6 and return this form to the Census Bureau – you are not required to complete the rest of this survey.

☐ No

1-5 **Did your company have discontinued operations in 2011?**

☐ Yes → **Include data for these operations in your survey responses**

☐ No

1-6 **Who is the survey coordinator?**

The survey coordinator is the person at your company responsible for gathering all requested information, ensuring instructions are followed, and submitting the completed survey. The survey coordinator may not be able to personally complete the entire survey and may need to request information from other knowledgeable resources concerning your company's R&D, accounting, human resources, and legal matters.

Name

Title

Telephone: Area code, Number; Extension

Fax: Area code, Number

E-mail address

11011061

SECTION 1

Business Codes

1-7 Do the business code(s) listed below reflect all applicable codes from the list on pages 42-43 in which your company operated worldwide during 2011?

NOTE: These codes will be used to describe both business activities and R&D activities and may differ from industry codes used by other government surveys and reports.

If no business codes are printed below, please write in the codes from pages 42-43 that apply to your company.

☐ Yes → **Continue with Question 1-8**

☐ No → **Draw a line through codes that are incorrect. As needed, enter additional codes and descriptions from pages 42-43 in the spaces provided.**

Business code (see pages 42-43)	Description

11011079

- Relative Standard Error =1.64%
- Collected on both forms.
- Unit non-response weight adjustment applied.
- Item imputation was done by HQ analysts for significant cases or by setting equal to sales for companies with equal employment.

/hat was the amount of your company's worldwide net sales and revenues uring 2011?

nclude:
Sales and operating revenues from discontinued operations

:xclude:
Non-operating income (i.e., dividends, interest)

Millions US $
12,900,545

SECTION 1

1-9 How much of the amount reported in Question 1-8 was attributable to or originated m domestic operations?

- Relative Standard Error = 2.23%
- Collected on both forms.
- Unit non-response weight adjustment applied.
- Item imputation was done by HQ analysts for significant cases or by using substitution from Business Register sales or by NALOB Ratio of Identicals (ROI, henceforth)

:lude:
Operating revenues and sales to foreign customers, including foreign subsidiaries

Millions US $
9,107,507

w much of the 2011 net sales and operating revenue amounts was for each siness code listed or amended in Question 1-7:
) Worldwide net sales and operating revenues reported in Question 1-8
) Domestic net sales and operating revenues reported in Question 1-9

	Business code (see page 6)	(1) Worldwide net sales and operating revenues			(2) Domestic net sales and operating revenues		
		$Bil.	Mil.	Thou.	$Bil.	Mil.	Thou.
a.							
b.							
c.							
d.							
e.							
f.							
g.							
h.							
i. Less eliminations – the sales and revenues that are eliminated in order to consolidate the business codes. . . .		(		)	(		)
j. **Total**							
		Total equals Question 1-8			*Total equals Question 1-9*		

- Collected only the largest code on the BRDI-1A form.

Form BRDI-1

11011087

SECTION 1

Product (good or service) innovation

A product innovation is the market introduction of a **new** or **significantly** improved good or service with respect to its capabilities, user friendliness, components, or sub-systems.

- Product innovations (new or improved) must be new to your company, but they do not need to be new to your market.
- Product innovations could have been originally developed by your company or by other companies.

-Collected on BOTH forms
-No item imputation
-estimates are for all respondents regardless of R&D performance.

1-11 During the three years 2009 to 2011, did your company introduce:

		Weighted Counts YES	Weighted Counts NO	
a. New or significantly improved goods (Exclude the simple resale of new goods purchased from other companies and changes of a solely aesthetic nature)?	RSE =4.40%	66,427	1,122,590	RSE =0.89%
b. New or significantly improved services?	RSE =4.57%	82,546	1,101,820	RSE = 0.91%

1-12 If you answered "yes" to either 1-11, line a, or 1-11, line b, were any of your product innovations during the three years 2009 to 2011:

		Weighted Counts YES	Weighted Counts NO	
a. New to your market? Your company introduced a new or significantly improved good or service to your market before your competitors. (It may be available in other markets.)	RSE = 4.67%	66,838	187,574	RSE = 3.12%
b. New only to your company? Your company introduced a new or significantly improved good or service that was already available from your competitors in your market.	RSE = 4.06%	93,788	160,114	RSE = 3.42%

1-13 Using the definitions above, please give the percentage of your total sales in 2011 from:

	Millions US $	
a. New or significantly improved goods and services introduced during 2009 to 2011 that were **new to your market**	873,851	RSE = 2.27%
b. New or significantly improved goods and services introduced during 2009 to 2011 that were **new only to your company**	679,612	RSE = 1.08%
c. Goods and services that were **unchanged or only marginally modified** during 2009 to 2011 (include the resale of new goods or services purchased from other companies)	11,347,082	RSE = 1.81%
d. **Total sales in 2011**		

-Collected on BOTH forms
-No item imputation
-NOTE: data are for R&D companies only

11011095

Process innovation

A process innovation is the implementation of a **new** or **significantly** improved production process, distribution method, or support activity for your goods or services.

- Process innovations must be new to your company, but they do not need to be new to your market.
- The innovation could have been originally developed by your company or by other companies.
- Exclude purely organizational innovations.

-Collected on BOTH forms
-No item imputation
-estimates are for all respondents regardless of R&D performance.

1-14 During the three years 2009 to 2011, did your company introduce:

		YES	NO	
a. New or significantly improved methods of manufacturing or producing goods or services?	RSE = 5.20%	46,326	1,136,978	RSE = 0.88%
b. New or significantly improved logistics, delivery or distribution methods for your inputs, goods, or services?	RSE = 6.35%	38,664	1,142,041	RSE = 0.88%
c. New or significantly improved supporting activities for your processes, such as maintenance systems or operations for purchasing, accounting, or computing?	RSE = 4.36%	87,968	1,092,244	RSE =0.92%

Form BRDI-1

11011103

SECTION 2
Financial Schedule A

Who should answer this section?
Persons familiar with accounting concepts and with access to financial records related to your company's R&D activities should complete this section.

What does this section cover?
This section requests financial information about your company's R&D expenses and capital expenditures for R&D. This section requests information about your company's R&D at multiple levels of detail: for your worldwide consolidated enterprise, for units or parts of your company defined by geography (countries, states, specific location), and for parts of your company defined by business code.

SECTION 2

2-1 What was the total worldwide R&D expense for your company in 2011?

- Relative Standard Error = 0.81%
- Collected on BOTH forms
- No item imputation
-NOTE: Q2.1 - Q2.3 does not equal Q2.4. (difference = $2.183 bil)

If your company is publicly traded, this amount is equivalent to that disclosed on SEC Form 10-K as defined in FASB ASC Topic 730, Research and Development (FASB Statement No. 2, "Accounting for Research and Development Costs.")
If your company is foreign-owned, refer to the instructions on page 4. Additional guidance, such as for **privately owned companies,** is available online at econhelp.census.gov/brdis.

NOTE: Report your company's R&D expense even if the amount is not considered material for your company's financial statements.

Millions US $
364,641

2-2 Does the amount reported in Question 2-1 include any of the following costs?

FOR 2.3:
- Relative Standard Error = 1.51%
- Collected on BOTH forms
- No item imputation

a. Collaborative R&D that was reimbursed by business partners, such as through cost-sharing agreements . . . . ☐ Yes ☐ No

b. R&D paid for by government or private foundation grants . . . . . . ☐ Yes ☐ No

c. Technical services not an integral part of an R&D project (such as product support provided by R&D employees) . . . . . . ☐ Yes ☐ No

d. Bid and proposal costs . . . . . . ☐ Yes ☐ No

e. Expense your company claimed resulting from the acquisition of another company with unfinished R&D projects (in-process R&D) . . . . . ☐ Yes ☐ No

2-3 If you answered "Yes" to any of the costs in Question 2-2, what was the amount of these costs that was included in your response to Question 2-1?

Millions US $
25,489

2-4 Subtract Question 2-3 from Question 2-1 and enter the result here. This is the total R&D paid for by your company in 2011.

Millions US $
341,335

2-5 Is the amount entered in Question 2-4 greater than zero?

☐ Yes → **Continue with Question 2-6**

☐ No → **Skip to Section 3 on page 20**

FOR 2.4:
- Relative Standard Error = 0.85%
- Collected on BOTH forms
- Unit non-response weight adjustment applied.
- Item imputation was done by HQ analysts for selected cases where 10K values or prior year data was available

Form BRDI-1

11011111

SECTION 2

R&D paid for by your company

2-6 **Of the amount reported in Question 2-4, what were the costs for each business code listed or amended on page 6 of this form?**

Allocate R&D that is applicable to more than one business code on a reasonable basis. Allocation in proportion to operating revenues is acceptable unless some alternative allocation basis is more appropriate.

	Business code (see page 6)	$Bil.	Mil.	Thou.
a.				
b.				
c.				
d.				
e.				
f.				
g.				
h.				
i. **Total** (equals Question 2-4)				

- Collected only the largest code on the BRDI-1A form.

2-7 **Of the amount reported in Question 2-4, what costs were incurred by your company in the following locations?**

-Collected on BOTH forms
-Unit non-response weight adjustment applied
-Item imputation: 2.7a by company trend and NALOB ROI and 2.7b by subtraction

	Millions US $	
a. Domestic United States (50 states and D.C.) [Include R&D performed by domestic operations that is paid for by foreign subsidiaries]	268,376	RSE = 1.07%
b. All other countries (also, Puerto Rico) [Include R&D performed by foreign subsidiaries that is paid for by domestic operations]	72,958	RSE = 0.31%
c. **Total** (equals Question 2-4)	341,335	

2-8 **Copy the amount from Question 2-7, line a. This is the total <u>domestic R&D paid for by your company</u> in 2011.**

Millions US $: 268,376

2-9 **Copy the amount reported from Question 2-7, line b. This is the total <u>foreign R&D paid for by your company</u> in 2011.**

Millions US $: 72,958

Form BRDI-1

11011129

SECTION 2

2-10 How much of the (1) domestic, (2) foreign, and (3) total worldwide R&D paid for by your company in 2011 was for each of the following types of costs?

	(1) Domestic Millions US $	(2) Foreign Millions US $	(3) Total worldwide Millions US $
a. Salaries, wages, and fringe benefits	141,487 (RSE = 1.26%)	37,128 (RSE = 0.49%)	178,616 (RSE = 1.01%)
b. Stock-based compensation	9,945 (RSE = 4.37%)	931 (RSE = 0.22%)	10,876 (RSE = 4.00%)
c. Temporary staffing, including on-site consulta[...]	10,145 (RSE = 1.20%)	2,405 (RSE = 0.88%)	12,550 (RSE = 1.00%)
d. Expensed equipment	2,938 (RSE = 5.26%)	929 (RSE = 0.15%)	3,867 (RSE = 4.00%)
e. Materials and supplies	15,961 (RSE = 0.69%)	3,776 (RSE = 0.08%)	19,737 (RSE = 0.56%)
f. Leased facilities and equipment	3,677 (RSE = 4.11%)	1,101 (RSE = 0.34%)	4,779 (RSE = 3.17%)
g. Depreciation and amortization on R&D propert[...]ment	9,796 (RSE = 2.54%)	2,370 (RSE = 0.22%)	12,166 (RSE = 2.05%)
h. Payments to business partners for collaborat[...]	5,617 (RSE = 0.81%)	1,401 (RSE = 0.06%)	7,018 (RSE = 0.65%)
i. Purchased R&D services (if your company is f[...], include payments to [...] owner for R&D)	23,991 (RSE = 0.43%)	6,520 (RSE = 0.69%)	30,511 (RSE = 0.37%)
j. All other purchased services except R&D	5,959 (RSE = 1.47%)	2,105 (RSE = 0.08%)	8,064 (RSE = 1.09%)
k. All other costs	38,860 (RSE = 1.98%)	14,291 (RSE = 0.18%)	53,152 (RSE = 1.45%)
l. **Total**	268,376	72,958	341,335
	[...]n 2-8	[...]9	*Total equals Question 2-4*

-2.10a, 2.10c, 2.10h, 2.10i, 2.10k collected on BOTH forms
-2.10b, 2.10d-2.10g, 2.10j collected on BRDI-1 form ONLY
-Unit non-response weight adjustment applied to all items

Item Imputation (cols 1 & 2) - Use company trend first. If company trend not available, use NALOB ROI
(col 3) - Imputed using the sum of columns 1 and 2

FOR 2.12:
-Collected on BOTH forms
-Unit non-response weight adjustment applied
-Item imputation for 2.11(1), (2) and (3): by subtraction

FOR 2.13:
-Collected on BOTH forms
-Unit non-response weight adjustment applied
-Item imputation for 2.12(1): Use company trend; if not available, NALOB ROI; for 2.12(2): subtract 2.12(1) from 2.12(3); for 2.12 (3): Use company trend; if not available, NALOB ROI

2-11 Add 2-10, lines h and i for each column, and enter the result here. This is R&D performed by others.

	(1) Domestic	(2) Foreign	(3) Total worldwide
	Millions US $	Millions US $	Millions US $
	29,608	7,921	37,529
	RSE = 0.38%	RSE = 0.56%	RSE = 0.33%

2-12 Subtract 2-11 from 2-10, line l, for each column and enter the result here. This is R&D performed by your company.

	(1) Domestic	(2) Foreign	(3) Total worldwide
	Millions US $	Millions US $	Millions US $
	238,768	65,038	303,806
	RSE = 1.20%	RSE = 0.34%	RSE = 0.95%

2-13 Copy the amount from Question 2-12, column 2. This is the foreign R&D paid for and performed by your company in 2011.

$Bil. Thou.

65,038

2-14 Of the amount reported in Question 2-13, how much R&D was performed in the following locations?

Location	Millions US $	RSE
a. Canada	5,082	RSE = 1.00%
b. Puerto Rico	95	RSE = 0%
Europe		
a. Austria	176	RSE = 0%
b. Belgium	1,613	RSE = 0%
c. Czech Rep.	151	RSE = 0%
d. Denmark	512	RSE = 0%
e. Finland	312	RSE = 0%
f. France	2,670	RSE = 0.32%
g. Germany	8,583	RSE = 0.41%
h. Ireland	1,115	RSE = 0%
i. Italy	1,311	RSE = 0%
j. Netherlands	1,526	RSE = 0.10%
k. Poland	345	RSE = 1.60%
l. Russia	429	RSE = 0%
m. Spain	436	RSE = 0%
n. Sweden	777	RSE = 0%
o. Switzerland	2,112	RSE = 0.28%
p. United Kingdom	9,567	RSE = 0.12%
q. Other Europe	2,803	RSE = 3.98%
Latin America		
a. Argentina	172	RSE = 0%
b. Brazil	1,751	RSE = 0%
c. Chile	45	RSE = 0%

Question continues on next page

Form BRDI-1

SECTION 2

11011145

2-14 (Continued)

		$Bil. Mil. Thou.
RSE = 0.53%	d. Mexico	865
RSE = 2.70%	Other Latin America	394
	Asia and Pacific	
RSE = 0%	a. Australia	1,187
RSE = 0.58%	b. China	4,141
RSE = 3.09%	c. Hong Kong	104
RSE = 0.31%	d. India	4,668
RSE = 0%	e. Indonesia	23
RSE = 0.60%	f. Japan	3,014
RSE = 0%	g. Malaysia	645
RSE = 11.29%	h. New Zealand	47
RSE = 0.30%	i. Singapore	1,632
RSE = 0%	j. South Korea	1,132
RSE = 1.32%	k. Taiwan	481
RSE = 28.55%	l. Thailand	344
RSE = 0.84%	m. Other Asia/Pacific	551

	$Bil. Mil. Thou.	
Middle East		
a. Israel	3,316	RSE = 0.65%
b. Other Middle East	169	RSE = 0%
Africa		
a. South Africa	126	RSE = 0%
b. Other Africa	53	RSE = 0%
Total (equals Question 2-13)	56,635	RSE = 0.03%

FOR 2.14:
-Collected on BRDI-1 ONLY; For BRDI-1A - Value of company foreign R&D performance moved to undistributed foreign.
-Item imputation - Use company trend (if available). Some data may be undistributed when using company trend as the former "other" column was split into several cells for 2011.
-Unit non-response weight adjustment applied

2-15 Copy the amount from Question 2-12, column 1. This is the domestic R&D paid for and performed by your company in 2011.

Millions US $: 238,768

Form BRDI-1

FOR 2.16:
-Collected only largest state on BRDI-1A
-Unit non-response weight adjustment applied
-Item imputation - Use company trend (if available). Use mail state for single units. Sum of states less than total, moved to undistributed.

11011152

2-16 How much of the amount reported in Question 2-15 was performed in each state (including D.C.) in 2011?

RSE	State	Millions US $	State	Millions US $	RSE
RSE = 3.67%	Alabama	835	Maine	264	RSE = 3.08%
RSE = 11.50%	Alaska	52	Maryland	2,867	RSE = 5.36%
RSE = 1.15%	Arizona	3,776	Massachusetts	12,712	RSE = 0.91%
RSE = 3.44%	Arkansas	307	Michigan	12,156	RSE = 1.02%
RSE = 0.92%	California	64,104	Minnesota	5,592	RSE = 0.89%
RSE = 1.61%	Colorado	3,642	Mississippi	189	RSE = 4.11%
RSE = 0.59%	Connecticut	6,272	Missouri	2,818	RSE = 1.32%
RSE = 0.79%	Delaware	1,453	Montana	118	RSE = 4.11%
RSE = 16.55%	District of Columbia	162	Nebraska	601	RSE = 2.69%
RSE = 2.95%	Florida	3,855	Nevada	554	RSE = 3.05%
RSE = 2.63%	Georgia	3,303	New Hampshire	943	RSE = 1.92%
RSE = 5.34%	Hawaii	183	New Jersey	11,977	RSE = 1.30%
RSE = 0.87%	Idaho	871	New Mexico	250	RSE = 3.36%
RSE = 1.11%	Illinois	10,764	New York	9,141	RSE = 1.99%
RSE = 0.82%	Indiana	5,484	North Carolina	5,157	RSE = 1.16%
RSE = 1.15%	Iowa	1,736	North Dakota	236	RSE = 2.01%
RSE = 2.02%	Kansas	1,037	Ohio	5,403	RSE = 1.57%
RSE = 3.13%	Kentucky	1,017	Oklahoma	507	RSE = 3.96%
RSE = 6.61%	Louisiana	382	Oregon	4,415	RSE = 0.60%

Question continues on next page

11011160

2-16 (Continued)

RSE	State	Millions US $	State	Millions US $	RSE
RSE = 1.16%	Pennsylvania	9,018	Vermont	329	RSE = 2.07%
RSE = 2.02%	Rhode Island	451	Virginia	3,138	RSE = 8.95%
RSE = 2.11%	South Carolina	949	Washington	13,659	RSE = 0.49%
RSE = 3.89%	South Dakota	112	West Virginia	213	RES = 3.04%
RSE = 2.34%	Tennessee	1,279	Wisconsin	3,548	RSE = 1.17%
RSE = 1.68%	Texas	12,920	Wyoming	33	RSE = 12.07%
RSE = 1.43%	Utah	1,874	**Total** (equals Question 2-15)	238,768	

2-17 At what domestic location did your company perform the largest dollar amount of R&D in 2011?

Address 1

Address 2

City **State** **ZIP**

2-18 How much of the amount reported in Question 2-15 was from the location identified in Question 2-17?

Millions US $: 142,897

-Relative Standard Error = 1.99%
-Collected on BOTH forms
-Unit non-response weight adjustment applied
-Item imputation was done for single units

11011178

2-19 **At what domestic location did your company perform the second largest dollar amount of R&D in 2011?**

Address 1

Address 2

City

-Relative Standard Error = 0.22%
-Collected only on BRDI-1
-Unit non-response weight adjustment applied
-There was no item imputation performed on this question.

2-20 **How much of the amount reported in Question 2-15 was from the location identified in Question 2-19?**

Millions US $
19,042

2-21 **Of the domestic R&D performed by your company reported in Question 2-15, how much was for each business code reported in Question 2-6?**

	Business code (see page 6)	(1) Domestic R&D performed by your company $Bil.	Mil.	Thou.
a.				
b.				
c.				
d.				
e.				
f.				
g.				
h.				
i. **Total**				

- Collected only the largest code on the BRDI-1A form.

Total equals Question 2-15

SECTION 2

11011186

SECTION 2

Domestic R&D performance paid for by foreign subsidiaries

-Relative Standard Error = 0%
-Collected on BRDI-1 ONLY
-Unit non-response weight adjustment applied
-No item imputation

2-22 How much of the amount reported in Question 2-15 was paid for by your company's foreign subsidiaries?

Example: Company Y owns a subsidiary in France. In order to complete the development of a product in 2011, the French subsidiary paid for R&D performed at Company Y's U.S. R&D center. The cost of the U.S. R&D that was paid for by the French subsidiary would be included in this item.

Millions US $
3,342

Foreign R&D performance paid for by domestic operations

-Relative Standard Error = 0%
-Collected on BRDI-1 ONLY
-Unit non-response weight adjustment applied
-No item imputation

2-23 How much of the amount reported in Question 2-13 was paid for by your company's domestic operations?

Example: Company Z owns a subsidiary in France. In order to complete the development of a product in 2011, the domestic operations paid for R&D performed at its subsidiary's R&D center in France. The cost of the French subsidiary's R&D that was paid for by the domestic operations would be included in this item.

Millions US $
9,689

R&D performed by others

2-24 Copy the amount from Question 2-11, column 1. This is the domestic R&D paid for by your company in 2011 that was performed by others.

Millions US $
29,608

2-25 How much of the amount reported in Question 2-24, was performed by the following types of organizations?

	Millions US $	
a. Companies located inside the United States	22,747	RSE = 0%
b. Your company's foreign owner (if your company is foreign-owned)	440	RSE = 0%
c. Other companies located outside the United States	3443	RSE = 2.43%
d. U.S. federal government agencies or laboratories	175	RSE = 0%
e. U.S. state and local government agencies or laboratories	10	RSE = 0%
f. Foreign government agencies or laboratories	8	RSE = 0%
g. All other organizations inside the United States	688	RSE = 0.07%
h. All other organizations outside the United States	144	RSE = 0%
i. **Total domestic R&D paid for by your company that was performed by others** (equals Question 2-24)	29,608	

- Collected on BRDI-1 ONLY
- Unit non-response weight adjustment applied
-Item imputation: for BRDI-1 - Use company trend first. If not available, use NALOB ROI; for BRDI-1A no item imputation

Form BRDI-1

11011194

Projected R&D for 2012

2-26 What are your company's projected 2012 costs for (1) domestic, (2) foreign, and (3) total worldwide R&D paid for by your company?

NOTE: This amount is the 2012 projection for what is reported in Question 2-10, line l.

(1) Domestic	(2) Foreign	(3) Total worldwide
Millions US $	Millions US $	Millions US $
293,157	72,054	365,211
RSE = 1.03%	RSE = 0.53%	RSE = 0.84%

2-27 How much of the amount reported in Question 2-26, column 1, is for projected purchased R&D services and projected payments to business partners for collaborative R&D?

Domestic
Millions US $
35,114
RSE = 1.02%

FOR 2.26 & 2.27:
- Collected on BOTH forms
-Unit non-response weight adjustment applied
-Item imputation was done using Company Trend, if available, otherwise NALOB ROI was used.

FOR 2.28 & 2.29
-Collected on BOTH forms
-Unit non-response weight adjustment applied
-Item imputation was done using NALOB ROI

Capital

2-28 Wh... your company's capital ex... estic United States in 2011?

Millions US $
477,001
RSE = 2.26%

2-29 How much of the amount reported in Question 2-28 was for R&D operations?

Millions US $
23,183
RSE = 1.67%

Reporting information

2-30 Is the information in this section reported for the 2011 calendar year?

☐ Yes

☐ No → **Enter time period covered below:**

From (MM) (YYYY) to (MM) (YYYY)

SECTION 2

11011202

SECTION 3

SECTION 3
Financial Schedule B

Who should answer this section?
Persons familiar with accounting concepts and with access to financial records related to your company's R&D activities should complete this section.

What does this section cover?
This section requests financial information about your company's costs for work that was funded, paid for, or reimbursed by others. This section requests information about these costs at multiple levels of detail: for your worldwide consolidated enterprise, for units or parts of your company defined by geography (countries, states), and for parts of your company defined by business code.

3-1 What were your company's total worldwide costs (both direct and indirect) in 2011 for the following that were funded, paid for, or reimbursed by others not owned by your company?

FOR 3.1a - 3.1i:
-No item imputation
-NOTE: sum details (Q3.1a - Q3.1i) does not equal Q3.1j because details edited based on revised Q3.1j values (difference = $3,148m)

Exclude:
- Costs that were paid for by your company, such as those reported in Question 2-4
- Payments in excess of the actual cost of the work performed (such as profit or fees)

	Millions US $	
a. R&D that was reimbursed by your company's foreign owner, if your company was foreign-owned	9,377	RSE = 4.21%
b. Collaborative R&D that was reimbursed by business partners, such as through cost-sharing agreements	6,763	RSE = 2.62%
c. R&D paid for by government or private foundation grants	11,852	RSE = 2.27%
d. Defense R&D services provided to the government and/or government contractors	27,573	RSE = 0.71%
e. Medical nonclinical R&D services provided to others not owned by your company	2,808	RSE = 1.17%
f. Medical clinical trial Phase I-III services provided to others not owned by your company (include pass-through costs)	5,588	RSE = 1.91%
g. Nondefense custom software development and/or computer systems designed for others not owned by your company	450	RSE = 34.97%

Exclude:
- Software development that does not depend on a scientific or technological advance, such as adding functionality to existing application programs, debugging systems, and adapting existing software

	Millions US $	
h. Developing, producing, and testing prototypes of customer's products prior to their introduction to the market (excluding defense-related prototyping reported in line d)	1,416	RSE = 7.27%
i. All other R&D services provided to others not owned by your company	3,210	RSE = 5.13%
j. **Total**	72,185	RSE = 0.87%

FOR 3.1j:
-Collected on BOTH forms
-Unit non-response weight adjustment applied
-Item imputation was done by analyst; substituted 10-k data or hand imputed using prior year data for the large R&D cases that were delinquent. Additionally, data from select companies were imputed using 2009 R&D contract information from usaspending.gov.

11011210

3-2 **Copy the amount from Question 3-1, line j. This is the total R&D paid for by others in 2011.**

Millions US $: 72,185

3-3 **Is the amount entered in Question 3-2 greater than zero?**

☐ Yes → **Continue with Question 3-4**

☐ No → **Skip to Section 4 on page 32**

R&D paid for by others

3-4 **Of the amount reported in Question 3-2, what costs were incurred by your company in the following locations?**

-Collected on BOTH forms
-Unit non-response weight adjustment applied
-Item imputation: 3.4a by NALOB ROI and 3.4b by subtraction

	Millions US $	
a. Domestic United States (50 states and D.C.)	66,847	RSE = 0.93%
b. All other countries (also, Puerto Rico)	5,340	RSE = 0.4%
c. **Total** (equals Question 3-2)	72,185	

SECTION 3

3-5 **Copy the amount from Question 3-4, line a. This is the total domestic R&D paid for by others in 2011.**

Millions US $: 66,847

3-6 **Copy the amount from Question 3-4, line b. This is the total foreign R&D paid for by others in 2011.**

Millions US $: 5,340

Form BRDI-1

11011228

SECTION 3

3-7 How much of the (1) domestic, (2) foreign, and (3) total worldwide R&D paid for by others in 2011 was for each of the following types of costs?

	(1) Domestic	(2) Foreign	(3) Total worldwide
	Millions US $	Millions US $	Millions US $
a. Salaries, wages, and fringe benefits	33,003 (RSE = 1.25%)	2,470 (RSE = 0.61%)	35,473 (RSE = 1.17%)
b. Stock-based compensation	228 (RSE = 0.95%)	4 (RSE = 0%)	233 (RSE = 0.93%)
c. Temporary staffing, including on-site consultants	1,345 (RSE = 3.10%)	108 (RSE = 0.05%)	1,453 (RSE = 2.87%)
d. Expensed equipment	346 (RSE = 1.88%)	26 (RSE = 0%)	371 (RSE = 1.75%)
e. Materials and supplies	6,340 (RSE = 1.44%)	361 (RSE = 0.11%)	6,701 (RSE = 1.37%)
f. Leased facilities and equipment	723 (RSE = 0.95%)	110 (RSE = 0.55%)	833 (RSE = 0.83%)
g. Depreciation and amortization on R&D property and equipment	1,049 (RSE= 0.88%)	88 (RSE= 0.38%)	1,138 (RSE= 0.82%)
h. Payments to business partners for collaborative R&D	1,477 (RSE = 7.28%)	54 (RSE = 0.85%)	1,531 (RSE = 7.02%)
i. Purchased R&D services (if your company is fo[...] include payments to yo[...]ner for R&D)	10,043 (RSE = 1.43%)	1,093 (RSE = 0.22%)	11,136 (RSE = 1.29%)
j. All other purchased services except R&D	925 (RSE = 1.57%)	26 (RSE = 0.24%)	951 (RSE = 1.53%)
k. All other costs	11,364 (RSE = 0.73%)	999 (RSE = 0.54%)	12,363 (RSE = 0.68%)
l. **Total**	66,847	5,340	72,185
	[...]-5		*Total equals Question 3-2*

-3.7a, 3.7c, 3.7h, 3.7i, 3.7k collected on BOTH forms
-3.7b, 3.7d-3.7g, 3.7j collected on BRDI-1 ONLY
-Unit non-response weight adjustment applied to all items

Item Imputation (cols 1 & 2) - Use company trend first. If company trend not available, use NALOB ROI
(col 3) - Imputed using the sum of columns 1 and 2

FOR 3.8:
-Collected on BOTH forms
-Unit non-response weight adjustment applied
-Item imputation for 3.8(1), (2) and (3): by subtraction

FOR 3.9:
-Collected on BOTH forms
-Unit non-response weight adjustment applied
-Item imputation for 3.9(1): Use company trend; if not available, NALOB ROI; for 3.9(2); subtract 3.9(1) from 3.9(3); for 3.9(3): Use company trend; if not available, NALOB ROI

3-8 **Add 3-7, lines h and i for each column, and enter the result here. This is R&D performed by others (e.g., subcontracted/passed-through R&D costs).**

(1) Domestic	(2) Foreign	(3) Total worldwide
Millions US $	Millions US $	Millions US $
11,520	1,147	12,667
RSE = 1.57%	RSE = 0.21%	RSE = 1.43%

3-9 **Subtract 3-8 from 3-7, line l, for each column and enter the result here. This is R&D performed by your company that was paid for by others.**

(1) Domestic	(2) Foreign	(3) Total worldwide
Millions US $	Millions US $	Millions US $
55,324	4,193	59,517
RSE = 1.02%	RSE = 0.51%	RSE = 0.95%

3-10 **Copy the amount from Question 3-9, column 2. This is the foreign R&D performed by your company that was paid for by others.**

Millions US $: 4,193

3-11 **Of the amount reported in Question 3-10, how much R&D was performed in the following locations?**

Location	Millions US $	RSE
a. Canada	386	RSE = 1.83%
b. Puerto Rico	19	RSE = 0%
Europe		
a. Austria	31	RSE = 0%
b. Belgium	89	RSE = 0%
c. Czech Rep.	29	RSE = 0%
d. Denmark	11	RSE = 0%
e. Finland	15	RSE = 0%
f. France	301	RSE = 0%
g. Germany	374	RSE = 0%
h. Ireland	46	RSE = 0%
i. Italy	114	RSE = 0%
j. Netherlands	84	RSE = 11.31%
k. Poland	54	RSE = 0%
l. Russia	47	RSE = 0%
m. Spain	108	RSE = 0%
n. Sweden	53	RSE = 0%
o. Switzerland	269	RSE = 0%
p. United Kingdom	814	RSE = 0%
q. Other Europe	207	RSE = 0%
Latin America		
a. Argentina	58	RSE = 0%
b. Brazil	74	RSE = 0%
c. Chile	10	RSE = 0%

($Bil. Mil. Thou.)

Question continues on next page

SECTION 3

Form BRDI-1

11011244

3-11 (Continued)

	$Bil. Mil. Thou.	
d. Mexico	46	RSE = 0%
e. Other Latin America	22	RSE = 0%
Asia and Pacific		
a. Australia	111	RSE = 2.02%
b. China	107	RSE = 0%
c. Hong Kong	5	RSE = 0%
d. India	115	RSE = 0%
e. Indonesia	2	RSE = 0%
f. Japan	215	RSE = 0%
g. Malaysia	7	RSE = 0%
h. New Zealand	9	RSE = 0%
i. Singapore	79	RSE = 0%
j. South Korea	51	RSE = 0%
k. Taiwan	13	RSE = 0%
l. Thailand	12	RSE = 0%
m. Other Asia/ Pacific	13	RSE = 0%
Middle East		
a. Israel	54	RSE = 0.16%
b. Other Middle East	14	RSE = 0%
Africa		
a. South Africa	69	RSE = 0%
b. Other Africa	5	RSE = 0%
Total (equals Question 3-10)	4,193	

FOR 3-11:
-Collected on BRDI-1 ONLY; For BRDI-1A - Value of company foreign R&D performance moved to undistributed foreign.
-Item imputation - Use company trend (if available). Some data may be undistributed when using company trend as the former "other" column was split into several cells for 2011.
-Unit non-response weight adjustment applied

Domestic R&D performed by your company t

3-12 Copy the amount from Question 3-9, column 1. This is the domestic R&D performed by your company that was paid for by others.

3-13 How much of the domestic R&D performed by your company that was paid for by others reported in Question 3-12 was for each business code listed or amended on page 6 of this form?

Allocate R&D that is applicable to more than one business code on a reasonable basis. Allocation in proportion to operating revenues is acceptable unless some alternative allocation basis is more appropriate.

Business code (see page 6)	$Bil. Mil. Thou.
a.	
b.	

Question continues on next page

11011251

3-13 (Continued)

Business code (see page 6)		$Bil.	Mil.	Thou.
c.	....			
d.	....			
e.	....			
f.	....			
g.	....			
h.	....			
i. **Total** (equals Question 3-12)........				

SECTION 3

3-14 How much of the amount reported in Question 3-12 was paid for by each of the following?

If your company is a subcontractor or subgrantee, report the original source of funds.

Example: Company Sub Inc. performs custom software development for a large defense company as a subcontractor on a contract with the U.S. Dept. of Defense. Even though Sub Inc. is working directly for the defense company, it reports the cost of this development in line d because the Dept. of Defense was the original source of funds.

-Collected on BOTH forms
-Unit non-response weight adjustment applied
-Item Imputation: Use company trend first. If not available, impute using NALOB X BEA Link Indicator ROI.

	Millions US $	
a. Other companies located inside the United States.........	11,124	RSE = 2.51%
b. Your company's foreign owner (if your company is foreign-owned)	7,438	RSE = 5.23%
c. Other companies located outside the United States	4,569	RSE = 1.10%
d. U.S. federal government agencies or laboratories.........	31,309	RSE = 0.75%
e. U.S. state government agencies or laboratories	321	RSE = 24.54%
f. Foreign government agencies or laboratories............	63	RSE = 18.79%
g. All other organizations inside the United States	369	RSE = 14.32%
h. All other organizations located outside the United States ...	129	RSE = 22.84%
i. **Total** (equals Question 3-12)........	55,324	

11011269

SECTION 3

3-15 **Copy the amount from Question 3-14, line d. This is domestic R&D performed by your company that was paid for by the U.S. federal government.**

	Millions US $
	31,309

3-16 **How much of the amount reported in Question 3-15 was paid for by the following agencies?**

	Millions US $	
a. Department of Defense	26,350	RSE = 0.66%
b. Department of Energy	881	RSE = 6.68%
c. National Aeronautics and Space Administration	1,580	RSE = 1.91%
d. National Institutes of Health	656	RSE = 2.39%
e. All other	1,843	RSE = 5.68%
f. **Total** (equals Question 3-15)	31,309	

3-17 **Subtract Question 3-15 from Question 3-12 and enter the result here. This is the domestic R&D performed by your company that was paid for by nonfederal sources.**

	Millions US $	
	24,015	RSE = 2.07%

3-18 **How much of the following three amounts was performed in each state (including D.C.):**
(1) Domestic R&D paid for by the U.S. federal government reported in Question 3-15
(2) Domestic R&D paid for by nonfederal sources reported in Question 3-17
(3) Total domestic R&D performed by your company that was paid for by others reported in Question 3-12

	(1) Federal Millions US $	(2) Nonfederal Millions US $	(3) Total Millions US $
Alabama	RSE = 0.74%	RSE = 2.58%	RSE = 1.00%
	697	348	1,045
Alaska	RSE = 6.07%	RSE = 6.82%	RSE = 4.97%
	14	19	33
Arizona	RSE = 0.57%	RSE = 1.36%	RSE = 0.67%
	678	477	1,155
Arkansas	RSE = 5.24%	RSE = 13.32%	RSE = 7.73%
	17	21	38

Question continues on next page

FOR 3.18:
-Collected only largest state on BRDI-1A
-Unit non-response weight adjustment applied
-Item imputation - Use company trend (if available). Use mail state for single units. Sum of states less than total, moved to undistributed.

11011277

3-18 (Continued)

	(1) Millions US $	(2) Millions US $	(3) Millions US $
California	RSE = 0.55% 6,968	RSE = 1.92% 3,963	RSE = 0.80% 10,931
Colorado	RSE = 1.08% 426	RSE = 3.80% 242	RSE = 1.59% 668
Connecticut	RSE = 0.37% 928	RSE = 2.54% 304	RSE = 0.69% 1,232
Delaware	RSE = 12.20% 18	RSE = 0.19% 625	RSE = 0.40% 643
District of Colu	RSE = 0.50% 230	RSE = 6.85% 23	RSE = 0.83% 253
Florida	RSE = 0.65% 1,680	RSE = 4.74% 453	RSE = 1.15% 2,133
Georgia	RSE = 3.94% 190	RSE = 3.27% 347	RSE = 2.64% 536
Hawaii	RSE = 14.4% 19	RSE = 5.54% 49	RSE = 6.05% 69
Idaho	RSE = 19.26% 7	RSE = 0.52% 294	RSE = 0.69% 301
Illinois	RSE = 4.15% 167	RSE = 1.88% 1,108	RSE = 1.76% 1,275
Indiana	RSE = 4.08% 82	RSE = 1.57% 593	RSE = 1.48% 674
Iowa	RSE = D D	RSE = D D	RSE = 0.68% 578
Kansas	RSE = 9.18% 17	RSE = 0.95% 455	RSE = 0.98% 472

Question continues on next page

Form BRDI-1

27

SECTION 3

11011285

SECTION 3

3-18 (Continued)

	(1) Millions US $	RSE	(2) Millions US $	RSE	(3) Millions US $	RSE
Kentucky	220	RSE = 0.93%	41	RSE = 12.51%	261	RSE = 2.12%
Louisiana	37	RSE = 7.81%	40	RSE = 13.46%	77	RSE = 8.33%
Maine	18	RSE = 5.74%	12	RSE = 16.43%	31	RSE = 7.54%
Maryland	1,677	RSE = 0.71%	558	RSE = 3.10%	2,235	RSE = 0.99%
Massachusetts	919	RSE = 1.06%	2,091	RSE = 0.87%	3,010	RSE = 0.70%
Michigan	292	RSE = 1.97%	1,212	RSE = 1.16%	1,504	RSE = 1.03%
Minnesota	240	RSE = 1.88%	342	RSE = 3.04%	582	RSE = 1.98%
Mississippi	31	RSE = 2.02%	15	RSE = 14.03%	46	RSE = 4.77%
Missouri	4,743	RSE = 0.07%	D	RSE = D	D	RSE = D
Montana	8	RSE = 10.36%	10	RSE = 18.03%	18	RSE = 10.98%
Nebraska	8	RSE = 11.22%	27	RSE = 7.58%	35	RSE = 6.59%
Nevada	23	RSE = 9.10%	62	RSE = 6.48%	85	RSE = 5.39%
New Hampshir	D	RSE = 0.41%	D	RSE = 2.56%	1,126	RSE = 0.54%

Question continues on next page

11011293

3-18 (Continued)

	(1) Millions US $		(2) Millions US $		(3) Millions US $	
	RSE	Value	RSE	Value	RSE	Value
New Jersey	RSE = 3.93%	246	RSE = 1.12%	1,706	RSE = 1.14%	1,952
New Mexico	RSE = 1.50%	182	RSE = 12.05%	41	RSE = 2.52%	223
New York	RSE = 0.62%	2,119	RSE = 4.65%	812	RSE = 1.39%	2,931
North Carolina	RSE = 6.30%	128	RSE = 1.49%	909	RSE = 1.53%	1,036
North Dakota	RSE = 6.58%	7	RSE = 6.31%	17	RSE = 4.95%	25
Ohio	RSE = 0.95%	956	RSE = 3.20%	634	RSE = 1.48%	1,590
Oklahoma	RSE = 2.62%	58	RSE = 11.58%	39	RSE = 5.00%	97
Oregon	RSE = 3.88%	48	RSE = 3.48%	168	RSE = 2.88%	216
Pennsylvania	RSE = 4.55%	237	RSE = 4.73%	463	RSE = 3.56%	700
Rhode Island	RSE = 8.93%	40	RSE = 14.83%	51	RSE = 9.38%	91
South Carolina	RSE = 2.30%	116	RSE = 1.44%	334	RSE = 1.24%	450
South Dakota	RSE = 22.00%	3	RSE = 8.65%	21	RSE = 8.07%	24
Tennessee	RSE = 8.09%	35	RSE = 6.36%	119	RSE = 5.31%	155

Question continues on next page

SECTION 3

11011301

SECTION 3

3-18 (Continued)

	(1) Millions US $	(2) Millions US $	(3) Millions US $
Texas	RSE = 1.3% 1,169	RSE = 2.81% 1,219	RSE = 1.62% 2,388
Utah	RSE = 0.69% 482	RSE = 7.66% 83	RSE = 1.29% 565
Vermont	RSE = 3.86% 24	RSE = 8.67% 20	RSE = 4.51% 44
Virginia	RSE = 0.98% 2,085	RSE = 8.87% 339	RSE = 1.58% 2,424
Washington	RSE = 1.05% 578	RSE = 7.04% 321	RSE = 2.62% 899
West Virginia	RSE = 1.89% 22	RSE = 6.80% 13	RSE = 2.92% 35
Wisconsin	RSE = 7.66% 38	RSE = 3.94% 467	RSE = 3.74% 505
Wyoming	RSE = 6.39% 7	RSE = 13.55% 6	RSE = 7.36% 13
Total	31,309	24,015	55,324
	Total equals Question 3-15	*Total equals Question 3-17*	*Total equals Question 3-12*

3-19 At what domestic location did your company perform the largest dollar amount of R&D that was paid for by others in 2011?

Address 1

Address 2

City **State** **ZIP**

11011319

3-20 **How much of the amount reported in Question 3-12 was from the location identified in Question 3-19?**

Millions US $: 21,420 RSE = 0.05%

3-21 **At what domestic location did your company perform the second largest dollar amount of R&D that was paid for by others in 2011?**

Address 1

Address 2

City **State** **ZIP**

3-22 **How much of the amount reported in Question 3-12 was from the location identified in Question 3-21?**

Millions US $: 4,758 RSE = 0%

Projected R&D paid for by others in 2012

3-23 **What are your company's projected 2012 costs for R&D that will be paid for by others?**

NOTE: This amount is the 2012 projection for what is reported in Question 3-2.

Millions US $: 70,807 RSE = 0.80%

3-24 **How much of the projected costs in 2012 for R&D that will be paid for by others reported in Question 3-23 will be performed by your company in the United States?**

NOTE: This amount is the 2012 projection for what is reported in Question 3-12.

Millions US $: 60,820 RSE = 0.91%

3-25 **How much of the projected costs in 2012 for domestic R&D performed by your company that will be paid for by others reported in Question 3-24 will be paid for by the U.S. federal government?**

NOTE: This amount is the 2012 projection for what is reported in Question 3-15.

Millions US $: 36,005 RSE = 0.78%

FOR 3.23-3.25:
- Collected on BOTH forms
- Unit non-response weight adjustment applied
- Item imputation: Company trend was used, if available, otherwise NALOB ROI was used.

SECTION 3

Form BRDI-1

11011327

SECTION 4
Management and Strategy of R&D

Who should answer this section?
Persons familiar with the technical, managerial, and strategic aspects of your company's R&D should complete this section.

What does this section cover?
This section requests information about the characteristics of the R&D reported in Sections 2 and 3. This section requests information about your company's worldwide consolidated R&D and the R&D your company performs in the domestic United States.

4-1 **Copy the amount from Question 2-4. This is the total R&D paid for by your company in 2011.**

Millions US $: 341,335

4-2 **Is the amount entered in Question 4-1 greater than zero?**

☐ Yes → **Continue with Question 4-3**

☐ No → **Skip to Question 4-12 on page 33**

Characteristics of domestic R&D paid for and performed by your company

4-3 **Copy the amount from Question 2-15. This is the domestic R&D paid for and performed by your company.**

$Bil. Mil. Thou.: 238,768

4-4 **How much of the amount reported in Question 4-3 was for the following categories?**

-Collected on BOTH forms
-Unit non-response weight adjustment applied
-Item imputation: Use company trend first. If not available, use NALOB ROI.

	Millions US $	
a. Research–the planned, systematic pursuit of new knowledge or understanding.	48,309	RSE = 1.18%
b. Development–the systematic use of research and practical experience to produce new or significantly improved goods, services, or processes	190,460	RSE = 1.38%
c. **Total** (equals Question 4-3).	238,768	

4-5 **If you reported any research in Question 4-4, line a, how much of that research was for the following categories?**

-Collected on BOTH forms
-Unit non-response weight adjustment applied
-Item imputation was done using NALOB ROI.

	Millions US $	
a. Applied research–the activity aimed at solving a specific problem or meeting a specific commercial objective.	37,573	RSE = 1.41%
b. Basic research–the activity aimed at acquiring new knowledge or understanding without specific immediate commercial application or use.	10,735	RSE = 0.72%
c. **Total** (equals Question 4-4, line a)	48,309	

11011335

Areas of application for domestic R&D paid for and performed by your company

NOTE: You may report the same R&D in multiple areas for Questions 4-6 to 4-7.

FOR 4.6 - 4.11
-Collected on BOTH forms
-Unit non-response weight adjustment applied
-Item imputation done using Industry ROI

4-6 **What percentage of the amount reported in Question 4-3 had energy applications, including energy production, distribution, stora[ge] and efficiency (excluding exploration and prospecting)?**

Example: Company B is a semiconductor manufacturer. Its products are not designed specifically for energy applications. In 2011, 10% of the domestic R&D performed by the company was focused on improving the energy efficiency of its products. Based on this, Company B reports "10%" for this question.

Millions US $: 18,417 — RSE = 1.37%

4-7 **What percentage of the amount reported in Question 4-3 had environmental protection applications, including pollution abatement?**

Millions US $: 7,246 — RSE = 3.40%

Technology focus of domestic R&D paid for and performed by your company

NOTE: You may report the same R&D in multiple areas for Questions 4-8 to 4-11.

4-8 **What percentage of the amount reported in Question 4-3 was for software products?**

Millions US $: 55,228 — RSE = 3.60%

4-9 **What percentage of the amount reported in Question 4-3 was for software embedded in other projects or products?**

Millions US $: 21,570 — RSE = 3.94%

4-10 **What percentage of the amount reported in Question 4-3 was for biotechnology—the use of cellular and bio-molecular processes to solve problems or make useful products?**

Millions US $: 20,933 — RSE = 1.13%

4-11 **What percentage of the amount reported in Question 4-3 was for nanotechnology—the science and technology involving work at the nanometer scale?**

Millions US $: 11,149 — RSE = 1.02%

SECTION 4

Domestic R&D performed by your company that was paid for by others

4-12 **Copy the amount from Question 3-12. This is the domestic R&D performed by your company that was paid for by others.**

$Bil.	Mil.	Thou.
		55,324

4-13 **Is the amount entered in Question 4-12 greater than zero?**

☐ Yes → **Continue with Question 4-14**

☐ No → **Skip to Section 5 on page 36**

Form BRDI-1

11011343

SECTION 4

4-14 How much of the amount reported in Question 4-12 was for the following categories?

-Collected on BOTH forms
-Unit non-response weight adjustment applied
-Item imputation: Use company trend first. If not available, use NALOB ROI.

	Millions US $	
a. Research–the planned, systematic pursuit of new knowledge or understanding	11,899	RSE = 1.83%
b. Development–the systematic use of research and practical experience to produce new or significantly improved goods, services, or processes	43,426	RSE = 1.05%
c. **Total** (equals Question 4-12)	55,324	

4-15 If you reported any research in Question 4-14, line a, how much of that research was for the following categories?

-Collected on BOTH forms
-Unit non-response weight adjustment applied
-Item imputation was done using NALOB ROI.

	Millions US $	
a. Applied research–the activity aimed at solving a specific problem or meeting a specific commercial objective	9,613	RSE = 2.22%
b. Basic research–the activity aimed at acquiring new knowledge or understanding without specific immediate commercial application or use	2,285	RSE = 1.98%
c. **Total** (equals Question 4-14, line a)	11,899	

NOTE: You may report the same R&D in multiple areas for Questions 4-16 to 4-21.

	Millions US $	
4-16 What percentage of the amount reported in Question 4-12 had energy applications, including energy production, distribution, storage, and efficiency (excluding exploration and prospecting)?	3,339	RSE = 4.76%
4-17 What percentage of the amount reported in Question 4-12 had environmental protection applications, including pollution abatement?	1,926	RSE = 6.02%
4-18 What percentage of the amount reported in Question 4-12 was for biotechnology—the use of cellular and bio-molecular processes to solve problems or make useful products?	5,205	RSE = 1.10%
4-19 What percentage of the amount reported in Question 4-12 was for nanotechnology—the science and technology involving work at the nanometer scale?	1,957	RSE = 3.46%
4-20 What percentage of the amount reported in Question 4-12 was for software products?	5,028	RSE = 4.20%
4-21 What percentage of the amount reported in Question 4-12 was for software embedded in other projects or products?	3,334	RSE = 5.11%

Form BRDI-1

11011350

Domestic R&D performed by your company that was paid for by the U.S. federal government

4-22 **Copy the amount from Question 3-15. This is domestic R&D performed by your company that was paid for by the U.S. federal government.**

$Bil.	Mil.	Thou.
		31,309

4-23 **Is the amount entered in Question 4-22 greater than zero?**

☐ Yes → **Continue with Question 4-24**

☐ No → **Skip to Section 5 on page 36**

4-24 **How much of the amount reported in Question 4-22 was for the following categories?**

-Collected on BRDI-1 form ONLY
-Unit non-response weight adjustment applied
-Item imputation - For BOTH forms, use company trend first. If not available, use NALOB ROI.

	Millions US $	
a. Research–the planned, systematic pursuit of new knowledge or understanding	5,082	RSE = 2.38%
b. Development–the systematic use of research and practical experience to produce new or significantly improved goods, services, or processes	26,228	RSE = 0.62%
c. **Total** (equals Question 4-22)	31,309	

4-25 **If you reported any research in Question 4-24, line a, how much of that research was for the following categories?**

-Collected on BRDI-1 form ONLY
-Unit non-response weight adjustment applied
-Item imputation: For BOTH forms - use NALOB ROI

	Millions US $	
a. Applied research–the activity aimed at solving a specific problem or meeting a specific commercial objective	4,404	RSE = 2.53%
b. Basic research–the activity aimed at acquiring new knowledge or understanding without specific immediate commercial application or use	677	RSE = 6.38%
c. **Total** (equals Question 4-22, line a)	5,082	

SECTION 4

Form BRDI-1

11011368

SECTION 5
Human Resources

Who should answer this section?
Persons familiar with human resources concepts and with access to records related to your company's employees should complete this section.

What does this section cover?
This section requests information about your company's employees, focusing on those who worked on R&D activities either full-time or part-time. Include employment data for operations or subsidiaries for which your company owned more than 50 percent.

-RSE = 2.80%
-Collected on BOTH forms
-Unit non-response weight adjustment applied
- Item imputation was done by HQ analysts primarily using 10-k

5-1 What was the total number of worldwide employees working at your company for the pay period that included March 12, 2011?

Include:
- Full- and part-time employees

Exclude:
- Leased or temporary employees and on-site consultants

Thousands
29,327

5-2 How many of the employees reported in Question 5-1 were employees of your company's <u>domestic</u> operations and <u>foreign</u> operations?

Domestic operations employees include all employees whose payroll was reported on the first quarter filing of IRS Form 941, Employer's Quarterly Tax Return.

	(1) Domestic	(2) Foreign	(3) Total
	Thousands	Thousands	Thousands
Employees	19,285	10,042	29,327
	RSE = 2.94%	RSE = 4.93%	*Total equals Question 5-1*

5-3 How many employees reported in Question 5-2 were <u>R&D employees</u>?

R&D employees include all employees who work on R&D or who provide direct support to R&D, such as researchers, R&D managers, technicians, clerical staff, and others assigned to R&D groups.

All other employees include employees who provide indirect support to R&D, such as corporate personnel, security guards, and cafeteria workers.

	(1) Domestic	(2) Foreign	(3) Total
	Thousands	Thousands	Thousands
a. R&D employees	1,471	554	2,025
b. All other employees . . .	17,814	9,487	27,302
c. **Total employees**.	19,285	10,042	29,327

Total line equals Question 5-2

R&D Employees	RSE = 1.32%	RSE = 0.98%	RSE = 1.00%
All other employees	RSE = 3.16%	RSE = 5.21%	RSE = 3.00%

SECTION 5

11011376

R&D employees

5-4 **Copy the numbers from Question 5-3, line a. These are your company's R&D employees.**

	(1) Domestic Thousands	(2) Foreign Thousands	(3) Total R&D Thousands
R&D employees	1,471	554	2,025

5-5 **How many of the R&D employees reported in Question 5-4 were female employees and male employees?**

	(1) Domestic Thousands	(2) Foreign Thousands	(3) Total R&D Thousands
a. Female R&D employees	352	151	504
b. Male R&D employees	1,118	403	1,521
c. **Total R&D employees**	1,471	554	2,025

Total line equals Question 5-4

-Collected on BOTH forms
-Unit non-response weight adjustment applied
-Item imputation was done using NALOB ROI using domestic R&D employee type and sex. If not possible, use company trend and if no company trend, then use NALOB ROI using worldwide R&D employees.

	Domestic	Foreign	Total R&D
Female R&D Employees	RSE = 1.35%	RSE = 1.07%	RSE = 1.00%
Male R&D Employees	RSE = 1.43%	RSE = 0.98%	RSE = 1.09%

5-6 **How many of the R&D employees reported in Question 5-4 worked in the occupations listed below?**

	(1) Domestic Thousands	(2) Foreign Thousands	(3) Total R&D Thousands
a. R&D scientists, engineers, and managers	1,003	395	1,398
b. R&D technicians and technologists	308	95	403
c. R&D support staff (clerical and other)	160	65	225
d. **Total R&D employees**	1,471	554	2,025

Total line equals Question 5-4

	Domestic	Foreign	Total R&D
R&D S&Es	RSE = 1.46%	RSE = 1.18%	RSE = 1.11%
R&D Technicians	RSE = 3.02%	RSE = 0.57%	RSE = 2.31%
R&D Support	RSE = 2.39%	RSE = 2.13%	RSE = 1.81%

SECTION 5

11011384

Domestic full-time equivalents (FTEs)

5-7 Of the domestic R&D employees reported in Question 5-4, column 1, what was the number of full-time equivalents (FTEs) for R&D activity for full-time R&D employees, other full-time employees not working solely on R&D, and part-time employees?

	Number	
a. FTEs for full-time R&D employees Count the number of full-time employees who work only on R&D.	1,061	RSE = 1.61%

Example:
50 full-time R&D employees worked only on R&D = 50 FTEs

b. FTEs for other full-time employees not working solely on R&D Use the proportion of the time they work on R&D to calculate the number of FTEs.	154	RSE = 2.77%

Example:
60 full-time employees averaged one-fourth of their time on R&D = 15 FTEs

c. FTEs for part-time employees working on R&D Use the portion of a full-time week (such as 40 hours) that they work on R&D to calculate the FTEs.	17	RSE = 4.98%

Example:
20 part-time employees averaged 20 hours a week on R&D activities = 10 FTEs

d. **Total FTEs**	1,231	RSE = 1.48%

Total FTEs should not exceed Question 5-4, column 1.

-Collected on BOTH forms
-Unit non-response weight adjustment applied
-Imputation using NALOB ROI except for line d, which employed company trend, if available.

SECTION 5

11011392

5-8 Of the domestic R&D scientists, engineers, and managers reported in Question 5-6, row a, column 1, what was the number of full-time equivalents (FTEs) for R&D activity for full-time R&D employees, other full-time employees not working solely on R&D, and part-time employees?

	Number	
a. FTEs for full-time R&D scientists, engineers, and managers Count the number of full-time employees who work only on R&D.	764	RSE = 1.78%

Example:
50 full-time R&D scientists worked only on R&D = 50 FTEs

b. FTEs for other full-time scientists, engineers, and managers not working soley on R&D Use the proportion of the time they work on R&D to calculate the number of FTEs.	81	RSE = 2.37%

Example:
60 full-time managers averaged one-fourth of their time on R&D = 15 FTEs

c. FTEs for part-time scientists, engineers, and managers working on R&D Use the portion of a full-time week (such as 40 hours) that they work on R&D to calculate the FTEs.	7	RSE = 4.76%

Example:
20 part-time employees averaged 20 hours a week on R&D activities = 10 FTEs

d. **Total FTEs**	853	RSE = 1.62%

Total FTEs should not exceed Question 5-6, row a, column 1.

-Collected on BOTH forms
-Unit non-response weight adjustment applied
-Imputation using NALOB ROI except for line d, which employed company trend, if available.

SECTION 5

Form BRDI-1

11011400

SECTION 6
Intellectual Property and Technology Transfer

Who should answer this section?
Persons with an understanding of your company's general business strategy and knowledge of its patenting, licensing, and other activities related to intellectual property should complete this section.

What does this section cover?
This section requests information about intellectual property and technology transfer activities such as:
- Patents
- Patent licensing
- Protection of intellectual property
- Transfer of intellectual property

Are responses to this survey confidential?
Yes. Your responses are completely confidential under Title 13, United States Code, and are seen only by persons sworn to uphold the confidentiality of Census Bureau information. Data provided will be used only to publish summary statistics that do not identify individual companies. Title 13 also provides that copies of reports retained in your files are immune from legal process. In addition, reported data are exempt from requests made under the Freedom of Information Act.

FOR 6.1 - 6.7:
-Collected on BOTH forms
-Item imputation was performed for select companies using USPTO data
-NOTE: data are for R&D companies only

Patents

6-1 **How many patents did your company apply for in 2011 from the U.S. Patent and Trademark Office (USPTO)?**
Number: 119,366 RSE = 1.70%

6-2 **What percentage of the patent applications reported in Question 6-1 has your company applied for or plans to apply for in foreign jurisdictions?**
Number: 59,457 RSE = 1.12%

6-3 **What percentage of the patent applications reported in Question 6-1 was for inventions that originated within your company's organized R&D activities?**
Number: 99,042 RSE = 1.50%

6-4 **How many patents were issued to your company in 2011 by the USPTO?**
Number: 79,214 RSE = 1.20%

6-5 **What percentage of your company's inventions considered for patenting in 2011 resulted in patent applications?**
%

Patent licensing

6-6 **What was the amount of your company's patent licensing revenue in 2011?**
Millions US $: 24,452 RSE = 1.41%

SECTION 6

11011418

6-7 **How many new agreements did your company enter into during 2011 to license patents to others not owned by your company?**

Number: 4,098 RSE = 7.73%

Intellectual property transfer activities

6-8 **Did your company perform the following activities in 2011?**

FOR 6.8:
-Collected on BOTH forms
-No item imputation
-estimates are for all respondents regardless of R&D performance.

	RSE	Weighted Counts YES	Weighted Counts NO	RSE
a. Transferred intellectual property (IP) to others not owned by company through participation in technical assistance or "know how" agreements	12.00%	5,618	1,074,568	0.94%
b. Received IP from others not owned by your company throug participation in technical assistance or "know how" agreeme	11.65%	8,259	1,069,431	0.94%
c. Transferred IP to a spin-off or spin-out of your company.	24.68%	1,303	1,078,241	0.93%
d. Received IP from a parent company as part of a spin-off or s	39.74%	736	1,075,892	0.94%
e. Acquired more than 50% ownership in another company for primary purpose of acquiring their IP.	33.53%	1,516	1,076,415	0.94%
f. Acquired any financial interest in another company in order gain access to their IP	31.06%	1,680	1,077,176	0.93%
g. Participated in cross-licensing agreements—the agreements in which two or more parties grant a license to each other for the use of the subject matter claimed in one or more of the patents ow by each party	18.19%	2,122	1,076,865	0.93%
h. Allowed free use of patents or other IP owned by your company (for example, allowing free use of software patents by the o source community)	24.07%	1,618	1,076,695	0.93%
i. Made use of open source patents or other freely available IP owned by your company	10.13%	11,678	1,065,448	0.95%

Intellectual property protection

6-9 **During 2011, how important to your company were the following types of intellectual property protection?**

FOR 6.9:
-Collected on BOTH forms
-No item imputation
-estimates are for all respondents regardless of R&D performance.

	RSE-VI	RSE-SI	Weighted Counts Very important	Weighted Counts Somewhat important	Weighted Counts Not important	RSE-NI
a. Utility patents (patents for invention)	6.50%	7.25%	20,940	27,288	987,971	1.02%
b. Design patents (patents for appearan	7.37%	6.28%	20,782	36,217	977,617	1.03%
c. Trademarks	4.63%	4.23%	65,034	87,817	883,364	1.14%
d. Copyrights	5.41%	4.65%	48,027	71,571	913,627	1.10%
e. Trade secrets	4.27%	4.52%	69,427	77,656	887,112	1.13%
f. Mask works (copyright protection for semiconductor products	14.11%	9.93%	6,526	16,271	1,010,361	1.00%

Form BRDI-1

11011426

Business codes
(used in Sections 1, 2, and 3)

Aerospace and Defense

33642	Aircraft engine and engine parts manufacturing
33641	Aircraft manufacturing
33644	Guided missiles, space vehicles, and parts manufacturing
33692	Military armored vehicle, tank, and tank components manufacturing
33643	Other aircraft parts and auxiliary equipment manufacturing
33452	Search, detection, navigation, guidance, aeronautical, and nautical system and instruments manufacturing
33660	Ship and boat building

Automobiles, Motorcycles, and Components

33620	Motor vehicle body and trailer manufacturing
33630	Motor vehicle parts manufacturing
33610	Motor vehicles manufacturing
33691	Motorcycle, bicycle, and parts manufacturing
33651	Railroad rolling stock manufacturing
33660	Ship and boat building
33699	All other transportation equipment manufacturing

Capital Equipment

33311	Agricultural machinery and equipment manufacturing
33332	Commercial, service industry, temperature control, and airflow control machinery manufacturing
33312	Construction machinery manufacturing
33500	Electrical equipment, appliances, and components manufacturing
33360	Engine, turbine, and power transmission equipment manufacturing
33322	Industrial machinery manufacturing, except semiconductor machinery
33390	Metalworking and other general purpose machinery manufacturing
33319	Mining, oil, and gas field machinery and equipment manufacturing
33331	Photographic and photocopying equipment manufacturing

Chemicals and Materials

32402	Asphalt paving, roofing, and saturated materials manufacturing
32510	Basic chemicals manufacturing
32790	Cement, concrete, lime, gypsum, and other nonmetallic mineral product manufacturing
32710	Clay and glass products manufacturing
21200	Mining
32403	Other petroleum and coal products manufacturing, including motor oil, hydraulic fluid, and charcoal
32592	Paint, adhesive, and other chemical manufacturing
32200	Paper manufacturing
32530	Pesticide, fertilizer, and other agricultural chemical manufacturing
32600	Plastics and rubber products manufacturing
33100	Primary metal manufacturing
32520	Resin, synthetic rubber, and artificial synthetic fibers and filaments manufacturing
32591	Soap, cleaning compound, and toilet preparations manufacturing
32100	Wood products manufacturing

Consumer Goods

33430	Audio and video equipment manufacturing
31210	Beverage manufacturing
33200	Fabricated metal products manufacturing
31100	Food manufacturing
33700	Furniture and related products manufacturing
33990	Miscellaneous manufacturing not listed elsewhere (games, office supplies, slot machines, etc.)
32300	Printing and related support activities
32591	Soap, cleaning compound, and toilet preparations manufacturing
31990	Textile, apparel, and leather products manufacturing
31220	Tobacco manufacturing

Energy and Mining

33360	Engine, turbine, and power transmission equipment manufacturing
21200	Mining
33319	Mining, oil, and gas field machinery and equipment manufacturing
21100	Oil and gas extraction
32401	Petroleum refineries
21300	Support activities for mining, including oil and gas

Finance, Insurance, and Real Estate

52200	Finance: banking and credit intermediation
52400	Insurance carriers and related activities
53100	Real estate
52310	Securities, commodity contracts, and other financial investments and related activities (including funds and trusts)

Healthcare

33451	Electromedical, electrotherapeutic, and irradiation apparatus manufacturing
62200	Hospitals and nursing care facilities
32542	In vitro diagnostic substances manufacturing
62150	Medical and diagnostic laboratories
33910	Medical equipment and supplies manufacturing
62110	Offices of physicians
62199	Other ambulatory health care services (ambulance, dental, home health care)
32541	Pharmaceutical, medicinal, botanical, and biological products (except diagnostic) manufacturing
54173	Research and development services in biotechnology
54174	Research and development services in physical, engineering, and life sciences (except biotechnology)

Form BRDI-1

11011434

Information Technology - Goods and Services

54150 Computer systems design and related services
33412 Computers and peripheral equipment manufacturing, including magnetic and optical media
51800 Data processing, hosting, and related services
33500 Electrical equipment, appliances, and components manufacturing
45411 Electronic shopping and electronic auctions
33459 Measuring and control instruments manufacturing (not listed elsewhere)
33429 Other communication equipment manufacturing (except radio, television, and wireless communication equipment)
51910 Other information services, including Internet publishing, broadcasting, and web search portals
33422 Radio, television, and wireless communication equipment manufacturing
33440 Semiconductor and other electronic components manufacturing
33321 Semiconductor machinery manufacturing
51120 Software publishers (except Internet)
33421 Telephone apparatus manufacturing including routers, modems, and gateways
42500 Wholesale electronic markets and agents and brokers (business to business)

Professional, Scientific, and Technical Services

54180 Advertising and related services
54130 Architectural, engineering, and related services
54150 Computer systems design and related services
54111 Legal, accounting, tax preparation, bookkeeping and payroll services
54160 Management, scientific, and technical consulting services
54190 Professional, scientific, and technical services (not listed elsewhere)
54173 Research and development services in biotechnology
54174 Research and development services in physical, engineering, and life sciences (except biotechnology)
54172 Research and development services in social sciences and humanities
54140 Specialized design services

Telecommunications and Utilities

51500 Broadcasting (except Internet)
51790 Other telecommunications (not listed elsewhere)
51740 Satellite telecommunications
22100 Utilities
56200 Waste management and remediation services
51710 Wired telecommunications carriers
51720 Wireless telecommunications carriers (except satellite)

Other Services

72000 Accommodation and food services
56100 Administrative and support services
71000 Arts, entertainment, and recreation
23000 Construction
53300 Lessors of nonfinancial intangible assets (including patent licensing)
55100 Management of companies and enterprises
42300 Merchant wholesalers, durable goods
42400 Merchant wholesalers, nondurable goods
51200 Motion picture and sound recording (except Internet)
51110 Newspaper, periodical, book, and directory publishers (except Internet)
81000 Other services (not listed elsewhere)
49200 Couriers, messengers, and express delivery services
53200 Rental and leasing services
44000 Retail trade, except electronic shopping and electronic auctions
62400 Social assistance services
21300 Support activities for mining, including oil and gas
48000 Transportation
49300 Warehousing and storage

Form BRDI-1

11011442

Remarks (Please use the space below for any explanations that may help us understand your reported data.)

Undistributed Lines (in millions):

2.14: (foreign perf total - foreign country total) 559 (RSE = 3.85%)
2.16: 6,111 (RSE = 0.38%)
2.25: 1,953 (RSE = 3.89%)
3.12 (foreign perf total - foreign country total) 52 (RSE = 33.91%)
3.18 (1): D
3.18 (2): D
3.18 (3): D

We estimate that it will take from .5 to 25 hours to complete this form, with 14.3 hours being the average. This includes time to read instructions, develop or assemble materials, conduct tests, organize and review the information, and maintain and report the information. Send comments regarding this burden estimate or any other aspect of this collection of information, including suggestions for reducing this burden, to:

Paperwork Project 0607-0912
U.S. Census Bureau
4600 Silver Hill Road
AMSD-3K138
Washington, D.C. 20233

You may e-mail comments to Paperwork@census.gov; use "Paperwork Project 0607-0912" as the subject.

~ **Thank you for completing your 2011 Business R&D and Innovation Survey** ~

PLEASE MAKE A COPY OF THIS FORM FOR YOUR RECORDS AND RETURN THE ORIGINAL

Appendix H

2011 BRDIS Table 46

TABLE 46 Number of Companies in the United States That Introduced New or Significantly Improved Products or Processes, by Industry and Company Size: 2009-2011

			New or Significantly Improved Product OR Process		New or Significantly Improved Product(s)					
					Any good/ service		New goods		New services	
Industry	NAICS Code	Number of Companies[a]	Yes	No	Yes	No	Yes	No	Yes	No
All industries	21-23, 31-33, 42-81	1,220,114	174,066	1,024,199	114,895	1,070,247	66,427	1,122,590	82,546	1,101,820
Manufacturing industries	31-33	103,220	30,396	71,475	22,365	78,250	19,755	81,060	9,058	90,889
Food	311	8,333	2,171	6,135	1,444	6,785	1,346	6,848	487	7,681
Beverage and tobacco products	312	1,245	437	799	264	943	200	1,013	125	1,067
Textile, apparel and leather	313-316	4,641	1,006	3,515	724	3,772	582	3,908	222	4,244
Wood products	321	4,876	583	4,266	331	4,440	259	4,509	157	4,598
Paper	322	1,218	323	888	201	1,001	167	1,035	72	1,094
Printing and related support activities	323	8,249	1,782	6,362	863	7,085	584	7,477	774	7,186
Petroleum and coal products	324	369	203	164	193	174	193	174	59	308
Chemicals	325	4,660	2,277	2,251	1,724	2,762	1,662	2,843	548	3,816
Basic chemicals	3251	867	508	354	418	443	413	449	181	658
Resin, synthetic rubber, fibers, and filament	3252	392	137	204	117	223	114	226	42	294
Pesticide, fertilizer, and other agricultural chemical	3253	429	77	352	67	361	57	372	33	396
Pharmaceuticals and medicines	3254	941	412	492	341	558	333	570	86	800
Soap, cleaning compound, and toilet preparation	3256	840	540	292	350	471	335	485	116	704
Paint, coating, adhesive, and other chemicals	3255, 3259	1,189	603	557	430	705	411	742	90	964
Plastics and rubber products	326	4,851	1,693	3,086	1,281	3,450	1,220	3,524	373	4,345
Nonmetallic mineral products	327	3,765	617	3,129	463	3,178	394	3,301	201	3,476
Primary metals	331	1,433	250	1,177	170	1,257	156	1,271	59	1,366
Fabricated metal products	332	24,026	5,642	18,166	3,504	19,927	2,605	20,846	1,938	21,561
Machinery	333	9,881	3,808	5,862	3,139	6,469	2,895	6,745	1,035	8,449
Agricultural implement	33311	375	189	181	165	200	163	207	38	322
Semiconductor machinery	333295	169	41	125	41	125	40	126	13	151
Engine, turbine, and power transmission equipment	3336	366	185	176	164	195	114	247	78	277
Other machinery	other 333	8,971	3,393	5,379	2,769	5,949	2,578	6,165	906	7,699
Computer and electronic products	334	5,637	3,063	2,465	2,779	2,711	2,511	2,978	1,186	4,267
Communications equipment	3342	674	399	250	380	268	371	276	99	533
Semiconductor and other electronic components	3344	1,782	913	834	737	1,000	635	1,099	320	1,412
Navigational, measuring, electromedical, and control instruments	3345	2,253	1,161	1,056	1,084	1,106	1,042	1,151	411	1,785
Electromedical, electrotherapeutic, and irradiation apparatus	334510, 334517	265	112	147	103	136	102	137	22	235
Search, detection, navigation, guidance, aeronautical, and nautical system and instrument	334511	218	160	57	157	60	156	61	48	168
Other measuring and controlling device	other 3345	1,770	888	852	824	910	784	953	341	1,382

Other computer and electronic products	other 334	928	590	326	578	337	463	452	356	537
Electrical equipment, appliances, and components	335	2,530	995	1,505	860	1,621	802	1,679	317	2,149
Transportation equipment	336	4,441	1,578	2,731	1,292	2,962	1,252	2,981	427	3,717
Automobiles, bodies, trailers, and parts	3361, 3362, 3363	2,718	890	1,761	706	1,917	674	1,921	267	2,357
Aerospace products and parts	3364	629	264	358	210	399	201	409	55	550
Aircraft, aircraft engine, and aircraft parts	336411-13	607	253	348	201	387	193	396	49	535
Guided missile, space vehicle, and related parts	336414-15, 336419	22	11	10	9	12	8	13	6	15
Military armored vehicle, tank, and tank component	336992	60	D	D	D	D	D	D	D	D
Other transportation	other 336	1,034	D	D	D	D	D	D	D	D
Furniture and related products	337	4,855	1,143	3,687	741	4,057	723	4,079	150	4,661
Miscellaneous manufacturing	339	8,208	2,825	5,287	2,394	5,656	2,204	5,849	930	6,904
Medical equipment and supplies	3391	3,194	1,250	1,915	1,117	2,023	992	2,128	486	2,490
Other miscellaneous manufacturing	3399	5,014	1,575	3,372	1,277	3,633	1,212	3,722	444	4,414
Nonmanufacturing industries	21-23, 42-81	1,116,895	143,670	952,724	92,529	991,997	46,673	1,041,530	73,488	1,010,931
Mining, extraction, and support activities	21	5,526	686	4,751	337	5,088	213	5,214	294	5,134
Utilities	22	660	103	554	36	621	17	636	30	627
Wholesale trade	42	81,652	15,581	64,891	9,550	70,312	7,032	73,230	4,868	74,683
Electronic shopping and electronic auctions	454111-12	2,322	744	1,564	510	1,798	334	1,924	385	1,922
Transportation and warehousing	48-49	35,067	2,643	32,170	1,385	32,925	78	34,481	1,362	33,198
Information	51	17,587	6,056	11,257	4,978	12,230	2,846	14,232	3,851	13,132
Publishing	511	7,152	2,874	4,234	2,319	4,704	1,736	5,247	1,374	5,440
Newspaper, periodical, book, and directory publishers	5111	4,945	1,431	3,497	929	3,941	655	4,174	494	4,176
Software publishers	5112	2,208	1,442	737	1,390	763	1,080	1,072	880	1,264
Telecommunications	517	2,065	554	1,449	388	1,606	80	1,915	374	1,620
Data processing, hosting, and related services	518	2,285	938	1,256	912	1,282	542	1,650	856	1,322
Other information	other 51	6,085	1,691	4,318	1,358	4,638	490	5,421	1,247	4,749
Finance and insurance	52	40,467	5,127	34,835	2,876	36,534	876	38,236	2,570	36,840
Real estate and rental and leasing	53	31,730	2,136	29,092	1,078	29,652	1,014	29,706	571	30,159
Lessors of nonfinancial intangible assets (except copyrighted works)	533	543	263	279	262	280	260	282	8	534
Other real estate and rental and leasing	other 53	31,187	1,873	28,814	816	29,372	754	29,424	563	29,625
Professional, scientific, and technical services	54	135,127	23,708	109,569	19,085	112,184	7,938	124,146	16,083	115,349
Architectural, engineering, and related services	5413	23,027	4,695	18,018	3,460	19,082	1,622	20,945	2,653	19,674
Computer systems design and related services	5415	15,507	6,893	8,425	5,491	9,591	3,213	11,884	4,288	10,638
Scientific research and development services	5417	2,670	1,146	1,492	1,005	1,585	767	1,849	564	2,015
Biotechnology research and development	541711	849	306	534	258	579	200	627	141	692
Physical, engineering, and life sciences (except biotechnology) research and development	541712	1,657	810	826	721	877	555	1,072	405	1,188
Social sciences and humanities research and development	541720	163	30	133	27	128	13	149	18	136

continued

TABLE 46 Continued

Industry	NAICS Code	Number of Companies[a]	New or Significantly Improved Product OR Process		New or Significantly Improved Product(s)					
					Any good/ service		New goods		New services	
			Yes	No	Yes	No	Yes	No	Yes	No
Other professional, scientific, and technical services	other 54	93,923	10,974	81,634	9,129	81,927	2,335	89,469	8,578	83,022
Health care services	621-23	160,040	25,286	130,943	15,648	139,012	4,747	149,913	15,142	139,533
Other nonmanufacturing	23, 44-45 (excluding 454111-12), 55-56, 624, 71-72, 81	606,717	61,600	533,097	37,046	551,641	21,578	569,811	28,332	560,352
All companies	-	1,220,114	174,066	1,024,199	114,895	1,070,247	66,427	1,122,590	82,546	1,101,820
Small companies (number of domestic employees)[b]										
5-499	-	1,210,429	172,091	1,016,640	113,423	1,062,216	65,151	1,114,355	81,836	1,093,046
5-99	-	1,155,721	161,950	973,139	106,952	1,015,712	60,266	1,065,724	78,394	1,043,648
5-49	-	1,072,528	149,659	904,395	98,681	944,515	54,839	991,153	73,131	969,597
5-9	-	468,836	58,088	401,335	39,289	415,503	18,896	436,863	31,430	422,872
10-24	-	438,543	60,751	370,602	40,523	385,455	23,676	404,092	29,357	397,310
25-49	-	165,150	30,820	132,458	18,869	143,557	12,266	150,197	12,343	149,416
50-99	-	83,193	12,291	68,743	8,271	71,198	5,427	74,571	5,263	74,052
100-249	-	42,412	7,606	34,140	5,035	36,055	3,833	37,797	2,701	38,319
250-499	-	12,297	2,536	9,361	1,437	10,448	1,051	10,834	742	11,080
Medium and large companies (number of domestic employees)										
500-999	-	4,326	722	3,574	580	3,702	484	3,804	244	4,030
1,000-4,999	-	4,068	920	3,077	592	3,392	527	3,459	279	3,697
5,000-9,999	-	692	144	529	128	544	115	558	68	603
10,000-24,999	-	397	109	265	96	276	87	286	57	312
25,000 or more	-	202	81	114	75	117	65	128	61	132

Industry	NAICS Code	Number of Companies[a]	New or Significantly Improved Process(es)							
			Any process		Manufacturing/ production methods		Logistics/delivery/ distribution methods		Support activities	
			Yes	No	Yes	No	Yes	No	Yes	No
All industries	21-23, 31-33, 42-81	1,220,114	115,958	1,062,204	46,326	1,136,978	38,664	1,142,041	87,968	1,092,244
Manufacturing industries	31-33	103,220	21,313	79,218	16,340	84,481	6,198	94,307	12,955	87,646
Food	311	8,333	1,566	6,642	1,257	6,951	553	7,650	922	7,289
Beverage and tobacco products	312	1,245	353	832	303	882	162	1,023	178	1,007
Textile, apparel and leather	313-316	4,641	650	3,801	522	3,981	188	4,263	404	4,067
Wood products	321	4,876	466	4,345	306	4,505	87	4,710	226	4,571
Paper	322	1,218	248	950	150	1,048	86	1,112	167	1,030
Printing and related support activities	323	8,249	1,504	6,581	995	7,089	429	7,597	794	7,235
Petroleum and coal products	324	369	82	285	76	291	63	305	73	294
Chemicals	325	4,660	1,645	2,865	1,030	3,479	430	4,078	1,234	3,272

Basic chemicals	3251	867	408	451	337	523	137	723	338	521
Resin, synthetic rubber, fibers, and filament	3252	392	92	247	63	272	22	316	59	280
Pesticide, fertilizer, and other agricultural chemical	3253	429	40	389	33	396	12	418	14	415
Pharmaceuticals and medicines	3254	941	270	632	210	692	49	850	143	757
Soap, cleaning compound, and toilet preparation	3256	840	397	429	127	701	86	741	346	480
Paint, coating, adhesive, and other chemicals	3255, 3259	1,189	438	717	260	895	125	1,031	334	819
Plastics and rubber products	326	4,851	1,174	3,535	1,035	3,677	323	4,373	669	4,032
Nonmetallic mineral products	327	3,765	455	3,147	342	3,260	175	3,554	246	3,483
Primary metals	331	1,433	197	1,229	160	1,265	51	1,375	74	1,352
Fabricated metal products	332	24,026	4,429	18,944	3,381	20,158	959	22,446	2,743	20,667
Machinery	333	9,881	2,476	7,130	2,053	7,578	717	8,858	1,526	8,077
Agricultural implement	33311	375	94	276	79	290	28	342	48	321
Semiconductor machinery	333295	169	18	148	17	149	6	160	10	156
Engine, turbine, and power transmission equipment	3336	366	139	214	124	230	74	279	89	264
Other machinery	other 333	8,971	2,225	6,492	1,833	6,909	609	8,078	1,378	7,337
Computer and electronic products	334	5,637	1,708	3,727	1,444	4,000	600	4,826	1,012	4,421
Communications equipment	3342	674	121	474	103	493	60	535	89	507
Semiconductor and other electronic components	3344	1,782	601	1,137	535	1,207	185	1,549	393	1,345
Navigational, measuring, electromedical, and control instruments	3345	2,253	648	1,553	560	1,645	178	2,018	280	1,921
Electromedical, electrotherapeutic, and irradiation apparatus	334510, 334517	265	50	206	46	210	10	245	19	237
Search, detection, navigation, guidance, aeronautical, and nautical system and instrument	334511	218	82	135	72	145	18	199	71	146
Other measuring and controlling device	other 3345	1,770	516	1,212	441	1,291	150	1,574	190	1,539
Other computer and electronic products	other 334	928	337	562	246	654	176	723	250	648
Electrical equipment, appliances, and components	335	2,530	550	1,904	357	2,101	142	2,316	371	2,092
Transportation equipment	336	4,441	1,154	3,079	1,018	3,216	458	3,744	758	3,450
Automobiles, bodies, trailers, and parts	3361, 3362, 3363	2,718	690	1,947	597	2,036	218	2,389	408	2,199
Aerospace products and parts	3364	629	189	431	170	450	40	580	132	488
Aircraft, aircraft engine, and aircraft parts	336411-13	607	181	418	164	435	D	D	125	474
Guided missile, space vehicle, and related parts	336414-15, 336419	22	8	13	6	15	D	D	7	14
Military armored vehicle, tank, and tank component	336992	60	6	4	6	4	D	D	3	7
Other transportation	other 336	1,034	269	696	244	727	D	D	215	756
Furniture and related products	337	4,855	811	4,006	517	4,302	272	4,545	495	4,322
Miscellaneous manufacturing	339	8,208	1,844	6,218	1,393	6,699	504	7,532	1,062	6,985
Medical equipment and supplies	3391	3,194	869	2,271	628	2,523	255	2,876	468	2,665
Other miscellaneous manufacturing	3399	5,014	975	3,947	764	4,176	249	4,656	595	4,320

continued

TABLE 46 Continued

Industry	NAICS Code	Number of Companies[a]	New or Significantly Improved Process(es): Any process		Manufacturing/ production methods		Logistics/delivery/ distribution methods		Support activities	
			Yes	No	Yes	No	Yes	No	Yes	No
Nonmanufacturing industries	21-23, 42-81	1,116,895	94,645	982,986	29,986	1,052,497	32,466	1,047,734	75,013	1,004,598
Mining, extraction, and support activities	21	5,526	463	4,918	268	5,112	128	5,226	379	4,975
Utilities	22	660	89	559	25	623	20	629	77	572
Wholesale trade	42	81,652	10,958	68,772	3,911	75,820	4,353	75,378	8,573	71,057
Electronic shopping and electronic auctions	454111-12	2,322	557	1,747	243	2,046	336	1,968	369	1,937
Transportation and warehousing	48-49	35,067	2,119	31,393	5	34,258	1,109	33,152	1,616	32,144
Information	51	17,587	3,569	13,348	1,295	15,829	1,985	15,157	2,045	14,997
Publishing	511	7,152	1,282	5,731	484	6,622	754	6,299	629	6,434
Newspaper, periodical, book, and directory publishers	5111	4,945	839	3,997	365	4,563	500	4,379	336	4,550
Software publishers	5112	2,208	443	1,734	120	2,059	255	1,921	293	1,884
Telecommunications	517	2,065	334	1,666	25	1,976	164	1,837	226	1,773
Data processing, hosting, and related services	518	2,285	676	1,509	232	1,957	165	2,019	533	1,653
Other information	other 51	6,085	1,277	4,442	554	5,274	903	5,001	657	5,137
Finance and insurance	52	40,467	3,447	36,510	763	39,193	690	39,265	3,328	36,629
Real estate and rental and leasing	53	31,730	1,626	28,601	309	29,919	613	29,366	1,619	28,857
Lessors of nonfinancial intangible assets (except copyrighted works)	533	543	7	535	5	537	D	D	4	538
Other real estate and rental and leasing	other 53	31,187	1,619	28,066	304	29,382	D	D	1,615	28,319
Professional, scientific, and technical services	54	135,127	13,405	117,910	5,569	126,300	4,331	126,801	9,992	121,505
Architectural, engineering, and related services	5413	23,027	2,922	19,481	1,480	20,935	1,462	21,022	1,994	20,221
Computer systems design and related services	5415	15,507	4,442	10,794	1,417	13,810	1,702	13,516	3,757	11,551
Scientific research and development services	5417	2,670	768	1,849	533	2,088	163	2,458	364	2,251
Biotechnology research and development	541711	849	222	609	179	653	82	751	80	751
Physical, engineering, and life sciences (except biotechnology) research and development	541712	1,657	535	1,088	350	1,276	75	1,551	278	1,343
Social sciences and humanities research and development	541720	163	11	152	3	159	6	157	6	156
Other professional, scientific, and technical services	other 54	93,923	5,273	85,786	2,139	89,468	1,004	89,805	3,877	87,483
Health care services	621-23	160,040	16,966	137,704	5,519	148,906	6,307	148,368	12,629	142,042
Other nonmanufacturing	23, 44-45 (excluding 454111-12), 55-56, 624, 71-72, 81	606,717	41,446	541,523	12,078	574,491	12,595	572,425	34,385	549,883
All companies	-	1,220,114	115,958	1,062,204	46,326	1,136,978	38,664	1,142,041	87,968	1,092,244
Small companies (number of domestic employees)[b]										
5-499	-	1,210,429	114,497	1,054,213	45,484	1,128,357	38,075	1,133,193	86,882	1,083,888
5-99	-	1,155,721	106,743	1,008,801	42,008	1,078,614	35,953	1,082,114	80,825	1,036,795

5-49	-	1,072,528	98,316	937,793	38,277	1,002,592	33,321	1,005,333	74,283	963,595
5-9	-	468,836	36,647	412,341	13,905	437,052	12,691	436,791	27,489	422,420
10-24	-	438,543	40,900	384,425	16,799	410,554	14,486	412,454	30,706	395,216
25-49	-	165,150	20,768	141,027	7,573	154,986	6,144	156,088	16,089	145,959
50-99	-	83,193	8,427	71,008	3,730	76,022	2,632	76,781	6,542	73,200
100-249	-	42,412	5,727	35,629	2,578	38,780	1,666	39,677	4,444	36,895
250-499	-	12,297	2,026	9,782	898	10,963	456	11,401	1,613	10,198
Medium and large companies (number of domestic employees)										
500-999	-	4,326	503	3,751	336	3,928	212	4,036	320	3,930
1,000-4,999	-	4,068	710	3,266	333	3,643	215	3,754	576	3,397
5,000-9,999	-	692	105	562	72	596	61	606	76	590
10,000-24,999	-	397	74	288	51	311	51	310	59	302
25,000 or more	-	202	69	124	50	143	50	143	55	138

D = data withheld to avoid disclosing operations of individual companies.

[a]Statistics are based on companies in the United States that reported to the survey, regardless of whether they did or did not perform or fund R&D. These statistics do not include an adjustment to the weight to account for unit nonresponse.

[b]Upper bound based on U.S. Small Business Administration's definition of a small business; BRDIS does not include companies with fewer than 5 domestic employees.

NAICS = 2007 North American Industry Classification System.

NOTES: Detail may not add to total because of rounding. Industry classification based on dominant business code for domestic R&D performance where available. For companies that did not report business codes, classification used for sampling was assigned. Sum of yes and no responses may not add to number of companies due to item nonresponse to some innovation question items. The full set of detailed statistical tables from the 2011 BRDIS will be available in Business R&D and Innovation: 2011 scheduled for release in 2014. Innovation statistics are found in Tables 46-49. Relative standard errors for each cell in those tables are available from NCSES upon request. Selected final statistics are summarized in NCSES's *InfoBrief*: "Business R&D Performance in the United States Increased in 2011" (NSF 13-335) at http://www.nsf.gov/statistics/infbrief/nsf13335.

SOURCE: National Science Foundation/National Center for Science and Engineering Statistics and U.S. Census Bureau, Business R&D and Innovation Survey, 2011.

Appendix I

2011 BRDIS Table 47

TABLE 47 Number of Companies in the United States That Introduced New or Significantly Improved Products or Processes and the Proportion of Companies in Each Industry and Company Size Classification: 2009-2011(number and percentage)

Industry	NAICS Code	Number of Companies[a]	New or Significantly Improved Product OR Process		New or Significantly Improved Product(s)					
					Any good/ service		New goods		New services	
			Yes	No	Yes	No	Yes	No	Yes	No
All industries	21-23, 31-33, 42-81	1,220,114	14.3	83.9	9.4	87.7	5.4	92.0	6.8	90.3
Manufacturing industries	31-33	103,220	29.4	69.2	21.7	75.8	19.1	78.5	8.8	88.1
Food	311	8,333	26.1	73.6	17.3	81.4	16.2	82.2	5.8	92.2
Beverage and tobacco products	312	1,245	35.1	64.2	21.2	75.8	16.0	81.3	10.0	85.7
Textile, apparel and leather	313-316	4,641	21.7	75.7	15.6	81.3	12.5	84.2	4.8	91.5
Wood products	321	4,876	12.0	87.5	6.8	91.0	5.3	92.5	3.2	94.3
Paper	322	1,218	26.5	72.9	16.5	82.2	13.7	85.0	5.9	89.8
Printing and related support activities	323	8,249	21.6	77.1	10.5	85.9	7.1	90.6	9.4	87.1
Petroleum and coal products	324	369	55.0	44.4	52.2	47.2	52.2	47.2	16.0	83.5
Chemicals	325	4,660	48.9	48.3	37.0	59.3	35.7	61.0	11.8	81.9
Basic chemicals	3251	867	58.5	40.8	48.2	51.1	47.6	51.8	20.9	75.9
Resin, synthetic rubber, fibers, and filament	3252	392	34.8	51.9	29.9	56.8	29.1	57.6	10.7	75.0
Pesticide, fertilizer, and other agricultural chemical	3253	429	18.0	82.0	15.6	84.2	13.2	86.5	7.6	92.2
Pharmaceuticals and medicines	3254	941	43.8	52.2	36.3	59.2	35.4	60.6	9.2	85.0
Soap, cleaning compound, and toilet preparation	3256	840	64.3	34.8	41.7	56.1	39.8	57.7	13.8	83.7
Paint, coating, adhesive, and other chemicals	3255, 3259	1,189	50.7	46.9	36.1	59.3	34.5	62.3	7.5	81.1
Plastics and rubber products	326	4,851	34.9	63.6	26.4	71.1	25.1	72.7	7.7	89.6
Nonmetallic mineral products	327	3,765	16.4	83.1	12.3	84.4	10.5	87.7	5.3	92.3
Primary metals	331	1,433	17.4	82.1	11.8	87.7	10.9	88.7	4.1	95.3
Fabricated metal products	332	24,026	23.5	75.6	14.6	82.9	10.8	86.8	8.1	89.7
Machinery	333	9,881	38.5	59.3	31.8	65.5	29.3	68.3	10.5	85.5
Agricultural implement	33311	375	50.4	48.4	44.0	53.3	43.5	55.3	10.1	85.9
Semiconductor machinery	333295	169	24.3	74.0	24.3	74.0	23.7	74.6	7.7	89.3
Engine, turbine, and power transmission equipment	3336	366	50.5	48.1	44.9	53.2	31.2	67.4	21.2	75.6
Other machinery	other 333	8,971	37.8	60.0	30.9	66.3	28.7	68.7	10.1	85.8
Computer and electronic products	334	5,637	54.3	43.7	49.3	48.1	44.6	52.8	21.0	75.7
Communications equipment	3342	674	59.2	37.1	56.3	39.8	55.1	40.9	14.7	79.1
Semiconductor and other electronic components	3344	1,782	51.2	46.8	41.3	56.1	35.6	61.7	18.0	79.2
Navigational, measuring, electromedical, and control instruments	3345	2,253	51.5	46.9	48.1	49.1	46.3	51.1	18.2	79.2
Electromedical, electrotherapeutic, and irradiation apparatus	334510, 334517	265	42.4	55.3	39.1	51.4	38.7	51.8	8.2	88.8
Search, detection, navigation, guidance, aeronautical, and nautical system and instrument	334511	218	73.3	26.2	72.0	27.6	71.5	28.0	22.1	77.0

Other measuring and controlling device	other 3345	1,770	50.2	48.2	46.5	51.4	44.3	53.8	19.2	78.0
Other computer and electronic products	other 334	928	63.6	35.1	62.4	36.4	49.9	48.7	38.4	57.9
Electrical equipment, appliances, and components	335	2,530	39.3	59.5	34.0	64.0	31.7	66.3	12.5	84.9
Transportation equipment	336	4,441	35.5	61.5	29.1	66.7	28.2	67.1	9.6	83.7
Automobiles, bodies, trailers, and parts	3361, 3362, 3363	2,718	32.7	64.8	26.0	70.5	24.8	70.7	9.8	86.7
Aerospace products and parts	3364	629	41.9	57.0	33.3	63.5	31.9	65.0	8.8	87.5
Aircraft, aircraft engine, and aircraft parts	336411-13	607	41.6	57.4	33.1	63.8	31.8	65.2	8.1	88.2
Guided missile, space vehicle, and related parts	336414-15, 336419	22	50.0	45.5	40.9	54.5	36.4	59.1	27.3	68.2
Military armored vehicle, tank, and tank component	336992	60	D	D	D	D	D	D	D	D
Other transportation	other 336	1,034	D	D	D	D	D	D	D	D
Furniture and related products	337	4,855	23.5	75.9	15.3	83.6	14.9	84.0	3.1	96.0
Miscellaneous manufacturing	339	8,208	34.4	64.4	29.2	68.9	26.9	71.3	11.3	84.1
Medical equipment and supplies	3391	3,194	39.1	59.9	35.0	63.3	31.0	66.6	15.2	78.0
Other miscellaneous manufacturing	3399	5,014	31.4	67.3	25.5	72.5	24.2	74.2	8.9	88.0
Nonmanufacturing industries	21-23, 42-81	1,116,895	12.9	85.3	8.3	88.8	4.2	93.3	6.6	90.5
Mining, extraction, and support activities	21	5,526	12.4	86.0	6.1	92.1	3.8	94.3	5.3	92.9
Utilities	22	660	15.7	84.0	5.5	94.2	2.6	96.4	4.6	95.1
Wholesale trade	42	81,652	19.1	79.5	11.7	86.1	8.6	89.7	6.0	91.5
Electronic shopping and electronic auctions	454111-12	2,322	32.1	67.4	22.0	77.5	14.4	82.9	16.6	82.8
Transportation and warehousing	48-49	35,067	7.5	91.7	4.0	93.9	0.2	98.3	3.9	94.7
Information	51	17,587	34.4	64.0	28.3	69.5	16.2	80.9	21.9	74.7
Publishing	511	7,152	40.2	59.2	32.4	65.8	24.3	73.4	19.2	76.1
Newspaper, periodical, book, and directory publishers	5111	4,945	28.9	70.7	18.8	79.7	13.3	84.4	10.0	84.5
Software publishers	5112	2,208	65.3	33.4	63.0	34.6	48.9	48.6	39.8	57.3
Telecommunications	517	2,065	26.8	70.2	18.8	77.8	3.9	92.7	18.1	78.5
Data processing, hosting, and related services	518	2,285	41.0	55.0	39.9	56.1	23.7	72.2	37.5	57.9
Other information	other 51	6,085	27.8	71.0	22.3	76.2	8.0	89.1	20.5	78.0
Finance and insurance	52	40,467	12.7	86.1	7.1	90.3	2.2	94.5	6.3	91.0
Real estate and rental and leasing	53	31,730	6.7	91.7	3.4	93.5	3.2	93.6	1.8	95.0
Lessors of nonfinancial intangible assets (except copyrighted works)	533	543	48.4	51.4	48.3	51.6	47.9	51.9	1.5	98.3
Other real estate and rental and leasing	other 53	31,187	6.0	92.4	2.6	94.2	2.4	94.3	1.8	95.0
Professional, scientific, and technical services	54	135,127	17.5	81.1	14.1	83.0	5.9	91.9	11.9	85.4
Architectural, engineering, and related services	5413	23,027	20.4	78.2	15.0	82.9	7.0	91.0	11.5	85.4
Computer systems design and related services	5415	15,507	44.4	54.3	35.4	61.8	20.7	76.6	27.7	68.6
Scientific research and development services	5417	2,670	42.9	55.9	37.6	59.4	28.7	69.2	21.1	75.5
Biotechnology research and development	541711	849	36.1	62.8	30.3	68.2	23.5	73.9	16.6	81.5
Physical, engineering, and life sciences (except biotechnology) research and development	541712	1,657	48.9	49.8	43.5	52.9	33.5	64.7	24.4	71.7
Social sciences and humanities research and development	541720	163	18.1	81.2	16.3	78.5	8.0	91.4	10.8	83.4
Other professional, scientific, and technical services	other 54	93,923	11.7	86.9	9.7	87.2	2.5	95.3	9.1	88.4

continued

TABLE 47 Continued

Industry	NAICS Code	Number of Companies[a]	New or Significantly Improved Product OR Process		New or Significantly Improved Product(s)					
					Any good/ service		New goods		New services	
			Yes	No	Yes	No	Yes	No	Yes	No
Health care services	621-23	160,040	15.8	81.8	9.8	86.9	3.0	93.7	9.5	87.2
Other nonmanufacturing	23, 44-45 (excluding 454111-12), 55-56, 624, 71-72, 81	606,717	10.2	87.9	6.1	90.9	3.6	93.9	4.7	92.4
All companies	-	1,220,114	14.3	83.9	9.4	87.7	5.4	92.0	6.8	90.3
Small companies (number of domestic employees)[b]										
5-499	-	1,210,429	14.2	84.0	9.4	87.8	5.4	92.1	6.8	90.3
5-99	-	1,155,721	14.0	84.2	9.3	87.9	5.2	92.2	6.8	90.3
5-49	-	1,072,528	14.0	84.3	9.2	88.1	5.1	92.4	6.8	90.4
5-9	-	468,836	12.4	85.6	8.4	88.6	4.0	93.2	6.7	90.2
10-24	-	438,543	13.9	84.5	9.2	87.9	5.4	92.1	6.7	90.6
25-49	-	165,150	18.7	80.2	11.4	86.9	7.4	90.9	7.5	90.5
50-99	-	83,193	14.8	82.6	9.9	85.6	6.5	89.6	6.3	89.0
100-249	-	42,412	17.9	80.5	11.9	85.0	9.0	89.1	6.4	90.3
250-499	-	12,297	20.6	76.1	11.7	85.0	8.6	88.1	6.0	90.1
Medium and large companies (number of domestic employees)										
500-999	-	4,326	16.7	82.6	13.4	85.6	11.2	87.9	5.7	93.2
1,000-4,999	-	4,068	22.6	75.6	14.6	83.4	12.9	85.0	6.9	90.9
5,000-9,999	-	692	20.8	76.5	18.5	78.6	16.6	80.7	9.8	87.1
10,000-24,999	-	397	27.4	66.8	24.2	69.5	21.9	72.1	14.3	78.6
25,000 or more	-	202	40.1	56.4	37.2	58.2	32.2	63.6	30.2	65.3

Industry	NAICS Code	Number of Companies[a]	New or Significantly Improved Process(es)							
			Any process		Manufacturing/ production methods		Logistics/delivery/ distribution methods		Support activities	
			Yes	No	Yes	No	Yes	No	Yes	No
All industries	21-23, 31-33, 42-81	1,220,114	9.5	87.1	3.8	93.2	3.2	93.6	7.2	89.5
Manufacturing industries	31-33	103,220	20.6	76.7	15.8	81.8	6.0	91.4	12.6	84.9
Food	311	8,333	18.8	79.7	15.1	83.4	6.6	91.8	11.1	87.5
Beverage and tobacco products	312	1,245	28.4	66.8	24.3	70.8	13.0	82.2	14.3	80.8
Textile, apparel and leather	313-316	4,641	14.0	81.9	11.2	85.8	4.0	91.9	8.7	87.6
Wood products	321	4,876	9.6	89.1	6.3	92.4	1.8	96.6	4.6	93.7
Paper	322	1,218	20.4	78.0	12.3	86.1	7.0	91.3	13.7	84.6
Printing and related support activities	323	8,249	18.2	79.8	12.1	85.9	5.2	92.1	9.6	87.7
Petroleum and coal products	324	369	22.2	77.2	20.7	78.8	16.9	82.5	19.9	79.6
Chemicals	325	4,660	35.3	61.5	22.1	74.7	9.2	87.5	26.5	70.2

Basic chemicals	3251	867	47.1	52.0	38.8	60.3	15.8	83.3	39.0	60.1
Resin, synthetic rubber, fibers, and filament	3252	392	23.4	62.9	16.0	69.3	5.5	80.5	14.9	71.3
Pesticide, fertilizer, and other agricultural chemical	3253	429	9.4	90.6	7.8	92.2	2.8	97.2	3.4	96.6
Pharmaceuticals and medicines	3254	941	28.7	67.1	22.3	73.5	5.2	90.3	15.2	80.4
Soap, cleaning compound, and toilet preparation	3256	840	47.3	51.0	15.2	83.5	10.3	88.2	41.1	57.2
Paint, coating, adhesive, and other chemicals	3255, 3259	1,189	36.8	60.3	21.9	75.3	10.5	86.7	28.1	68.9
Plastics and rubber products	326	4,851	24.2	72.9	21.3	75.8	6.7	90.1	13.8	83.1
Nonmetallic mineral products	327	3,765	12.1	83.6	9.1	86.6	4.6	94.4	6.5	92.5
Primary metals	331	1,433	13.7	85.7	11.2	88.3	3.5	95.9	5.1	94.3
Fabricated metal products	332	24,026	18.4	78.8	14.1	83.9	4.0	93.4	11.4	86.0
Machinery	333	9,881	25.1	72.2	20.8	76.7	7.3	89.7	15.4	81.7
Agricultural implement	33311	375	25.0	73.5	21.1	77.5	7.4	91.2	12.9	85.6
Semiconductor machinery	333295	169	10.7	87.6	10.1	88.2	3.6	94.7	5.9	92.3
Engine, turbine, and power transmission equipment	3336	366	38.0	58.6	33.9	62.8	20.2	76.1	24.3	72.0
Other machinery	other 333	8,971	24.8	72.4	20.4	77.0	6.8	90.0	15.4	81.8
Computer and electronic products	334	5,637	30.3	66.1	25.6	71.0	10.6	85.6	18.0	78.4
Communications equipment	3342	674	18.0	70.4	15.3	73.2	8.9	79.5	13.2	75.2
Semiconductor and other electronic components	3344	1,782	33.7	63.8	30.0	67.7	10.4	86.9	22.1	75.5
Navigational, measuring, electromedical, and control instruments	3345	2,253	28.8	68.9	24.8	73.0	7.9	89.6	12.4	85.3
Electromedical, electrotherapeutic, and irradiation apparatus	334510, 334517	265	19.0	77.6	17.5	79.1	3.9	92.3	7.3	89.3
Search, detection, navigation, guidance, aeronautical, and nautical system and instrument	334511	218	37.5	62.0	33.1	66.5	8.3	91.3	32.7	66.8
Other measuring and controlling device	other 3345	1,770	29.2	68.5	24.9	72.9	8.5	88.9	10.7	86.9
Other computer and electronic products	other 334	928	36.3	60.6	26.5	70.5	19.0	78.0	26.9	69.9
Electrical equipment, appliances, and components	335	2,530	21.7	75.3	14.1	83.0	5.6	91.5	14.7	82.7
Transportation equipment	336	4,441	26.0	69.3	22.9	72.4	10.3	84.3	17.1	77.7
Automobiles, bodies, trailers, and parts	3361, 3362, 3363	2,718	25.4	71.6	22.0	74.9	8.0	87.9	15.0	80.9
Aerospace products and parts	3364	629	30.0	68.6	27.1	71.5	6.4	92.2	20.9	77.7
Aircraft, aircraft engine, and aircraft parts	336411-13	607	29.8	68.9	27.1	71.6	D	D	20.5	78.2
Guided missile, space vehicle, and related parts	336414-15, 336419	22	36.4	59.1	27.3	68.2	D	D	31.8	63.6
Military armored vehicle, tank, and tank component	336992	60	10.1	6.7	10.1	6.7	D	D	5.0	11.8
Other transportation	other 336	1,034	26.0	67.3	23.6	70.3	D	D	20.8	73.1
Furniture and related products	337	4,855	16.7	82.5	10.7	88.6	5.6	93.6	10.2	89.0
Miscellaneous manufacturing	339	8,208	22.5	75.7	17.0	81.6	6.1	91.8	12.9	85.1
Medical equipment and supplies	3391	3,194	27.2	71.1	19.7	79.0	8.0	90.0	14.6	83.5
Other miscellaneous manufacturing	3399	5,014	19.4	78.7	15.2	83.3	5.0	92.9	11.9	86.2

continued

TABLE 47 Continued

Industry	NAICS Code	Number of Companies[a]	New or Significantly Improved Process(es): Any process		Manufacturing/ production methods		Logistics/delivery/ distribution methods		Support activities	
			Yes	No	Yes	No	Yes	No	Yes	No
Nonmanufacturing industries	21-23, 42-81	1,116,895	8.5	88.0	2.7	94.2	2.9	93.8	6.7	89.9
Mining, extraction, and support activities	21	5,526	8.4	89.0	4.8	92.5	2.3	94.6	6.9	90.0
Utilities	22	660	13.5	84.8	3.8	94.5	3.0	95.3	11.7	86.7
Wholesale trade	42	81,652	13.4	84.2	4.8	92.9	5.3	92.3	10.5	87.0
Electronic shopping and electronic auctions	454111-12	2,322	24.0	75.2	10.5	88.1	14.5	84.8	15.9	83.4
Transportation and warehousing	48-49	35,067	6.0	89.5	0.0	97.7	3.2	94.5	4.6	91.7
Information	51	17,587	20.3	75.9	7.4	90.0	11.3	86.2	11.6	85.3
Publishing	511	7,152	17.9	80.1	6.8	92.6	10.5	88.1	8.8	90.0
Newspaper, periodical, book, and directory publishers	5111	4,945	17.0	80.8	7.4	92.3	10.1	88.6	6.8	92.0
Software publishers	5112	2,208	20.1	78.5	5.4	93.3	11.5	87.0	13.3	85.4
Telecommunications	517	2,065	16.2	80.7	1.2	95.7	7.9	89.0	10.9	85.9
Data processing, hosting, and related services	518	2,285	29.6	66.1	10.2	85.6	7.2	88.4	23.3	72.3
Other information	other 51	6,085	21.0	73.0	9.1	86.7	14.8	82.2	10.8	84.4
Finance and insurance	52	40,467	8.5	90.2	1.9	96.9	1.7	97.0	8.2	90.5
Real estate and rental and leasing	53	31,730	5.1	90.1	1.0	94.3	1.9	92.6	5.1	90.9
Lessors of nonfinancial intangible assets (except copyrighted works)	533	543	1.3	98.5	0.9	98.9	D	D	0.7	99.1
Other real estate and rental and leasing	other 53	31,187	5.2	90.0	1.0	94.2	D	D	5.2	90.8
Professional, scientific, and technical services	54	135,127	9.9	87.3	4.1	93.5	3.2	93.8	7.4	89.9
Architectural, engineering, and related services	5413	23,027	12.7	84.6	6.4	90.9	6.3	91.3	8.7	87.8
Computer systems design and related services	5415	15,507	28.6	69.6	9.1	89.1	11.0	87.2	24.2	74.5
Scientific research and development services	5417	2,670	28.8	69.2	19.9	78.2	6.1	92.1	13.6	84.3
Biotechnology research and development	541711	849	26.1	71.7	21.1	76.9	9.7	88.4	9.4	88.5
Physical, engineering, and life sciences (except biotechnology) research and development	541712	1,657	32.3	65.7	21.1	77.0	4.5	93.6	16.8	81.0
Social sciences and humanities research and development	541720	163	6.4	92.9	1.8	97.5	3.4	96.0	3.7	95.7
Other professional, scientific, and technical services	other 54	93,923	5.6	91.3	2.3	95.3	1.1	95.6	4.1	93.1
Health care services	621-23	160,040	10.6	86.0	3.4	93.0	3.9	92.7	7.9	88.8
Other nonmanufacturing	23, 44-45 (excluding 454111-12), 55-56, 624, 71-72, 81	606,717	6.8	89.3	2.0	94.7	2.1	94.3	5.7	90.6
All companies	-	1,220,114	9.5	87.1	3.8	93.2	3.2	93.6	7.2	89.5
Small companies (number of domestic employees)[b]										
5-499	-	1,210,429	9.5	87.1	3.8	93.2	3.1	93.6	7.2	89.5
5-99	-	1,155,721	9.2	87.3	3.6	93.3	3.1	93.6	7.0	89.7
5-49	-	1,072,528	9.2	87.4	3.6	93.5	3.1	93.7	6.9	89.8
5-9	-	468,836	7.8	88.0	3.0	93.2	2.7	93.2	5.9	90.1

10-24	-	438,543	9.3	87.7	3.8	93.6	3.3	94.1	7.0	90.1
25-49	-	165,150	12.6	85.4	4.6	93.8	3.7	94.5	9.7	88.4
50-99	-	83,193	10.1	85.4	4.5	91.4	3.2	92.3	7.9	88.0
100-249	-	42,412	13.5	84.0	6.1	91.4	3.9	93.6	10.5	87.0
250-499	-	12,297	16.5	79.6	7.3	89.2	3.7	92.7	13.1	82.9
Medium and large companies (number of domestic employees)										
500-999	-	4,326	11.6	86.7	7.8	90.8	4.9	93.3	7.4	90.8
1,000-4,999	-	4,068	17.4	80.3	8.2	89.5	5.3	92.3	14.1	83.5
5,000-9,999	-	692	15.2	81.2	10.4	86.1	8.8	87.6	11.0	85.3
10,000-24,999	-	397	18.6	72.6	12.8	78.3	12.8	78.1	14.9	76.1
25,000 or more	-	202	34.2	61.4	24.8	70.8	24.8	70.8	27.2	68.3

D = data withheld to avoid disclosing operations of individual companies.

[a]Statistics are based on companies in the United States that reported to the survey, regardless of whether they did or did not perform or fund R&D. These statistics do not include an adjustment to the weight to account for unit nonresponse.

[b]Upper bound based on U.S. Small Business Administration's definition of a small business; BRDIS does not include companies with fewer than 5 domestic employees.

NAICS = 2007 North American Industry Classification System.

NOTES: Detail may not add to total because of rounding. Industry classification based on dominant business code for domestic R&D performance where available. For companies that did not report business codes, classification used for sampling was assigned. Sum of yes and no percentages may not add to 100% due to item nonresponse to some innovation question items. The full set of detailed statistical tables from the 2011 BRDIS will be available in Business R&D and Innovation: 2011 scheduled for release in 2014. Innovation statistics are found in Tables 46-49. Relative standard errors for each cell in those tables are available from NCSES upon request. Selected final statistics are summarized in NCSES's *InfoBrief*: "Business R&D Performance in the United States Increased in 2011" (NSF 13-335) at http://www.nsf.gov/statistics/infbrief/nsf13335.

SOURCE: National Science Foundation/National Center for Science and Engineering Statistics and U.S. Census Bureau, Business R&D and Innovation Survey, 2011.

Appendix J

2011 BRDIS Table 48

TABLE 48 Number of R&D-Active and Non-R&D-Active Companies in the United States That Introduced New or Significantly Improved Products or Processes, by Size of R&D Program: 2009-2011

Company Type	Number of Companies	New or Significantly Improved Product OR Process		New or Significantly Improved Product(s): Any good/ service		New goods		New services	
		Yes	No	Yes	No	Yes	No	Yes	No
All companies[a]	1,220,114	174,066	1,024,199	114,895	1,070,247	66,427	1,122,590	82,546	1,101,820
With R&D activity[b]	57,103	36,655	19,224	32,473	23,075	27,111	28,252	16,052	38,205
< $10 million	55,119	35,425	18,582	31,376	22,310	26,169	27,338	15,490	36,928
≥ $10 but < $50 million	1,331	810	484	697	590	566	714	356	919
≥ $50 but < $100 million	235	167	57	158	65	145	79	73	146
≥ $100 million	418	253	102	242	110	231	122	134	212
Without R&D activity	1,163,011	137,411	1,004,975	82,422	1,047,172	39,316	1,094,337	66,494	1,063,615

Company Type	Number of Companies	New or Significantly Improved Process(es): Any process		Manufacturing/ production methods		Logistics/delivery/ distribution methods		Support activities	
		Yes	No	Yes	No	Yes	No	Yes	No
All companies[a]	1,220,114	115,958	1,062,204	46,326	1,136,978	38,664	1,142,041	87,968	1,092,244
With R&D activity[b]	57,103	22,110	33,229	13,988	41,688	8,761	46,550	14,526	41,014
< $10 million	55,119	21,318	32,184	13,424	40,412	8,355	45,124	13,926	39,778
≥ $10 but < $50 million	1,331	529	750	360	921	252	1,027	393	885
≥ $50 but < $100 million	235	105	116	84	136	57	161	81	139
≥ $100 million	418	158	179	120	220	97	238	126	211
Without R&D activity	1,163,011	93,848	1,028,975	32,337	1,095,290	29,903	1,095,491	73,442	1,051,231

[a]Statistics are based on companies in the United States that reported to the survey, regardless of whether they did or did not perform or fund R&D. These statistics do not include an adjustment to the weight to account for unit nonresponse.

[b]Statistics pertain to companies located in the United States that performed or funded R&D.

NOTES: Survey asked companies to identify innovations introduced in 2009-2011. Sum of yes and no responses may not add to number of companies due to item nonresponse to some innovation question items. Figures are preliminary and may later be revised. The full set of detailed statistical tables from the 2011 BRDIS will be available in Business R&D and Innovation: 2011 scheduled for release in 2014. Innovation statistics are found in Tables 46-49. Relative standard errors for each cell in those tables are available from NCSES upon request. Selected final statistics are summarized in NCSES's *InfoBrief*: "Business R&D Performance in the United States Increased in 2011" (NSF 13-335) at http://www.nsf.gov/statistics/infbrief/nsf13335.

SOURCE: National Science Foundation/National Center for Science and Engineering Statistics and U.S. Census Bureau, Business R&D and Innovation Survey, 2011.

Appendix K

2011 BRDIS Table 49

TABLE 49 Number of R&D-Active and Non-R&D-Active Companies in the United States That Introduced New or Significantly Improved Products or Processes and the Proportion of Companies in Each R&D Program Size Classification: 2009–2011 (number and percentage)

		New or Significantly Improved Product OR Process		New or Significantly Improved Product(s)					
				Any good/ service		New goods		New services	
Company Type	Number of Companies	Yes	No	Yes	No	Yes	No	Yes	No
All companies[a]	1,220,114	14.3	83.9	9.4	87.7	5.4	92.0	6.8	90.3
With R&D activity[b]	57,103	64.2	33.7	56.9	40.4	47.5	49.5	28.1	66.9
< $10 million	55,119	64.3	33.7	56.9	40.5	47.5	49.6	28.1	67.0
≥ $10 but < $50 million	1,331	60.9	36.3	52.3	44.3	42.6	53.6	26.7	69.1
≥ $50 but < $100 million	235	71.1	24.3	67.2	27.7	61.7	33.6	31.1	62.1
≥ $100 million	418	60.5	24.4	57.9	26.3	55.3	29.2	32.0	50.7
Without R&D activity	1,163,011	11.8	86.4	7.1	90.0	3.4	94.1	5.7	91.5

		New or Significantly Improved Process(es)							
		Any process		Manufacturing/ production methods		Logistics/delivery/ distribution methods		Support activities	
Company Type	Number of Companies	Yes	No	Yes	No	Yes	No	Yes	No
All companies[a]	1,220,114	9.5	87.1	3.8	93.2	3.2	93.6	7.2	89.5
With R&D activity[b]	57,103	38.7	58.2	24.5	73.0	15.3	81.5	25.4	71.8
< $10 million	55,119	38.7	58.4	24.4	73.3	15.2	81.9	25.3	72.2
≥ $10 but < $50 million	1,331	39.7	56.4	27.1	69.2	18.9	77.2	29.6	66.5
≥ $50 but < $100 million	235	44.7	49.4	35.7	57.9	24.3	68.5	34.5	59.2
≥ $100 million	418	37.8	42.8	28.7	52.6	23.2	56.9	30.2	50.5
Without R&D activity	1,163,011	8.1	88.5	2.8	94.2	2.6	94.2	6.3	90.4

[a]Statistics are based on companies in the United States that reported to the survey, regardless of whether they did or did not perform or fund R&D. These statistics do not include an adjustment to the weight to account for unit nonresponse.

[b]Statistics pertain to companies located in the United States that performed or funded R&D.

NOTES: Survey asked companies to identify innovations introduced in 2009-2011. Sum of yes and no percentages may not add to 100% due to item nonresponse to some innovation question items. The full set of detailed statistical tables from the 2011 BRDIS will be available in Business R&D and Innovation: 2011 scheduled for release in 2014. Innovation statistics are found in Tables 46-49. Relative standard errors for each cell in those tables are available from NCSES upon request. Selected final statistics are summarized in NCSES's *InfoBrief*: "Business R&D Performance in the United States Increased in 2011" (NSF 13-335) at http://www.nsf.gov/statistics/infbrief/nsf13335.

SOURCE: National Science Foundation/National Center for Science and Engineering Statistics and U.S. Census Bureau, Business R&D and Innovation Survey, 2011.

COMMITTEE ON NATIONAL STATISTICS

The Committee on National Statistics was established in 1972 at the National Academies to improve the statistical methods and information on which public policy decisions are based. The committee carries out studies, workshops, and other activities to foster better measures and fuller understanding of the economy, the environment, public health, crime, education, immigration, poverty, welfare, and other public policy issues. It also evaluates ongoing statistical programs and tracks the statistical policy and coordinating activities of the federal government, serving a unique role at the intersection of statistics and public policy. The committee's work is supported by a consortium of federal agencies through a National Science Foundation grant.

SCIENCE, TECHNOLOGY, AND ECONOMIC POLICY BOARD

The National Academies' Science, Technology, and Economic Policy (STEP) Board was established in 1991 at the National Research Council. Its mission is to advise federal, state, and local governments and to inform the public about economic and related public policies to promote the creation, diffusion, and application of new scientific and technical knowledge to enhance the productivity and competitiveness of the U.S. economy and foster economic prosperity for all Americans. The STEP Board and its committees marshal research and the expertise of scholars, industrial managers, investors, and former public officials in a wide range of policy areas that affect the speed and direction of scientific and technological change and their contributions to the growth of the U.S. and global economies. STEP activities and products are described at http://www.nationalacademies.org/step/.